AF346941

Nutrition Support for Athletic Performance

Nutrition Support for Athletic Performance

Special Issue Editors

Mark Russell
Jill Parnell

MDPI • Basel • Beijing • Wuhan • Barcelona • Belgrade • Manchester • Tokyo • Cluj • Tianjin

Special Issue Editors

Mark Russell
Leeds Trinity University
UK

Jill Parnell
Mount Royal University
Canada

Editorial Office
MDPI
St. Alban-Anlage 66
4052 Basel, Switzerland

This is a reprint of articles from the Special Issue published online in the open access journal *Nutrients* (ISSN 2072-6643) (available at: https://www.mdpi.com/journal/nutrients/special_issues/Nutrition_Athletic_Performance).

For citation purposes, cite each article independently as indicated on the article page online and as indicated below:

LastName, A.A.; LastName, B.B.; LastName, C.C. Article Title. *Journal Name* **Year**, *Article Number*, Page Range.

ISBN 978-3-03928-362-0 (Pbk)
ISBN 978-3-03928-363-7 (PDF)

© 2020 by the authors. Articles in this book are Open Access and distributed under the Creative Commons Attribution (CC BY) license, which allows users to download, copy and build upon published articles, as long as the author and publisher are properly credited, which ensures maximum dissemination and a wider impact of our publications.

The book as a whole is distributed by MDPI under the terms and conditions of the Creative Commons license CC BY-NC-ND.

Contents

About the Special Issue Editors

Mark Russell is based at Leeds Trinity University and focuses his research on topics allied with performance nutrition and applied exercise physiology for team sport athletes. As a result of this research, Mark has published over 75 peer-reviewed articles, presented at international conferences, and led multiple industry-funded contract research projects from inception to completion. Professor Russell presently supervises a number of Doctoral candidates at Leeds Trinity University, which includes leading projects related to profiling the performance and physiological responses of soccer goalkeepers and team sport substitutes, optimising half-time strategies, and facilitating recovery in Rugby League players. Professor Russell also works with a range of professional rugby and football teams and has consulted for a number of English Premier League football clubs and National and International Rugby squads. Professor Russell was also the National Lead for Applied Exercise Physiology with U.K. Deaf Sport between 2010 and 2017 and was responsible for the co-ordination of the sports science support services for DeaflympicsGB at the 2013 Summer Deaflympics held in Sofia, Bulgaria.

Jill Parnell is an Associate Professor at Mount Royal University in the Department of Health and Physical Education. Jill obtained her Ph.D. in medical sciences at the University of Calgary, researching novel nutritional therapies for obesity and associated co-morbidities. Upon completion of her Ph.D., she turned her focus to performance nutrition. Consequently, Dr. Parnell has published a wide range of studies in the field of nutrition in relation to obesity therapies and exercise performance. Presently, she maintains an active research portfolio supporting undergraduate and graduate level research. Dr. Parnell's research is multifaceted as she works with young athletes, Paralympic athletes, and endurance runners. Her research focuses on diet quality, ergogenic aids, and nutritional interventions to minimize exercise-induced gastrointestinal symptoms.

Preface to "Nutrition Support for Athletic Performance"

Athletes and their support personnel are constantly seeking evidence-informed recommendations to enhance athletic performance during competition and to optimize training-induced adaptations. Accordingly, nutritional and supplementation strategies are commonplace when seeking to achieve these aims, with such practices being implemented before, during, or after competition and/or training in a periodized manner. Performance nutrition is becoming increasingly specialized and needs to consider the diversity of athletes and the nature of the competitions. This Special Issue, Nutrition Support for Athletic Performance, describes recent advances in these areas.

Mark Russell, Jill Parnell
Special Issue Editors

Article

Influence of Equimolar Doses of Beetroot Juice and Sodium Nitrate on Time Trial Performance in Handcycling

Joelle Leonie Flueck [1,*], Alessandro Gallo [2], Nynke Moelijker [3], Nikolay Bogdanov [3], Anna Bogdanova [3] and Claudio Perret [1]

1   Institute of Sports Medicine, Swiss Paraplegic Centre Nottwil, CH-6207 Nottwil, Switzerland
2   Department of Health and Technology, ETH Zurich (Swiss Federal Institute of Technology Zurich), CH-8003 Zurich, Switzerland
3   Institute of Veterinary Physiology, Vetsuisse Faculty and the Zurich Center for Integrative Human Physiology (ZIHP), University of Zurich, CH-8057 Zurich, Switzerland
*   Correspondence: joelle.flueck@paraplegie.ch; Tel.: +41-41-939-66-17

Received: 7 June 2019; Accepted: 16 July 2019; Published: 18 July 2019

Abstract: This study aimed to investigate the influence of a single dose of either beetroot juice (BR) or sodium nitrate (NIT) on performance in a 10 km handcycling time trial (TT) in able-bodied individuals and paracyclists. In total, 14 able-bodied individuals [mean ± SD; age: 28 ± 7 years, height: 183 ± 5 cm, body mass (BM): 82 ± 9 kg, peak oxygen consumption (VO_{2peak}): 33.9 ± 4.2 mL/min/kg] and eight paracyclists (age: 40 ± 11 years, height: 176 ± 9cm, BM: 65 ± 9 kg, VO_{2peak}: 38.6 ± 10.5 mL/min/kg) participated in the study. All participants had to perform three TT on different days, receiving either 6 mmol nitrate as BR or NIT or water as a placebo. Time-to-complete the TT, power output (PO), as well as oxygen uptake (VO_2) were measured. No significant differences in time-to-complete the TT were found between the three interventions in able-bodied individuals ($p = 0.80$) or in paracyclists ($p = 0.61$). Furthermore, VO_2 was not significantly changed after the ingestion of BR or NIT in either group ($p < 0.05$). The PO to VO_2 ratio was significantly higher in some kilometers of the TT in able-bodied individuals ($p < 0.05$). The ingestion of BR or NIT did not increase handcycling performance in able-bodied individuals or in paracyclists.

Keywords: Paralympic; sports nutrition; supplementation

---

1. Introduction

Sports nutrition and performance enhancing supplements are prominently discussed topics in today's sports science. Currently, dietary nitrate is one of the latest trends in sports nutrition [1]. The main sources of dietary nitrate are green leafy and root vegetables, including beetroot, spinach, and rocket [2]. After ingestion, nitrate (NO_3^-) is reduced to nitrite (NO_2^-) and then to nitric oxide (NO), leading to a temporary increase in plasma and muscle NO [3]. Production of NO from nitrate and nitrite is enhanced under hypoxic and acidic conditions [4]. The role of NO as a signaling molecule is that it interacts with multiple targets (e.g., heme and thiol groups of proteins), triggering numerous downstream events, among which stimulation of protein kinase G and control over mitochondrial function are just a few [5]. At the organism level, NO controls heart and muscle function [6,7], vascular tone [7,8], O_2 transport by red blood cells, and other processes important for sports performance [4]. It has been proven that dietary nitrate supplementation reduces blood pressure in healthy individuals as well as in patients suffering from hypertension, diabetes, or ischemia [9–11]. Besides its beneficial therapeutic application, dietary nitrate supplementation, either as nitrate-rich beetroot juice (BR) or as sodium nitrate (NIT), has become a popular sport supplement based on its ergogenic effects (for review, see [12–14]).

Several studies have investigated the use of dietary nitrate for athletes in different disciplines, predominantly focusing on lower-body or whole-body exercises, including cycling [15,16] and running [17,18]. It has been demonstrated that dietary nitrate supplementation can improve both exercise capacity (e.g., time-to-exhaustion) [19,20] as well as exercise performance (e.g., time-to-completion) [16,21]. In addition, nitrate supplementation has been shown to reduce the oxygen cost (VO_2) of exercise [19,22]. It is assumed that these observed effects of nitrate supplementation are a consequence of an improved mitochondrial efficiency (reduced oxygen cost of ATP production) [23] and/or due to enhanced muscle contractile functions [24,25]. Complementary nitrate supplementation might promote vasorelaxation and elevate skeletal muscle oxygen delivery during exercise, as shown in rodents [26]. However, it seems unclear whether vasodilatation might improve performance [27,28]. Noteworthy targeted effects of nitrate supplementation on fast-twitch muscle fibers have been demonstrated previously [29,30].

Fat-twitch muscle fibers are more abundant in the upper body compared to the lower body muscles [31,32], and they exhibit different cardiovascular regulation mechanisms during exercise [33]. Furthermore, previous investigations reported that the cardiovascular strain in arm cranking was much higher compared to leg pedaling [34], and that arm cranking relies more on glycolytic metabolism [35]. All these physiological features of the upper body suggest that it should be sensitive to nitrate supplementation. However, the effects of dietary nitrate on upper body exercise performance have never been investigated in detail.

To date, only few studies have examined the effect of dietary nitrate supplementation on upper body exercise performance such as kayaking [36,37] or rowing [38,39]. These studies reported inconsistent results. Some investigations reported that BR supplementation improved repeated sprint and 2 km rowing performance in well trained athletes [38,39]. Furthermore, Peeling et al. [37] demonstrated that 500 m kayaking performance was significantly improved after the intake of BR. On the other hand, Muggeridge et al. [36] could not observe any ergogenic effects of BR supplementation on repeated sprint or 1 km time trial (TT) performance in trained kayakers.

To our knowledge, the effects of dietary nitrate supplementation on handcycling performance were never examined. Therefore, we aimed to investigate whether a single dose of 6 mmol nitrate as BR or NIT supplement improved 10 km handcycling TT performance in recreationally active individuals compared to paracycling athletes. Furthermore, we aimed to assess the effects of dietary nitrate supplementation on VO_2 during the TT. We hypothesized that, compared to placebo (PLA), a single intake of dietary nitrate would either (1) improve 10 km TT performance with similar VO_2 or (2) reduce VO_2 (whereas TT performance would remain unchanged).

2. Materials and Methods

2.1. Participants

Upper body trained able-bodied individuals as well as paracyclists with a spinal cord injury were recruited for this study. Athletes with a minimal training duration of 3 h per week and 3 sessions per week were admitted into the study. All participants were non-smokers. Participants using medication that might influence performance were excluded from the study. Furthermore, participants with nitrate hypersensitivity, diabetes, cardiovascular disease, pulmonary disease, or any other disease were not allowed to participate. Participants were aware of the study aim and the effects of nitrate. Before the start of the study, all participants provided written informed consent. The study was approved by the local ethical committee (EC-No. 2015-209, SNCTP-No. NCT02454049, Ethikkomission Nordwest- und Zentralschweiz, EKNZ, Basel, Switzerland) and met all ethical standards of the institutional and national research committee and the 1964 Helsinki Declaration [40].

2.2. Study Design

A randomized, placebo-controlled, single-blind, and cross-over study was performed at the Institute of Sports Medicine Nottwil, Switzerland. Each participant visited the exercise lab on five separate occasions. On the first visit, medical history was checked to fulfill all inclusion criteria, and participants were asked to carefully read and sign the written informed consent. On the second visit, a screening questionnaire was completed, and a subsequent incremental exercise test was performed to determine peak oxygen consumption (VO_{2peak}) and maximal power output (PO_{max}). Shortly after, a familiarization TT was performed. On the following three occasions, the participants performed a 10 km handcycling TT after the ingestion of BR, NIT, or PLA. Diet, sleep, training load, and acute sickness were recorded by a questionnaire before each trial to ensure equal testing conditions. All experimental trials were conducted within a maximum of six weeks, and TTs were at minimum 48 h apart for recovery purposes.

For the three experimental trials, participants arrived at the laboratory at approximately the same time of day ($\pm$ 2 h). After a 5 min resting period and the completion of the questionnaire, resting blood pressure and heart rate were measured in triplets in a sitting position at the wrist using an automated cuff (Boso medistars, Bosch + Sohn GmbH, Jungingen, Germany). The forearm was placed at heart level. Subsequently, participants received a non-transparent flask containing the supplement, which they ingested immediately. Afterwards, a 3 h resting period followed, where participants could leave the laboratory but were asked to refrain from any strenuous physical activity. During this time, participants were allowed to drink water ad libitum but were not allowed to eat. Following the 3 h resting period, blood pressure and heart rate were measured again. Thereafter, participants started with the 10 km TT.

2.3. Physical Activity and Dietary Standardization

Participants were asked to maintain their regular training schedule as consistent as possible over the time course of the study and were instructed to avoid high-intensity training the last two days before each experimental trial. They were allowed to maintain their habitual diet but were requested to abstain from nitrate-enriched foods (e.g., beetroot, spinach, rocket, lettuce, etc.) the last two days before each test. Furthermore, participants were requested to abstain from caffeine and alcohol the last 12 h before each test and were advised to eat a standardized meal 1 h before the arrival at the laboratory. They were instructed to maintain their sleep constant at a minimum of 7 h. To prevent any modification in the degradation process of nitrate by commensal bacteria in the oral cavity, participants were instructed to abstain from anti-bacterial mouthwash on all testing days [41].

2.4. Supplementation

Supplements were ingested in a fluid form. The supplements were either BR (6 mmol nitrate), water with sodium nitrate (NIT) (6 mmol nitrate), or plain water (PLA). The BR was a special production by an external company and was delivered in the form of a standardized shot (Biotta AG, Tägerwilen, Switzerland). For the preparation of the 6 mmol NIT supplements, the amount of 510 mg of sodium nitrate (Pure sodium nitrate, POCH S.A., supplier: Pharmaserv LTD, Stansstad, Switzerland) was weighted and dissolved in 85 mL of plain water. All supplements were bottled in 85 mL non-transparent flasks. Shortly after each exercise test, participants were asked about the supplement they received. Optically, they could not distinguish between NIT and PLA but could set PLA/NIT apart from BR because of taste and color. Only 60% of the athletes chose the right supplement after the NIT and PLA test. All athletes were able to distinguish BR apart from the other supplements.

2.5. Exercise Testing

2.5.1. Maximal Oxygen Uptake and Familiarization

Participants performed an incremental exercise ramp test on the handcycle ergometer Cyclus 2 (Avantronic, Leipzig, Germany) to detect VO_{2peak} and PO_{max}. After a baseline measurement of 1 min at rest, a standardized warm-up of 2 min cycling at 20 W was performed. Subsequently, the test started at 20 W. The workload increased continuously by 10 W per minute. Participants cycled until exhaustion and were verbally encouraged. Participants cycled at a self-selected pedal rate (60–100 rpm) and were instructed to keep this frequency constant during the test. Pedal rates < 50 rpm were regarded as test termination. During the test, heart rate was monitored (S610i, Polar Electro Oy, Kempele, Finland). These data were analyzed with the Polar Pro Trainer 5 software (Polar Electro Oy, Kempele, Finland). The breath-by-breath pulmonary gas exchange data were continuously measured during the test and averaged over 15 s periods (Oxycon Pro, Jaeger GmbH, Würzburg, Germany). VO_{2peak} was determined as the highest 15 s VO_2 value during the test. Ergometer configurations were recorded and maintained equally for all following exercise tests. After a 20 to 30 min break, participants performed a 10 km familiarization TT identical to the following experimental TT.

2.5.2. Time Trial

Three hours after the ingestion of the supplement, participants performed a simulated 10 km TT (0.5% incline) on the Cyclus 2 with a provided handbike (able-bodied individuals) or with their own handbike (paracyclists). All TTs were performed with the same fixed gearing. Participants performed a 5 min warm-up at self-selected pedal rate and with the fixed gearing. After a 1 min measurement at rest for baseline parameters, the 10 km TT began. After the warm-up and at the end of the TT, a capillary blood sample was taken from the earlobe and analyzed for blood lactate concentrations ([Lac]) using an enzymatic amperometric chip sensor system (Biosen C-Line Clinic, EKF diagnostic GmbH, Cardiff, UK), and rating of perceived exertion (RPE) was recorded using a Borg scale ranging from 6–20 [42]. During the test, a computer screen was placed in front of participants, which displayed the distance they had already cycled. No information was given about completion time, and participants received no feedback on performance during or after the TT. Furthermore, no verbal encouragement was provided during the TT. Heart rate was monitored during the warm-up and the TT. Respiratory gas exchange parameters were measured continuously during the TT using the above mentioned metabolic cart. Shortly before each measurement, the metabolic cart was calibrated by automatic volume calibration and by gas calibration using a standardized gas containing 16% O_2 and 5% CO_2 (Pangas, Dagmarsellen, Switzerland).

2.6. Plasma Nitrate and Nitrite Measurements

Venous blood samples were drawn into lithium-heparin tubes (2.7 mL Monovette Lithium Heparin; Sarstedt, Sevelen, Switzerland). All samples were immediately centrifuged at 3000 r·min^{-1} (1549 g) and 4 °C for 10 min within 1 min of collection. Blood plasma was harvested and stored at −80 °C in 1.5 mL tubes (Eppendorf, Wesseling–Berzdorf, Germany). Plasma nitrate ([NO_3^-]) and nitrite ([NO_2^-]) concentrations were assessed by chemiluminescence NO analyzer (CLD 88 sp NO Analyzer, ECO Medics AG, Duernten, Switzerland). [NO_2^-] was reduced to NO in Brown's solution (0.57 g 99.99% iodine; 1.62 g KI, 15 mL bi-distilled water, 200 mL acetic acid), and subsequent photons produced as NO interacted with O_3 were detected. Detection of [NO_3^-] included one more step in which [NO_3^-] was quantitatively reduced to [NO_2^-] using World Precisions Instrument's Nitralyzer kit (World Precision Instruments, Sarasota, FL, USA). Reduction of [NO_3^-] to [NO_2^-] was performed on copper-coated cadmium beads. Thereafter, total NO metabolites level (NO_x) = ([NO_2^-] + [NO_3^-]) was measured, and [NO_3^-] concentrations were calculated as a difference between [NO_2^-] levels before and after the reduction step. Calibration solutions of sodium nitrite ($NaNO_2^-$) (50, 100, 150, and 200 nmol·L^{-1}) and sodium nitrate ($NaNO_3^-$) (20, 40, 60, 80, and 100 $\mu mol \cdot L^{-1}$) were prepared in bi-distilled water to produce calibration plots.

2.7. Data Analysis and Statistics

The statistical analysis was performed using the IBM SPSS Statistics software for Windows (Version 24, IBM Corp., Armonk, NY, USA). Statistical significance was set at an α-level of 0.05. The analysis with the Q-Q plot and the Shapiro-Wilk test showed that data for the able-bodied individuals were normally distributed, and data from paracycling athletes were not normally distributed. The data were presented as mean ± standard deviation (SD). The data of the able-bodied participants were analyzed using parametric testing. To analyze changes in blood pressure and heart rate from pre- to post ingestion between NIT, BR, and PLA, a one-way repeated-measures ANOVA was performed. Differences in completion time, mean PO, mean VO_2, mean PO/VO_2 ratio, and other cardiorespiratory parameters were assessed using a one-way repeated-measures ANOVA. Pairwise t-tests were used for post hoc analysis, and Bonferroni corrections were applied for multiple testing. The PO and PO/VO_2 ratio data were graphed for the 10 km distance and analyzed using a two-way (supplement x distance) repeated-measures ANOVA with Bonferroni correction. In case of violation of sphericity, Greenhouse–Geisser correction was applied. The effect of order on TT performance was analyzed using a one-way repeated-measures ANOVA and pairwise t-tests with Bonferroni correction.

The data from paracyclists were analyzed using non-parametric testing. The Friedman test was applied to detect differences between the three groups for all parameters. If significant differences occurred, the Wilcoxon post-hoc test was applied with Bonferroni correction. For comparison between able-bodied individuals and paracyclists, the Mann-Whitney-U test was used to detect any differences.

3. Results

Fourteen healthy, recreationally active men [mean ± SD; age: 28 ± 7 years, height: 183 ± 5 cm, body mass (BM): 82 ± 9 kg, VO_{2peak}: 33.9 ± 4.2 mL/min/kg, PO_{max}: 152 ± 20 W] who were used to upper body exercises volunteered to participate in this study. They trained 4 ± 2 times per week with a total weekly duration of 6 ± 3 h. Additionally, eight paracyclists from the national team (data presented in Table 1) participated in this study. They trained 11 ± 4 h and 7 ± 2 times per week.

Table 1. Characteristics of the eight paracyclists.

ID	Age [year]	Height [cm]	Body Mass [kg]	VO_{2peak} [ml/min/kg]	PO_{max} [W]	Lesion Level	Category
1	42	188	66	41.5	200	Th5	MH3
2	34	169	72	29.7	144	L4	MH4
3	29	170	47	42.8	151	C4	MH2
4	54	178	75	49.7	222	Th12	MH5
5	32	173	60	36.9	180	Th3	MH3
6	61	178	61	45.9	208	Th4	MH3
7	40	165	64	44.8	220	Th10	MH4
8	35	190	76	17.3	98	C6	MH1
	40 ± 11	176 ± 9	65 ± 9	38.6 ± 10.5	178 ± 44		

Note: Data presented as mean ± standard deviation, PO_{max} = maximal power output in the ramp test, VO_{2peak} = peak oxygen uptake measured during the ramp test.

3.1. Blood Pressure and Heart Rate at Rest

No significant differences ($p > 0.05$) in systolic and diastolic blood pressure pre and post ingestion of the supplements between the interventions were found in able-bodied participants or paracyclists. Neither in able-bodied individuals nor in paracyclists did the change in blood pressure from pre to post ingestion significantly differ ($p > 0.05$) in all three interventions. No significant difference in systolic and diastolic blood pressure was found between able-bodied individuals and paracyclists ($p >$

0.05). No significant differences between the three interventions in able-bodied individuals and in paracyclists were observed for the change of heart rate from pre to post ingestion of the supplement ($p < 0.05$). No significant difference between able-bodied individuals and paracyclists was found ($p < 0.05$). Data are presented in Table 2.

Table 2. Blood pressure and heart rate before and after ingestion of the supplements.

	Able-Bodied Individuals			Paracyclists		
	PLA	**NIT**	**BR**	**PLA**	**NIT**	**BR**
Systolic BP [mmHg]						
Pre	124 ± 9	125 ± 9	124 ± 10	129 ± 16	130 ± 19	128 ± 17
Post	125 ± 9	125 ± 13	125 ± 10	134 ± 14	131 ± 18	135 ± 17
Diastolic BP [mmHg]						
Pre	77 ± 7	78 ± 10	79 ± 9	79 ± 15	78 ± 18	79 ± 15
Post	80 ± 8	80 ± 9	82 ± 9	84 ± 13	83 ± 16	84 ± 14
HR [bpm]						
Pre	61 ± 6	62 ± 6	61 ± 7	70 ± 17	70 ± 14	67 ± 17
Post	56 ± 7	55 ± 7	55 ± 9	61 ± 10	62 ± 10	59 ± 12

Note: data presented as mean $\pm$ SD, HR = heart rate, BP = blood pressure, PLA = placebo, NIT = sodium nitrate, BR = beetroot juice, bpm = beats per minute.

3.2. Plasma Nitrate and Nitrite Concentrations Pre and Post Supplement Ingestion

Plasma $[NO_2^-]$ and $[NO_3^-]$ concentrations before and after the ingestion of NIT, BR, or PLA are illustrated in Table 3. No significant difference was found in plasma $[NO_2^-]$ and $[NO_3^-]$ pre supplement ingestion in able-bodied individuals ($p = 0.99$) or in paracyclists ($p = 0.19$). A significantly higher $[NO_2^-]$ and $[NO_3^-]$ concentration was found in able-bodied individuals (NO_2^-: $p = 0.011$; NO_3^-: $p = 0.001$) and in paracyclists (NO_2^-: $p = 0.002$; NO_3^-: $p = 0.001$) in the BR intervention compared to PLA. No significant difference was found post ingestion between plasma $[NO_3^-]$ and $[NO_2^-]$ concentration in BR and NIT in either group ($p > 0.05$). Able-bodied individuals showed significantly higher $[NO_2^-]$ concentrations after BR consumption compared to paracyclists ($p < 0.05$). Paracyclists showed significantly higher $[NO_3^-]$ concentrations after BR and NIT consumption compared to able-bodied individuals ($p < 0.05$).

Table 3. Plasma $[NO_3^-]$ and $[NO_2^-]$ in able-bodied individuals and paracycling pre and post ingestion of BR, NIT, and PLA.

		Able-Bodied Individuals		Paracyclists	
		Pre	**Post**	**Pre**	**Post**
Plasma $[NO_3^-]$ **in uM**	PLA	68.5 ± 8.4	67.9 ± 8.6	37.9 ± 18.8	38.1 ± 18.9
	NIT	64.8 ± 10.9	278.7 ± 154.9 *[†]	36.8 ± 19.9	85.9 ± 73.2 *[†]
	BR	63.5 ± 8.9	273.7 ± 82.8 *[†]	38.8 ± 17.9	125.8 ± 99.3 *[†]
Plasma $[NO_2^-]$ **in nM**	PLA	44.7 ± 21.5	41.2 ± 28.7	66.9 ± 27.9	92.3 ± 109.9
	NIT	56.7 ± 22.6	141.2 ± 75.7 *[†]	57.7 ± 11.6	136.7 ± 69.6 *[†]
	BR	61.5 ± 53.8	121.2 ± 57.3 *[†]	108.7 ± 134.3 [†]	263.3 ± 159.2 *[†]

Note: PLA = placebo, NIT = sodium nitrate, BR = beetroot juice, * significant difference ($p < 0.05$) compared to pre ingestion, [†] = significant difference ($p < 0.05$) to PLA, pre = before ingestion of the supplement, post = 3 h after ingestion of the supplement.

3.3. Performance Parameters during Warm-up and TT

The warm-up parameters were not significantly different in the three intervention groups in paracyclists or in able-bodied individuals ($p < 0.05$). [Lac] in the warm-up was significantly lower in paracyclists compared to able-bodied individuals ($p = 0.029$). Maximal and average heart rate (HR) as well as PO during the warm-up were all significantly different between the two groups of participants ($p < 0.05$). Data are presented in Table 4.

Table 4. Warm-up data for the two groups.

	Able-Bodied Individuals			Paracyclists		
	PLA	**NIT**	**BR**	**PLA**	**NIT**	**BR**
$PO_{average}$ [W]	55 ± 19	58 ± 16	55 ± 18	90 ± 31 *	95 ± 31 *	88 ± 29 *
$HR_{average}$ [bpm]	93 ± 19	95 ± 15	94 ± 16	119 ± 20 *	115 ± 18 *	111 ± 19 *
HR_{max} [bpm]	105 ± 21	107 ± 15	107 ± 16	140 ± 25 *	137 ± 23 *	137 ± 26 *
RPE [6; 20]	11 [7; 13]	11 [7; 12]	11 [7; 13]	12 [8; 14]	11 [7; 12]	12 [8; 13]
[Lac] [mmol/L]	2.57 ± 1.43	2.67 ± 1.34	2.68 ± 1.13	1.89 ± 0.86 *	1.74 ± 0.77 *	1.87 ± 0.73 *

Note: Data presented as mean ± standard deviation and mean [min; max] for RPE, PLA = placebo, NIT = sodium nitrate, BR = beetroot juice, [Lac] = plasma lactate concentration after the time trial, RPE = rated perceived exertion, HR = heart rate, PO = power output, * = significant difference ($p < 0.05$) compared to able-bodied individuals.

No significant difference in time to complete the 10 km handcycling TT (Table 5) was found between the three interventions in able-bodied individuals ($p = 0.61$). Furthermore, no significantly different time to complete the TT in paracyclists was found ($p = 0.80$). Paracyclists completed the 10 km TT significantly faster in all three trials compared to able-bodied individuals ($p < 0.05$) and therefore showed a significantly higher $PO_{average}$ compared to able-bodied individuals ($p < 0.05$).

Table 5. Data from the 10 km time trail (TT) performance in both groups of participants.

	Able-Bodied Individuals			Paracyclists		
	PLA	**NIT**	**BR**	**PLA**	**NIT**	**BR**
Time to complete [s]	1182 ± 84	1194 ± 102	1195 ± 94	1106 ± 247 *	1091 ± 235 *	1071 ± 199 *
$PO_{average}$ [W]	114 ± 16	113 ± 20	112 ± 17	142 ± 49	144 ± 44	145 ± 42
$HR_{average}$ [bpm]	152 ± 13	153 ± 9	150 ± 11	162 ± 27 *	158 ± 30 *	158 ± 31 *
HR_{max} [bpm]	174 ± 10	175 ± 8	175 ± 9	176 ± 24	170 ± 31	173 ± 28
RPE [6; 20]	18 [16; 20]	19 [17; 20]	18.5 [17; 20]	19 [17; 20]	19 [17; 20]	19 [17; 20]
[Lac] [mmol/L]	11.94 ± 2.40	12.28 ± 2.84	11.66 ± 2.15	10.13 ± 5.98	10.05 ± 6.37	9.92 ± 5.99

Note: Data presented as mean ± standard deviation and mean [min; max] for RPE, RPE = rated perceived exertion, $HR_{average}$ = average heart rate during the test, HR_{max} = maximal heart rate during the test, $PO_{average}$ = average power output during the test, [Lac] = lactate concentration after the test, NIT = sodium nitrate, BR = beetroot juice, PLA = placebo, * = significant ($p < 0.05$) difference between the two groups of participants.

Neither [Lac] ($p = 0.31$) after the time trial nor rated perceived exertion (RPE) ($p = 0.41$) for the trials were significantly different between the three interventions in able-bodied individuals. The same was true in paracyclists between the three interventions (RPE: $p = 0.43$; [Lac]: $p = 0.68$). No significant differences were found in RPE ($p = 0.08$) and [Lac] ($p = 0.86$) concentration between able-bodied individuals and paracyclists. No significant differences between the three interventions were found in $HF_{average}$ in able-bodied individuals ($p = 0.25$) or in paracyclists ($p = 0.69$) during the TT, and there was no difference between the two groups of participants ($p = 0.11$). Data are presented in Table 5.

Out of all 22 participants, only one suffered from gastrointestinal side effects. The participant reported nausea. No significant order effect was found comparing time to complete the first, the second, and the third TT regardless of the supplement in able-bodied individuals ($p = 0.07$) and in paracyclists ($p = 0.14$). $PO_{average}$ was significantly greater in the third TT compared to the second in able-bodied individuals ($p = 0.039$, mean difference + 4.4 W, 95%-CI [0.2; 8.85]) but not in paracycling athletes ($p = 0.11$).

3.4. Oxygen Uptake and PO to VO$_2$ Ratio

No significant difference was found in $VO_{2average}$ between the three interventions in able-bodied individuals ($p = 0.15$) or in paracyclists ($p = 0.07$). $VO_{2average}$ was found to be significantly higher in paracyclists ($VO_{2average}$: NIT = 35.4 ± 9.5 min/kg, BR = 34.7 ± 10.3 mL/min/kg, PLA = 35.2 ± 9.9 mL/min/kg) compared to able-bodied ($VO_{2average}$: NIT = 29.0 ± 4.2 mL/min/kg,

BR = 27.8 ± 4.6 mL/min/kg, PLA = 29.3 ± 4.4 mL/min/kg) individuals (p = 0.042). Furthermore, no significant difference was found in PO to VO_2 ratio between the three interventions in able-bodied individuals (p = 0.62) or in paracyclists (p = 0.88). PO to VO_2 ratio was significantly different in some kilometers over the course of the TT in able-bodied individuals ($p < 0.05$). No significant differences in the same parameter were found in the paracyclists ($p > 0.05$) during the TT (Figure 1).

(a)

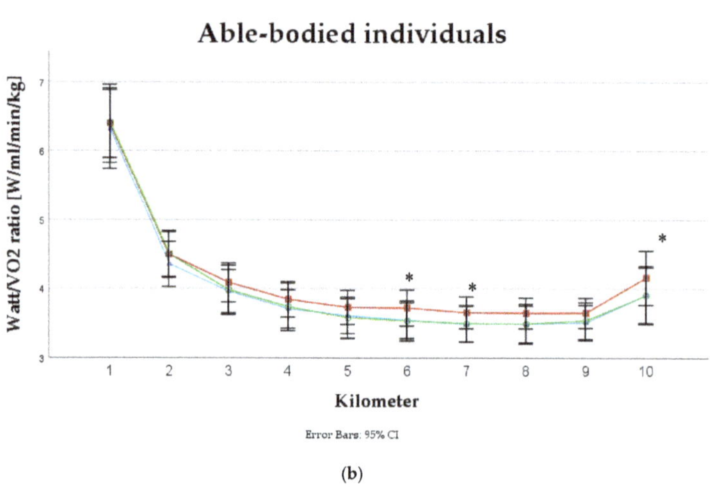

(b)

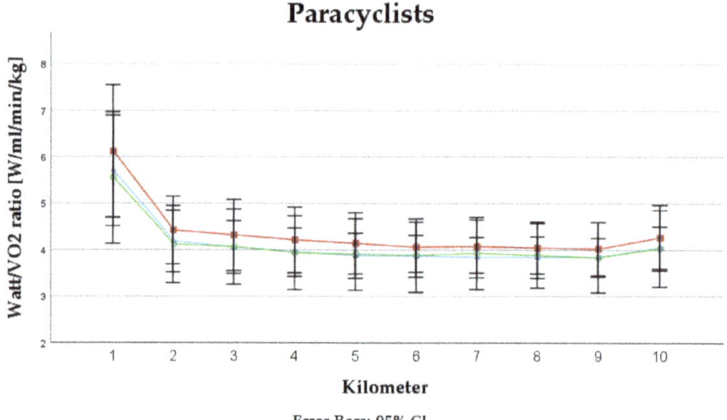

Figure 1. PO to VO_2 ratio in (**a**) able-bodied individuals and (**b**) in paracyclists during a 10 km handcycling TT. Note: * = significant difference ($p < 0.05$), red squares (■) = beetroot juice intervention, blue triangles (▲) = sodium nitrate intervention and green circles (○) = placebo intervention.

4. Discussion

To our knowledge, this is the first study examining the effects of acute nitrate supplementation on 10 km handcycling TT performance. The principal finding of the current investigation was that acute nitrate supplementation did not improve 10 km handcycling TT performance in recreationally active men or in trained paracyclists by BR or by NIT ingestion.

4.1. Performance

A single dose of 6 mmol nitrate (BR or NIT) did not improve 10 km handcycling TT performance in able-bodied, upper body trained individuals or in trained paracyclists with a spinal cord injury (Table 5). To date, only a few studies examined the effects of nitrate supplementation on upper body exercise performance [36–39,43,44]. It is worth mentioning that rowing is a whole body exercise; most of its performance is produced by the lower extremities. Furthermore, these studies focused on short-term exercises, including 4 min maximal effort exercises [37], single [43] or repeated sprint tests [36,38], and 500 m to 2 km TT [36–38]. In the study of Peeling et al. [37], the ingestion of 9.6 mmol BR significantly enhanced 500 m kayaking TT performance by 1.7%. However, in the same study, a 4 min TT performance remained unchanged after the intake of 4.8 mmol BR. Comparable results were shown by Hoon et al. [39], where the authors concluded that the ingestion of 8.4 mmol but not 4.2 mmol BR may improve 2 km rowing TT performance [39]. Furthermore, Bond et al. [38] reported improved repeated sprint rowing performance after six days of BR supplementation. The fact that these studies did not reveal any ergogenic effects after the ingestion of a small nitrate dose (e.g., 4.2 to 5 mmol BR) may lead to the assumption that the dose used in the present study (6 mmol BR or NIT) was insufficient to induce any significant effects specifically on upper body exercise performance. Therefore, we hypothesize that a higher dosage of BR or NIT is needed for an acute supplementation protocol to improve 10 km handcycling TT performance. However, of course, a lack of effect of BR supplementation on performance cannot be ruled out.

In the review of Jones [45], it was concluded that nitrate supplementation appears to be more effective in exercises within the duration from 5 to 30 min compared to exercises lasting more than 40 min. Short duration exercises are typically performed at high intensity with predominant fast-twitch muscle fiber contribution and are likely to result in hypoxic and acidic skeletal muscle state. Recalling that the reduction from $[NO_3^-]$ to NO is enhanced in hypoxic and acidic conditions [4] and that nitrate has targeted effects on fast-twitch muscle fibers [29,30], it becomes clear that short duration exercises may benefit more from nitrate supplementation. Nevertheless, type II muscle fibers are also recruited during prolonged low-intensity exercise. In the present study, the exercise duration was around 20 min, and the mean intensity was about 85% of VO_{2peak}, which indicates that the 10 km handcycling TT would provide optimal conditions for effective nitrate supplementation, particularly in view of the fact that upper body muscles possess a higher contribution of fast-twitch muscle fibers compared to the muscles of the lower body in healthy humans [31,32]. Furthermore, individuals with a spinal cord injury, M. deltoideus, seem to have a higher portion of type I muscle fibers [46]. Thus, nitrate would be expected to be less beneficial in those individuals compared to healthy, able-bodied controls. However, our study showed no improvement in performance in either group of athletes.

To date, only one study examined the effect of chronic nitrate supplementation on whole body exercise performance [38]. They reported an improved repeated 500 m sprint rowing performance after six days of BR supplementation (5.5 mmol per day). The dosage was therefore very similar to the 6 mmol used in our study but, interestingly, the authors revealed beneficial effects on performance. Therefore, it may be suggested that a chronic supplementation protocol may be more beneficial in terms of improving exercise performance compared to an acute supplementation protocol. This assumption is in line with the findings of Vanhatalo et al. [47], who reported that a chronic 15 day BR supplementation (5.2 mmol per day) resulted in a greater effect on incremental cycling exercise performance compared to a single dose of BR [47]. These findings are supported by the two studies of Cermak et al. [15], [48]. In the first study, the authors reported significant improvements in 10 km cycling TT performance after six days of BR supplementation (8 mmol per day) [15]. Conversely, in the second study, acute BR ingestion (8.7 mmol) did not show any positive effects on 1 h cycling performance [48]. The authors assumed that a longer supplementation period (chronic) is necessary to potentiate any ergogenic effects of BR. It is worth noting that, in these two studies, exercise duration and participant characteristics were different; therefore, comparison of the two studies should be taken with caution. A recent investigation by Jo et al. [49] compared the effects of acute (i.e., 2.5 h before exercise) and chronic (i.e., 15 days) nitrate supplementation (8 mmol per day) on 8 km cycling TT performance. The authors reported significant improvements in TT performance after chronic supplementation, whereas acute nitrate ingestion was unable to induce any ergogenic effects. Taking all these findings into account, we assume that the missing ergogenic effect of nitrate supplementation in the present study is partly attributable to the acute supplementation regimen. To date, no study investigated the effect of chronic BR supplementation on upper body exercise performance.

Recent studies reporting no improvement of performance after nitrate supplementation concluded that the fitness level of their participants (VO_{2peak} > 6 min/kg) might play an important role [36,50–53]. It has been shown that highly trained athletes exhibit increased NO synthase activity (NOS) and a better muscle capillarization compared to untrained individuals [54]. Furthermore, highly trained endurance athletes showed higher plasma nitrite levels [55] and fewer type II muscle fibers compared to recreationally active individuals [56]. All these adaptations to training may result in limited or attenuated effects of nitrate supplementation on performance in trained individuals [45]. In the study of Porcelli et al. [21], nitrate supplementation improved 3 km running TT performance in low to moderate trained participants but not in highly trained athletes (VO_{2peak} > 63 mL/min/kg). Therefore, it was concluded that highly trained athletes are less likely to benefit from nitrate supplementation [21]. In the present study, we tested recreationally active participants (VO_{2peak} = 33.7 ± 4 mL/min/kg) and paracyclists (VO_{2peak} = 38.6 ± 10.5 mL/min/kg). To note, the recorded VO_{2peak} may not reflect the true VO_{2max} value, as during our specific test, only upper body muscles were active. According to the literature [57], we expect the real VO_{2peak} values to be in a range of 45 to 55 mL/min/kg for the able-bodied individuals. Calculating the according value for the paracyclists produces a range between 50 to 60 mL/min/kg, but of course those athletes have secondary conditions such as a limited heart rate [58] or a reduced muscle mass in the core and the upper body [59]. Therefore, we assume that their VO_{2peak} would be even higher, as in able-bodied cyclists. Nevertheless, our able-bodied participants did not fall in the category of highly trained individuals. It seems that the fitness level of those participants would not be a major reason for missing performance enhancing effects.

It is possible that some of our athletes could have been identified as a so-called "responder" with a beneficial effect of either BR or NIT supplementation. Several studies have identified responders and non-responders in their participant samples [51,60–62]. The physiological background of these individual responses to nitrate supplementation is presently unknown and requires further investigation. It is suggested that habitual nitrate intake, baseline [NO_2^-] concentrations, activity of oral and gut bacteria, training status, and fiber type distribution may play a major role in the appearance of responders and non-responders [45]. Although the present study did not report an ergogenic effect of nitrate supplementation, some participants improved their 10 km TT performance after the ingestion of both BR and NIT. Therefore, we would recommend athletes trialing whether they benefit from nitrate ingestion or not. As long as it is well tolerated, we would not advise athletes against the use of nitrate supplementation. Furthermore, it might possibly be an advantage to follow a chronic supplementation protocol over several days or to use a higher dosage than 6 mmol nitrate.

4.2. Oxygen Consumption and PO to VO_2 Ratio

It is assumed that the ergogenic effect of nitrate supplementation is due to a reduction in VO_2, as a consequence of enhanced mitochondrial efficiency [23] and/or improved muscle contractile functions [24,25,29]. Larsen et al. [23] demonstrated that following three days of nitrate supplementation VO_2 at a given workload was significantly reduced by 3% during submaximal cycling exercise. This reduction in VO_2 was predominately related to the increased mitochondrial P/O ratio (ATP produced by O_2 consumed) [23]. Further investigations revealed that, after chronic nitrate supplementation, the expression of adenine nucleotide translocase (ANT) was reduced, a protein that is responsible for a substantial part of mitochondrial proton leakage, thus improving mitochondrial efficiency [23]. On the other hand, Bailey et al. [24] demonstrated that the reduction in VO_2 after six days of BR ingestion was related to a reduced ATP cost of muscle force production (i.e., higher muscle efficiency). However, the exact mechanisms remained unsolved. Furthermore, Hernandez et al. [29] reported that after seven days of nitrate supplementation in mice, force production of fast-twitch muscle fibers was increased. The authors suggested that the result was due to an increased expression of muscle Ca^{2+}-handling proteins in fast-twitch muscle fibers after nitrate supplementation [28]. In addition to these findings, Haider et al. [25] revealed that seven days of BR ingestion enhanced excitation–contraction coupling in human skeletal muscles. Thus, upper body muscles in general would benefit more, as they tend to have more type II muscle fibers [32]. However, a lesion of the spinal cord (and therefore a predominant use of upper body muscle) seems to lead to an increase in muscle type I fibers [46]. Therefore, we expected an increase in performance following nitrate ingestion in able-bodied individuals but possibly not in paracyclists with a spinal cord injury.

To date, it was shown that dietary nitrate supplementation either by BR or NIT reduces steady-state VO_2 (3 to 14%) for a given PO during submaximal exercises irrespective of exercise modality (cycling, running, kayaking, etc.) [20,47,63,64]. However, not all studies were able to detect such a reduction of VO_2 [65,66]. Our study (Figure 1) showed that PO to VO_2 ratio tended to be higher in the BR trial compared to NIT and PLA in able-bodied individuals with a significant difference in some km during the TT. However, VO_2 alone was not significantly lower in the BR trial. To summarize the most recent literature, it was noticed that the PO/VO_2 ratio increased to a larger extent during cycling (i.e., lower body: 7 to 11%) compared to handcycling and kayaking (i.e., upper body: 2 to 3%) [16,37]. It could be hypothesized that the smaller increase in the PO/VO_2 ratio is simply a consequence of a smaller active muscle mass during upper body exercise compared to cycling. On the other hand, upper body exercises rely largely on fast-twitch muscle fibers [67,68], and therefore it would be expected that upper body exercises may show greater benefits compared to cycling or running. However, the results of the present study do not support this assumption.

In the present study, we found an increase in PO/VO_2 ratio in some km of the TT after BR ingestion but not after NIT despite the same dosages (6 mmol) of the supplements (Figure 1a). This observation is contradictory, since several studies reported a reduced VO_2 during exercise following NIT supplementation [23,63]. In contrast to our study, these investigations used a chronic supplementation protocol. Thus far, few studies examined the effects of acute NIT ingestion on VO_2 [65,66]. Bescos et al. [66] reported that 10 mg/kg BM nitrate (~11 mmol) did not reduce VO_2 during submaximal cycling in trained endurance athletes. These results are in close agreement with those obtained by Flueck et al. [65], where different dosages of NIT (3, 6, and 12 mmol) did not reduce VO_2 response of moderate- and severe-intensity cycling. On the other hand, BR supplementation appeared to elicit greater effects on VO_2 due to the additional antioxidants, which facilitate nitrate reduction. Additionally, BR contains polyphenols with a potential for ergogenic effects in performance [69]. It is possible that BR might be more beneficial in reducing VO_2 compared to NIT, but more research is needed to confirm those assumptions.

4.3. Other Parameters

In the present study, neither BR nor NIT had any significant influence on systolic or diastolic blood pressure at rest (Table 2). Similar findings were shown in a similar study [65], whereas only the ingestion of a higher BR dosage seemed to result in a significantly lower blood pressure at rest. It seems that a dose-dependent decrease in blood pressure occurs after the ingestion of nitrate supplementation [10]. The mechanism by which dietary nitrate supplementation lowers blood pressure may be explained by the increased bioavailability of NO via the nitrate-nitrite-NO-pathway [70,71]. NO is known as a potent vasodilator that governs systemic blood pressure by reducing arterial pressure and peripheral resistance [4]. Therefore, a sufficient amount of nitrate needs to be ingested to increase the concentration of NO, thus inducing a reduction in blood pressure. Previous studies concluded that, compared to nitrate salts, BR has a greater potential to reduce systolic blood pressure due to the additional antioxidants in BR that facilitate the reduction of nitrite to NO [65]. However, the findings of the present study did not confirm this assumption.

Resting levels of [NO_3^-] seemed to be lower in the paracyclists (~38 uM) compared to able-bodied individuals (~66 uM). To our knowledge, no other study measured resting [NO_3^-] concentration in participants with a spinal cord injury. Therefore, we might only speculate about the reason for such a difference. Possible reasons might be the lower energy or the nitrate intake in general due to lower active muscle mass resulting in lower energy expenditure. Another explanation might involve the slower gastrointestinal transition time in those participants due to the impaired nervous system [72]. This might also affect the absorption of nutrients. Significant increases in [NO_2^-] and [NO_3^-] concentrations after NIT and BR were found in both groups of athletes. Interestingly, able-bodied individuals showed significantly higher [NO_2^-] concentrations after BR consumption compared to paracyclists, whereas paracyclists showed significantly higher [NO_3^-] concentrations after BR and NIT ingestion compared to able-bodied individuals. It seems possible that reduction from [NO_3^-] to [NO_2^-] was reduced in the paracyclists and therefore a higher [NO_3^-] was present 3 h after the ingestion. Such a hypothesis needs to be confirmed in future studies. No significant differences were found in [NO_2^-] and [NO_3^-] concentrations between NIT and BR ingestion in the same group. These findings seemed to be very similar to the previous study with a 6 mmol NIT and BR intervention in able-bodied individuals [65]. Thus, the results for [NO_2^-] and [NO_3^-] concentrations in this study seem to be consistent with the literature.

4.4. Limitations

One major limitation of this study was that we did not use a nitrate-depleted BR as a placebo supplement. The participants were possibly able to distinguish between BR and NIT or between BR and PLA. Therefore, we cannot exclude the possibility that our findings were influenced by a potential placebo effect in the BR group. However, the participants were not able to distinguish between PLA and NIT because of similar taste, smell, and appearance.

Additionally, there was a significant difference between the two groups. Able-bodied individuals showed a significantly higher body mass compared to paracyclists. This fact evolved mainly from the loss of muscle mass in the legs of athletes with a spinal cord injury due to immobilization. Furthermore, the age was significantly different between the two groups, as paracyclists were older. This is a known phenomenon in Paralympic sports, as most athletes start their career after the incident of the injury. Last but not least, both groups showed a different fitness level, as paracyclists appeared to be more highly trained than able-bodied individuals. All these issues might have influenced the outcome of the study and therefore needed to be acknowledged.

Finally, although our able-bodied participants were used to upper body exercise, we still reported a significant order effect on 10 km handcycling TT performance in this group of participants. Therefore, we assume that only one familiarization trial was insufficient in the present study. We cannot exclude the possibility that the effects of dietary nitrate on exercise performance and VO$_2$ in the present study were masked or blunted by the observed order effect.

5. Conclusions

Our study demonstrated that acute nitrate supplementation either by BR or NIT (6 mmol) did not improve 10 km handcycling TT performance in recreationally active men or in paracyclists. BR ingestion seemed to increase PO to VO$_2$ ratio in some km of the TT compared to PLA in able-bodied individuals but not in paracyclists. Overall, the present results do not support previous findings where acute nitrate supplementation enhanced short-term, high-intensity TT performance [36–38] in upper body exercise. Further research is required to examine optimal supplementation strategies and effects of nitrate supplementation on upper body exercise performance, especially in wheelchair athletes with physiological adaptations following the incident of the spinal cord injury.

Author Contributions: Individual contributions were as following: conceptualization, J.L.F. and C.P.; methodology, J.L.F., A.B., C.P.; validation, J.L.F., A.B. and C.P.; formal analysis, J.L.F., A.G., A.B.; investigation, A.G., J.L.F., N.M. and N.B.; resources, J.L.F.; data curation, J.L.F., A.G., A.B.; writing—original draft preparation, J.L.F. and A.G.; writing—review and editing, C.P. and A.B.; visualization, J.L.F. and A.G.; supervision, J.L.F.; project administration, J.L.F.

Funding: This project has received funding from the European Union's Horizon 2020 research and innovation program under grant agreement No. 675115—RELEVANCE—H2020-MSCA-ITN-2015/H2020-MSCA-ITN-2015 to A.B.

Acknowledgments: We are grateful for the participation of all athletes in this study. Furthermore, we would like to thank the medical staff, Eliane Arnold, Melanie Müller and Ruth Heller, for helping with the blood withdrawal. This may include administrative and technical support, or donations in kind (e.g., materials used for experiments).

Conflicts of Interest: The authors declare no conflict of interest.

References

1. Ahrendt, D.M. Ergogenic aids: Counseling the athlete. *Am. Fam. Physician* **2001**, *63*, 913–922. [PubMed]
2. Santamaria, P. Nitrate in vegetables: Toxicity, content, intake and ec regulation. *J. Sci. Food Agric.* **2006**, *86*, 10–17. [CrossRef]
3. Nyakayiru, J.; Kouw, I.W.K.; Cermak, N.M.; Senden, J.M.; van Loon, L.J.C.; Verdijk, L.B. Sodium nitrate ingestion increases skeletal muscle nitrate content in humans. *J. Appl. Physiol.* **2017**, *123*, 637–644. [CrossRef] [PubMed]
4. Lundberg, J.O.; Weitzberg, E.; Gladwin, M.T. The nitrate-nitrite-nitric oxide pathway in physiology and therapeutics. *Nat. Rev. Drug Discov.* **2008**, *7*, 156–167. [CrossRef] [PubMed]
5. Pacher, P.; Beckman, J.S.; Liaudet, L. Nitric oxide and peroxynitrite in health and disease. *Physiol. Rev.* **2007**, *87*, 315–424. [CrossRef] [PubMed]
6. Reid, M.B. Role of nitric oxide in skeletal muscle: Synthesis, distribution and functional importance. *Acta Physiol. Scand.* **1998**, *162*, 401–409. [CrossRef]
7. Stamler, J.S.; Meissner, G. Physiology of nitric oxide in skeletal muscle. *Physiol. Rev.* **2001**, *81*, 209–237. [CrossRef]
8. Moncada, S.; Palmer, R.M.; Higgs, E.A. Nitric oxide: Physiology, pathophysiology, and pharmacology. *Pharmacol. Rev.* **1991**, *43*, 109–142.
9. Gilchrist, M.; Shore, A.C.; Benjamin, N. Inorganic nitrate and nitrite and control of blood pressure. *Cardiovasc. Res. Cent. Bull.* **2011**, *89*, 492–498. [CrossRef]
10. Kapil, V.; Milsom, A.B.; Okorie, M.; Maleki-Toyserkani, S.; Akram, F.; Rehman, F.; Arghandawi, S.; Pearl, V.; Benjamin, N.; Loukogeorgakis, S.; et al. Inorganic nitrate supplementation lowers blood pressure in humans: Role for nitrite-derived no. *Hypertension* **2010**, *56*, 274–281. [CrossRef]
11. Kapil, V.; Weitzberg, E.; Lundberg, J.O.; Ahluwalia, A. Clinical evidence demonstrating the utility of inorganic nitrate in cardiovascular health. *Nitric Oxide* **2014**, *38*, 45–57. [CrossRef] [PubMed]
12. Hoon, M.W.; Johnson, N.A.; Chapman, P.G.; Burke, L.M. The effect of nitrate supplementation on exercise performance in healthy individuals: A systematic review and meta-analysis. *Int. J. Sport Nutr. Exerc. Metab.* **2013**, *23*, 522–532. [CrossRef] [PubMed]
13. Jones, A.M. Dietary nitrate supplementation and exercise performance. *Sports Med.* **2014**, *44* (Suppl. 1), S35–S45. [CrossRef]
14. McMahon, N.F.; Leveritt, M.D.; Pavey, T.G. The effect of dietary nitrate supplementation on endurance exercise performance in healthy adults: A systematic review and meta-analysis. *Sports Med.* **2017**, *47*, 735–756. [CrossRef] [PubMed]
15. Cermak, N.M.; Gibala, M.J.; van Loon, L.J. Nitrate supplementation's improvement of 10-km time-trial performance in trained cyclists. *Int. J. Sport Nutr. Exerc. Metab.* **2012**, *22*, 64–71. [CrossRef] [PubMed]
16. Lansley, K.E.; Winyard, P.G.; Bailey, S.J.; Vanhatalo, A.; Wilkerson, D.P.; Blackwell, J.R.; Gilchrist, M.; Benjamin, N.; Jones, A.M. Acute dietary nitrate supplementation improves cycling time trial performance. *Med. Sci. Sports Exerc.* **2011**, *43*, 1125–1131. [CrossRef] [PubMed]
17. Vasconcellos, J.; Henrique Silvestre, D.; Dos Santos Baiao, D.; Werneck-de-Castro, J.P.; Silveira Alvares, T.; Paschoalin, V.M. A single dose of beetroot gel rich in nitrate does not improve performance but lowers blood glucose in physically active individuals. *J. Nutr. Metab.* **2017**, *2017*, 7853034. [CrossRef]

18. Shannon, O.M.; Barlow, M.J.; Duckworth, L.; Williams, E.; Wort, G.; Woods, D.; Siervo, M.; O'Hara, J.P. Dietary nitrate supplementation enhances short but not longer duration running time-trial performance. *Eur. J. Appl. Physiol.* **2017**, *117*, 775–785. [CrossRef]

19. Bailey, S.J.; Winyard, P.; Vanhatalo, A.; Blackwell, J.R.; Dimenna, F.J.; Wilkerson, D.P.; Tarr, J.; Benjamin, N.; Jones, A.M. Dietary nitrate supplementation reduces the o2 cost of low-intensity exercise and enhances tolerance to high-intensity exercise in humans. *J. Appl. Physiol.* **2009**, *107*, 1144–1155. [CrossRef]

20. Lansley, K.E.; Winyard, P.G.; Fulford, J.; Vanhatalo, A.; Bailey, S.J.; Blackwell, J.R.; DiMenna, F.J.; Gilchrist, M.; Benjamin, N.; Jones, A.M. Dietary nitrate supplementation reduces the o2 cost of walking and running: A placebo-controlled study. *J. Appl. Physiol.* **2011**, *110*, 591–600. [CrossRef]

21. Porcelli, S.; Ramaglia, M.; Bellistri, G.; Pavei, G.; Pugliese, L.; Montorsi, M.; Rasica, L.; Marzorati, M. Aerobic fitness affects the exercise performance responses to nitrate supplementation. *Med. Sci. Sports Exerc.* **2015**, *47*, 1643–1651. [CrossRef] [PubMed]

22. Wylie, L.J.; Kelly, J.; Bailey, S.J.; Blackwell, J.R.; Skiba, P.F.; Winyard, P.G.; Jeukendrup, A.E.; Vanhatalo, A.; Jones, A.M. Beetroot juice and exercise: Pharmacodynamic and dose-response relationships. *J. Appl. Physiol.* **2013**, *115*, 325–336. [CrossRef] [PubMed]

23. Larsen, F.J.; Schiffer, T.A.; Borniquel, S.; Sahlin, K.; Ekblom, B.; Lundberg, J.O.; Weitzberg, E. Dietary inorganic nitrate improves mitochondrial efficiency in humans. *Cell Metab.* **2011**, *13*, 149–159. [CrossRef] [PubMed]

24. Bailey, S.J.; Fulford, J.; Vanhatalo, A.; Winyard, P.G.; Blackwell, J.R.; DiMenna, F.J.; Wilkerson, D.P.; Benjamin, N.; Jones, A.M. Dietary nitrate supplementation enhances muscle contractile efficiency during knee-extensor exercise in humans. *J. Appl. Physiol.* **2010**, *109*, 135–148. [CrossRef] [PubMed]

25. Haider, G.; Folland, J.P. Nitrate supplementation enhances the contractile properties of human skeletal muscle. *Med. Sci. Sports Exerc.* **2014**, *46*, 2234–2243. [CrossRef]

26. Ferguson, S.K.; Hirai, D.M.; Copp, S.W.; Holdsworth, C.T.; Allen, J.D.; Jones, A.M.; Musch, T.I.; Poole, D.C. Impact of dietary nitrate supplementation via beetroot juice on exercising muscle vascular control in rats. *J. Physiol.* **2013**, *591*, 547–557. [CrossRef] [PubMed]

27. Lundby, C.; Boushel, R.; Robach, P.; Moller, K.; Saltin, B.; Calbet, J.A. During hypoxic exercise some vasoconstriction is needed to match o2 delivery with o2 demand at the microcirculatory level. *J. Physiol.* **2008**, *586*, 123–130. [CrossRef] [PubMed]

28. Calbet, J.A.; Lundby, C.; Sander, M.; Robach, P.; Saltin, B.; Boushel, R. Effects of atp-induced leg vasodilation on vo2 peak and leg o2 extraction during maximal exercise in humans. *Am. J. Physiol. Regul. Integr. Comp. Physiol.* **2006**, *291*, R447–R453. [CrossRef]

29. Hernandez, A.; Schiffer, T.A.; Ivarsson, N.; Cheng, A.J.; Bruton, J.D.; Lundberg, J.O.; Weitzberg, E.; Westerblad, H. Dietary nitrate increases tetanic [ca2+]i and contractile force in mouse fast-twitch muscle. *J. Physiol.* **2012**, *590 Pt 15*, 3575–3583. [CrossRef]

30. Jones, A.M.; Ferguson, S.K.; Bailey, S.J.; Vanhatalo, A.; Poole, D.C. Fiber type-specific effects of dietary nitrate. *Exerc. Sport Sci. Rev.* **2016**, *44*, 53–60. [CrossRef]

31. Kiilerich, K.; Birk, J.B.; Damsgaard, R.; Wojtaszewski, J.F.; Pilegaard, H. Regulation of pdh in human arm and leg muscles at rest and during intense exercise. *Am. J. Physiol. Endocrinol. Metab.* **2008**, *294*, E36–E42. [CrossRef] [PubMed]

32. Schantz, P.; Randall-Fox, E.; Hutchison, W.; Tyden, A.; Astrand, P.O. Muscle fibre type distribution, muscle cross-sectional area and maximal voluntary strength in humans. *Acta Physiol. Scand.* **1983**, *117*, 219–226. [CrossRef] [PubMed]

33. Calbet, J.A.; Holmberg, H.C.; Rosdahl, H.; van Hall, G.; Jensen-Urstad, M.; Saltin, B. Why do arms extract less oxygen than legs during exercise? *Am. J. Physiol. Regul. Integr. Comp. Physiol.* **2005**, *289*, R1448–R1458. [CrossRef] [PubMed]

34. Calbet, J.A.; Gonzalez-Alonso, J.; Helge, J.W.; Sondergaard, H.; Munch-Andersen, T.; Saltin, B.; Boushel, R. Central and peripheral hemodynamics in exercising humans: Leg vs arm exercise. *Scand J. Med. Sci. Sports.* **2015**, *25* (Suppl. 4), 144–157. [CrossRef] [PubMed]

35. Van Hall, G.; Jensen-Urstad, M.; Rosdahl, H.; Holmberg, H.C.; Saltin, B.; Calbet, J.A. Leg and arm lactate and substrate kinetics during exercise. *Am. J. Physiol. Endocrinol. Metab.* **2003**, *284*, E193–E205. [CrossRef] [PubMed]

36. Muggeridge, D.J.; Howe, C.C.; Spendiff, O.; Pedlar, C.; James, P.E.; Easton, C. The effects of a single dose of concentrated beetroot juice on performance in trained flatwater kayakers. *Int. J. Sport Nutr. Exerc. Metab.* **2013**, *23*, 498–506. [CrossRef] [PubMed]

37. Peeling, P.; Cox, G.R.; Bullock, N.; Burke, L.M. Beetroot juice improves on-water 500 m time-trial performance, and laboratory-based paddling economy in national and international-level kayak athletes. *Int. J. Sport Nutr. Exerc. Metab.* **2015**, *25*, 278–284. [CrossRef]

38. Bond, H.; Morton, L.; Braakhuis, A.J. Dietary nitrate supplementation improves rowing performance in well-trained rowers. *Int. J. Sport Nutr. Exerc. Metab.* **2012**, *22*, 251–256. [CrossRef]

39. Hoon, M.W.; Jones, A.M.; Johnson, N.A.; Blackwell, J.R.; Broad, E.M.; Lundy, B.; Rice, A.J.; Burke, L.M. The effect of variable doses of inorganic nitrate-rich beetroot juice on simulated 2,000-m rowing performance in trained athletes. *Int. J. Sports Physiol. Perform.* **2014**, *9*, 615–620. [CrossRef]

40. World Medical Association. Declaration of Helsinki—Ethical principles for medical research involving human subjects. *J. Indian Med. Assoc.* **2009**, *107*, 403–405.

41. Govoni, M.; Jansson, E.A.; Weitzberg, E.; Lundberg, J.O. The increase in plasma nitrite after a dietary nitrate load is markedly attenuated by an antibacterial mouthwash. *Nitric Oxide* **2008**, *19*, 333–337. [CrossRef] [PubMed]

42. Borg, G.A. Psychophysical bases of perceived exertion. *Med. Sci. Sports Exerc.* **1982**, *14*, 377–381. [CrossRef] [PubMed]

43. Esen, O.; Nicholas, C.; Morris, M.; Bailey, S.J. No effect of beetroot juice supplementation on 100-m and 200-m swimming performance in moderately trained swimmers. *Int. J. Sports Physiol. Perform.* **2019**. [CrossRef] [PubMed]

44. Bruce, C.R.; Anderson, M.E.; Fraser, S.F.; Stepto, N.K.; Klein, R.; Hopkins, W.G.; Hawley, J.A. Enhancement of 2000-m rowing performance after caffeine ingestion. *Med. Sci. Sports Exerc.* **2000**, *32*, 1958–1963. [CrossRef] [PubMed]

45. Jones, A.M. Influence of dietary nitrate on the physiological determinants of exercise performance: A critical review. *Appl. Physiol. Nutr. Metab.* **2014**, *39*, 1019–1028. [CrossRef]

46. Schantz, P.; Sjoberg, B.; Widebeck, A.M.; Ekblom, B. Skeletal muscle of trained and untrained paraplegics and tetraplegics. *Acta Physiol. Scand.* **1997**, *161*, 31–39. [CrossRef]

47. Vanhatalo, A.; Bailey, S.J.; Blackwell, J.R.; DiMenna, F.J.; Pavey, T.G.; Wilkerson, D.P.; Benjamin, N.; Winyard, P.G.; Jones, A.M. Acute and chronic effects of dietary nitrate supplementation on blood pressure and the physiological responses to moderate-intensity and incremental exercise. *Am. J. Physiol. Regul. Integr. Comp. Physiol.* **2010**, *299*, R1121–R1131. [CrossRef] [PubMed]

48. Cermak, N.M.; Res, P.; Stinkens, R.; Lundberg, J.O.; Gibala, M.J.; van Loon, L.J. No improvement in endurance performance after a single dose of beetroot juice. *Int. J. Sport Nutr. Exerc. Metab.* **2012**, *22*, 470–478. [CrossRef]

49. Jo, E.; Fischer, M.; Auslander, A.T.; Beigarten, A.; Daggy, B.; Hansen, K.; Kessler, L.; Osmond, A.; Wang, H.; Wes, R. The effects of multi-day vs. Single pre-exercise nitrate supplement dosing on simulated cycling time trial performance and skeletal muscle oxygenation. *J. Strength Cond. Res.* **2019**, *33*, 217–224. [CrossRef]

50. Bescos, R.; Ferrer-Roca, V.; Galilea, P.A.; Roig, A.; Drobnic, F.; Sureda, A.; Martorell, M.; Cordova, A.; Tur, J.A.; Pons, A. Sodium nitrate supplementation does not enhance performance of endurance athletes. *Med. Sci. Sports Exerc.* **2012**, *44*, 2400–2409. [CrossRef]

51. Christensen, P.M.; Nyberg, M.; Bangsbo, J. Influence of nitrate supplementation on vo(2) kinetics and endurance of elite cyclists. *Scand. J. Med. Sci. Sports* **2013**, *23*, e21–e31. [CrossRef] [PubMed]

52. Peacock, O.; Tjonna, A.E.; James, P.; Wisloff, U.; Welde, B.; Bohlke, N.; Smith, A.; Stokes, K.; Cook, C.; Sandbakk, O. Dietary nitrate does not enhance running performance in elite cross country skiers. *Med. Sci. Sports Exerc.* **2012**, *44*, 2213–2219. [CrossRef] [PubMed]

53. Lane, S.C.; Hawley, J.A.; Desbrow, B.; Jones, A.M.; Blackwell, J.R.; Ross, M.L.; Zemski, A.J.; Burke, L.M. Single and combined effects of beetroot juice and caffeine supplementation on cycling time trial performance. *Appl. Physiol. Nutr. Metab.* **2014**, *39*, 1050–1057. [CrossRef] [PubMed]

54. Cocks, M.; Shaw, C.S.; Shepherd, S.O.; Fisher, J.P.; Ranasinghe, A.M.; Barker, T.A.; Tipton, K.D.; Wagenmakers, A.J. Sprint interval and endurance training are equally effective in increasing muscle microvascular density and enos content in sedentary males. *J. Physiol.* **2013**, *591*, 641–656. [CrossRef] [PubMed]

55. Totzeck, M.; Hendgen-Cotta, U.B.; Rammos, C.; Frommke, L.M.; Knackstedt, C.; Predel, H.G.; Kelm, M.; Rassaf, T. Higher endogenous nitrite levels are associated with superior exercise capacity in highly trained athletes. *Nitric Oxide* **2012**, *27*, 75–81. [CrossRef] [PubMed]

56. Gollnick, P.D.; Matoba, H. The muscle fiber composition of skeletal muscle as a predictor of athletic success. An overview. *Am. J. Sports Med.* **1984**, *12*, 212–217. [CrossRef] [PubMed]

57. Ade, C.J.; Broxterman, R.M.; Craig, J.C.; Schlup, S.J.; Wilcox, S.L.; Barstow, T.J. Upper body aerobic exercise as a possible predictor of lower body performance. *Aerosp. Med. Hum. Perform.* **2015**, *86*, 599–605. [CrossRef]

58. Bhambhani, Y. Physiology of wheelchair racing in athletes with spinal cord injury. *Sports Med.* **2002**, *32*, 23–51. [CrossRef]

59. Keil, M.; Totosy de Zepetnek, J.O.; Brooke-Wavell, K.; Goosey-Tolfrey, V.L. Measurement precision of body composition variables in elite wheelchair athletes, using dual-energy x-ray absorptiometry. *Eur. J. Sport Sci.* **2016**, *16*, 65–71. [CrossRef]

60. Boorsma, R.K.; Whitfield, J.; Spriet, L.L. Beetroot juice supplementation does not improve performance of elite 1500-m runners. *Med. Sci. Sports Exerc.* **2014**, *46*, 2326–2334. [CrossRef]

61. Wilkerson, D.P.; Hayward, G.M.; Bailey, S.J.; Vanhatalo, A.; Blackwell, J.R.; Jones, A.M. Influence of acute dietary nitrate supplementation on 50 mile time trial performance in well-trained cyclists. *Eur. J. Appl. Physiol.* **2012**, *112*, 4127–4134. [CrossRef] [PubMed]

62. Bourdillon, N.; Fan, J.L.; Uva, B.; Muller, H.; Meyer, P.; Kayser, B. Effect of oral nitrate supplementation on pulmonary hemodynamics during exercise and time trial performance in normoxia and hypoxia: A randomized controlled trial. *Front. Physiol.* **2015**, *6*, 288. [CrossRef] [PubMed]

63. Larsen, F.J.; Weitzberg, E.; Lundberg, J.O.; Ekblom, B. Effects of dietary nitrate on oxygen cost during exercise. *Acta Physiol.* **2007**, *191*, 59–66. [CrossRef] [PubMed]

64. Pawlak-Chaouch, M.; Boissiere, J.; Gamelin, F.X.; Cuvelier, G.; Berthoin, S.; Aucouturier, J. Effect of dietary nitrate supplementation on metabolic rate during rest and exercise in human: A systematic review and a meta-analysis. *Nitric Oxide* **2016**, *53*, 65–76. [CrossRef] [PubMed]

65. Flueck, J.L.; Bogdanova, A.; Mettler, S.; Perret, C. Is beetroot juice more effective than sodium nitrate? The effects of equimolar nitrate dosages of nitrate-rich beetroot juice and sodium nitrate on oxygen consumption during exercise. *Appl. Physiol. Nutr. Metab.* **2016**, *41*, 421–429. [CrossRef] [PubMed]

66. Bescos, R.; Rodriguez, F.A.; Iglesias, X.; Ferrer, M.D.; Iborra, E.; Pons, A. Acute administration of inorganic nitrate reduces vo(2peak) in endurance athletes. *Med. Sci. Sports Exerc.* **2011**, *43*, 1979–1986. [CrossRef] [PubMed]

67. Pimental, N.A.; Sawka, M.N.; Billings, D.S.; Trad, L.A. Physiological responses to prolonged upper-body exercise. *Med. Sci. Sports Exerc.* **1984**, *16*, 360–365. [CrossRef]

68. Pendergast, D.R. Cardiovascular, respiratory, and metabolic responses to upper body exercise. *Med. Sci. Sports Exerc.* **1989**, *21* (Suppl. 5), S121–S125. [CrossRef]

69. Braakhuis, A.J.; Hopkins, W.G. Impact of dietary antioxidants on sport performance: A review. *Sports Med.* **2015**, *45*, 939–955. [CrossRef]

70. Webb, A.J.; Patel, N.; Loukogeorgakis, S.; Okorie, M.; Aboud, Z.; Misra, S.; Rashid, R.; Miall, P.; Deanfield, J.; Benjamin, N.; et al. Acute blood pressure lowering, vasoprotective, and antiplatelet properties of dietary nitrate via bioconversion to nitrite. *Hypertension* **2008**, *51*, 784–790. [CrossRef]

71. Siervo, M.; Lara, J.; Ogbonmwan, I.; Mathers, J.C. Inorganic nitrate and beetroot juice supplementation reduces blood pressure in adults: A systematic review and meta-analysis. *J. Nutr.* **2013**, *143*, 818–826. [CrossRef] [PubMed]

72. Krogh, K.; Mosdal, C.; Laurberg, S. Gastrointestinal and segmental colonic transit times in patients with acute and chronic spinal cord lesions. *Spinal Cord* **2000**, *38*, 615–621. [CrossRef] [PubMed]

© 2019 by the authors. Licensee MDPI, Basel, Switzerland. This article is an open access article distributed under the terms and conditions of the Creative Commons Attribution (CC BY) license (http://creativecommons.org/licenses/by/4.0/).

Article

No Effect of Tart Cherry Juice or Pomegranate Juice on Recovery from Exercise-Induced Muscle Damage in Non-Resistance Trained Men

Kirstie L. Lamb [1], Mayur K. Ranchordas [2], Elizabeth Johnson [3], Jessica Denning [3], Faye Downing [1] and Anthony Lynn [1,*]

[1] Food and Nutrition Group, Sheffield Hallam University, Sheffield S1 1WB, UK
[2] Academy of Sport and Physical Activity, Sheffield Hallam University, Sheffield S10 2BP, UK
[3] Department of Oncology, Human Nutrition, University of Sheffield, Sheffield S10 2TN, UK
*   Correspondence: T.Lynn@shu.ac.uk

Received: 11 June 2019; Accepted: 11 July 2019; Published: 14 July 2019

Abstract: Tart cherry juice (TC) and pomegranate juice (POM) have been demonstrated to reduce symptoms of exercise-induced muscle damage (EIMD), but their effectiveness has not been compared. This randomized, double-blind, parallel study compared the effects of TC and POM on markers of EIMD. Thirty-six non-resistance trained men (age 24.0 (Interquartile Range (IQR) 22.0, 33.0) years, body mass index (BMI) 25.6 ± 4.0 kg·m^{-2}) were randomly allocated to consume 2×250 mL of: TC, POM, or an energy-matched fruit-flavored placebo drink twice daily for nine days. On day 5, participants undertook eccentric exercise of the elbow flexors of their non-dominant arm. Pre-exercise, immediately post-exercise, and at 24 h, 48 h, 72 h and 96 h post-exercise, maximal isometric voluntary contraction (MIVC), delayed onset muscle soreness (DOMS), creatine kinase (CK), and range of motion (ROM) were measured. The exercise protocol induced significant decreases in MIVC ($p < 0.001$; max decrease of 26.8%, 24 h post-exercise) and ROM ($p = 0.001$; max decrease of 6.8%, 72 h post-exercise) and significant increases in CK ($p = 0.03$; max increase 1385 U·L^{-1}, 96 h post-exercise) and DOMS ($p < 0.001$; max increase of 26.9 mm, 48 h post-exercise). However, there were no statistically significant differences between treatment groups (main effect of group $p > 0.05$ or group x time interaction $p > 0.05$). These data suggest that in non-resistance trained men, neither TC nor POM enhance recovery from high-force eccentric exercise of the elbow flexors.

Keywords: exercise-induced muscle damage; recovery; polyphenols; tart cherry; pomegranate

1. Introduction

Eccentric muscle contractions, which are a feature of many sports, can induce substantial muscle damage [1]. Typical symptoms of exercise-induced muscle damage (EIMD) include loss of force, a reduced range of motion (ROM), and the development of muscle soreness. Since these symptoms can impair subsequent performance, there is much interest in strategies that accelerate recovery from EIMD [1–3]. The development of EIMD is not fully understood, but it has been described as a two-phase process [2]. The initial phase is believed to involve structural disruption to sarcomeres and failure of the excitation-coupling process [4,5]. The secondary phase is characterized by a large increase in intracellular Ca^{2+}, which: (i) activates calcium-dependent proteases that degrade muscle proteins [6], and (ii) triggers an inflammatory response [7]; a feature of these is an elevation in reactive oxygen species [8]. Since inflammation and reactive species are involved in the process of EIMD, numerous dietary factors with antioxidant and anti-inflammatory effects have been examined to determine whether they accelerate recovery from EIMD [3].

Polyphenols are one dietary component with antioxidant and anti-inflammatory effects that have recently attracted substantial interest for their potential to promote recovery from EIMD [1]. Studies have explored the effects of isolated polyphenolic supplements such as quercetin [9], and various foods and drinks rich in polyphenols including beetroot [10], bilberry [11], blueberry [12], cocoa flavonoids [13], green tea extract [14], pomegranate [15], and tart cherry [16]. The results of these studies have been somewhat equivocal. This could reflect differences in study design, but also probably reflects differences in the total content and profile of polyphenols that are present in the dietary supplements employed in the various studies. Different plant foods and different cultivars of the same fruit or vegetable can vary substantially in their content of polyphenols [17,18]. Moreover, even the same cultivar of a fruit or vegetable can vary manifold in its content of polyphenols depending on factors such as growing conditions and ripeness [19].

Of the various polyphenol-rich foods studied so far, tart cherries and pomegranate have arguably shown the most potential for promoting recovery from EIMD. Connolly et al. [16] were the first to report that tart cherry juice (TC) could accelerate recovery from EIMD. In untrained young men, the consumption of two × 12 fl oz bottles of a tart cherry/apple juice blend daily for 4 days prior to and four days after a bout of maximal eccentric contractions of the elbow flexors (2 × 20 reps) reduced post-exercise muscle soreness and reduced the loss in isometric strength. Since then, tart cherry juice has been shown to promote the recovery of muscle strength/function after: (i) a marathon [20], (ii) an eccentric knee extensor protocol [21], (iii) cycling [22], and (iv) high-intensity shuttle running [23,24]. When exercise protocols have induced a significant increase in inflammation, tart cherry has typically attenuated the rise in one or more serum markers of inflammation [20,22,23,25,26]. Positive effects on markers of oxidative stress [20,21,25] and DOMS [26,27] have also sometimes been observed, whereas tart cherry has demonstrated little ability to lower serum markers of muscle damage, such as creatine kinase [20,21]. Although studies on tart cherry have typically reported beneficial effects on one or more markers of EIMD, a couple of studies have failed to report any beneficial effects [28,29]. McCormick et al. [28] found no effect of supplementation with tart cherry juice on DOMS and markers of inflammation and oxidative stress in male water polo players who undertook a simulation designed to mimic the demands of a game of water polo. Beals et al. [29] failed to find a beneficial effect of supplementation with a beverage containing freeze-dried tart cherry powder on the recovery of muscle strength or DOMS in recreationally active men and women who completed a fatiguing eccentric knee flexor protocol. However, in that study the exercise protocol failed to induce measurable muscle damage.

To our knowledge, four studies have reported on the ability of pomegranate to accelerate recovery from EIMD. In an initial cross-over study of 16 recreationally active young men, Trombold et al. (2010) reported that a daily supplement of pomegranate-extract taken for nine days was effective at enhancing the recovery of isometric strength after a bout of maximal eccentric contractions of elbow flexors (2 × 20 reps) undertaken on day 5 [15]. However, they found no difference between pomegranate extract and placebo for reported muscle soreness or serum markers of muscle damage and inflammation. In a subsequent study of resistance trained men, Trombold et al. (2011) found that a daily supplement of pomegranate juice (POM; taken for eight days before and seven days after a muscle damage protocol) enhanced the recovery of isometric strength of the elbow flexors. However, POM was ineffective at promoting the recovery of isometric strength of the knee extensors [30]. Then, the same group conducted a dose–response study in recreationally active young men, and found that one or two daily drinks of a POM concentrate were similarly effective at promoting the recovery of isometric strength of knee extensors following downhill running and of elbow flexors after 40 eccentric contractions [31]. In a small study of nine elite weightlifters, Ammar et al. (2016) reported that supplementation with POM for two days before a session of three Olympic weightlifting exercises seemed to have some beneficial effects on serum markers of muscle damage and reported muscle soreness measured acutely post-exercise and 48 h later [32]. However, the data are difficult to interpret, because all the participants consumed the placebo drink first and the two exercise sessions were separated by only 48 h. So, it seems that during the 48-h recovery period after the placebo drink, participants were consuming POM.

Collectively, these studies indicate that TC and POM seem to accelerate post-exercise recovery; however, it is unclear whether they are equally as effective. Although POM and TC are both rich sources of anthocyanins, especially cyanidin and its derivatives [33], their profile of other polyphenols differs. For example, POM is particularly rich in ellagitannins, especially punicalagin and its isomers [34]. Whereas tart cherry contains flavan-3-ols such as catechin and epicatechins [35], which are absent from POM [34]. Due to these and other differences in composition, the potency of TC and POM to enhance recovery may differ, but no studies have directly addressed this possibility. This is surprising, because it is of practical importance to know whether the efficacy of TC and POM varies, so that optimal advice can be provided to athletes. Therefore, the purpose of this study was to compare the efficacy of TC and POM to promote recovery from EIMD.

2. Materials and Methods

2.1. Participants

Participants were recruited via word of mouth, poster advertisements in Sheffield Hallam University and the University of Sheffield, and recruitment emails to staff and students. Thirty-six non-resistance trained men (age 24.0 (interquartile range (IQR) 22.0, 33.0) years, body mass index (BMI) 25.6 ± 4.0 kg·m^{-2}) volunteered to participate. Inclusion criteria included aged 18–65 years, non-smokers, no upper-body strength training in the last three months, injury-free, and not presently taking anti-inflammatories or antioxidants. The study was conducted in accordance with the Declaration of Helsinki. Ethical approval was received from Sheffield Hallam University Research Ethics Committee (SBS-244). Following written and verbal study briefings, participants provided written informed consent.

2.2. Study Design

This was a nine-day, randomized, double-blind, placebo-controlled, parallel study with three experimental treatment groups. Participants were block randomized to receive either: (i) pomegranate juice (POM), (ii) tart cherry juice (TC), or (iii) a fruit-flavored placebo drink (PLA). The order of the blocks (sizes 3 and 6) was determined using a computerized random number generator (www.random.org) by a researcher not involved in participant recruitment. At the start of the study, there was a familiarization session during which baseline measures of body composition were collected and participants practiced the protocols for inducing muscle damage and measuring strength (maximal voluntary isometric contractions (MIVC)). Participants were instructed to use minimal effort to prevent unintentional muscle damage and to ensure they did not induce a protective repeated bout effect [36]. Participants consumed 2 × 250 mL servings per day (one in the morning and one in the evening) of their allotted drink for nine days. On day 5, participants attended the research laboratory after consuming their normal breakfast and their morning test drink. Pre-exercise measures of MIVC, range of motion (ROM), delayed onset muscle soreness (DOMS), and blood creatine kinase (CK) were collected; then, participants completed a muscle damage protocol consisting of 5 × 10 sets of unilateral eccentric elbow flexions using their non-dominant arm. Participants consumed an extra 250 mL of their allotted drink and consumed a standardized meal immediately post exercise. The consumption of other food or drink, except water, was avoided for 3 h post-exercise. Further measures of MIVC, ROM, DOMS, and CK were collected immediately post-exercise and daily for four days (Figure 1). For the duration of the study, participants were asked to avoid: strenuous physical activity, other methods for promoting recovery from EIMD (including non-steroidal anti-inflammatory drugs, foods rich in polyphenols, and nutritional supplements). The physical activity diaries were used to check compliance with the instructions to avoid strenuous exercise, and the food diary was used to calculate the average number of portions of polyphenol-rich foods consumed daily across the duration of the study.

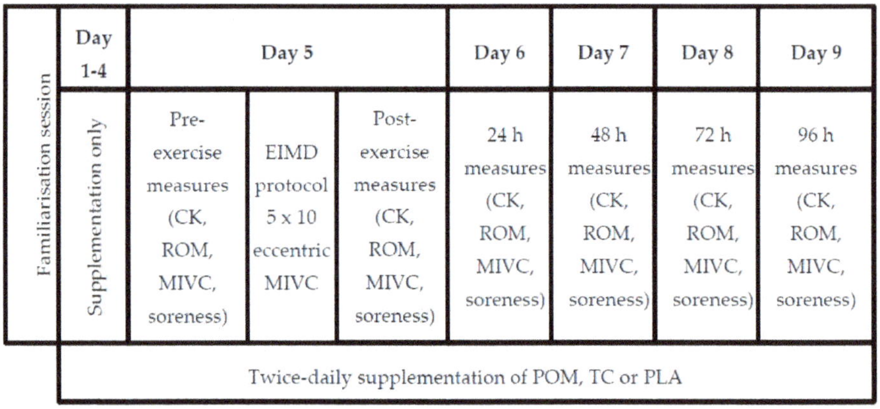

Familiarisation session	Supplementation only	Day 1-4	Day 5			Day 6	Day 7	Day 8	Day 9
			Pre-exercise measures (CK, ROM, MIVC, soreness)	EIMD protocol 5 x 10 eccentric MIVC	Post-exercise measures (CK, ROM, MIVC, soreness)	24 h measures (CK, ROM, MIVC, soreness)	48 h measures (CK, ROM, MIVC, soreness)	72 h measures (CK, ROM, MIVC, soreness)	96 h measures (CK, ROM, MIVC, soreness)
Twice-daily supplementation of POM, TC or PLA									

Figure 1. Timeline of study protocol from familiarization to 96-h post-exercise. POM = pomegranate; TC = tart cherry; PLA = placebo; EIMD = exercise-induced muscle damage; CK = creatine kinase; ROM = range of motion; MIVC = maximal isometric voluntary contraction.

2.3. Fruit Juices, Placebo Beverage and Post-Exercise Meal

Each serving of POM (POM Wonderful LLC, Los Angeles, CA, USA) consisted of 250 mL of undiluted juice, whilst a serving of TC (CherryActive Ltd., Hanworth, UK) contained 30 mL of concentrate diluted with 220 mL of water. The energy content of the TC was adjusted to match the energy content of POM by the addition of maltodextrin powder (My Protein Ltd., Northwhich, UK). The PLA beverage consisted of a blackcurrant-flavored maltodextrin sports drink (Science in Sport Energy Go, SiS Ltd., Nelson, Lancashire, UK), which was prepared to match the energy content of the POM and TC drinks. The serving sizes of POM and TC were based on standard serving sizes and were congruent with volumes shown to be effective in previous studies [21,25,30]. A technician not involved in data collection prepared all the drinks in brown bottles to mask the color from the investigators. At recruitment, participants were informed that the aim of the study was to explore the effect of different fruit-flavored drinks on EIMD, so participants were unaware of which drink was the placebo drink. Fresh drinks were prepared and delivered to participants every two days. Participants were advised to store the drinks in their refrigerators to minimize the degradation of bioactive compounds. Empty bottles were counted to assess compliance.

The total phenolic content of the test drinks was analyzed using the Folin–Ciocalteu method, and results were expressed as gallic acid equivalents [37]. The pH differential method was employed to determine the total content of anthocyanins present in each drink, and results were expressed as cyanidin-3-glucoside equivalents [38]. For both analyses, three bottles of each juice were analyzed in triplicate.

The post-exercise meal consisted of two white-bread rolls (8 g protein·100 g^{-1}; Kingsmeal, Maidenhead, UK) containing chicken breast (30.7 g protein·100 g^{-1}; Sainsbury's Supermarkets Ltd., London, UK) or mature cheddar cheese for vegetarians (25.4 g protein·100 g^{-1}; Cathedral City, Dairy Crest, Surrey, UK) with soya spread (Pure, Kerry Foods, Warrington, UK). Sandwiches were prepared to provide a total of 0.3 g protein·kg BW^{-1}.

2.4. Anthropometric Measures

Height was measured to the nearest 0.1 cm using a wall-mounted stadiometer (Harpenden, Holtain Model 602VR, UK). Weight and body composition were measured using an InBody 720 bioelectrical impedance scale (Biospace Co., Ltd., Leicester, UK).

2.5. Measurement of Maximal Isometric Voluntary Contractions (MIVC)

At the familiarization session, the Biodex System3 isokinetic dynamometer (Biodex Medical Systems, Inc., Shirley, NY, USA) was set up according to manufacturer settings [39] for elbow flexion with the dynamometer orientated at a 30° angle and the position of the seat adjusted for each individual. Settings were recorded to ensure that participants were in the same position for each test. To measure maximal isometric strength, participants completed 5 × 3 s MIVC at a 90° angle with a 10-s rest interval between repetitions. The mean of the five repetitions was recorded.

2.6. Eccentric (EIMD) Protocol

A two-minute rest period was provided between measuring MIVC and the EIMD protocol. The eccentric exercise protocol consisted of 50 (five sets of 10) maximal voluntary eccentric contractions of the non-dominant elbow flexors on the isokinetic dynamometer. With the wrist supinated, the elbow joint was extended forcibly through 90°, from 90° to 180°, at an angular velocity of $60°·s^{-1}$. To ensure maximal resistance throughout each extension, verbal encouragement was given. The eccentric contraction was followed by a 12-s rest period during which the arm was returned to the flexed 90° position at $10°·s^{-1}$. A two-minute rest period was provided between sets.

2.7. Measurement of Creatine Kinase

Blood samples were obtained via finger prick capillary sampling. A 30-µL sample of capillary blood was applied to a CK test strip (Reflotron®, HaB Direct, Warwickshire, UK). Blood samples were analyzed for CK concentration using a Roche Reflotron® Plus analyser (HaB Direct, Warwickshire, UK). When CK values exceeded the measurement range of the analyzer, samples were diluted with a control blood sample of a known low CK concentration and reanalyzed. To calculate the CK concentration of these samples, the reanalyzed values were multiplied by the dilution factor, and the concentration of CK in the control was subtracted. A hyper-response was defined as a CK measurement above $2000 \ U·L^{-1}$ at any time-point following the EIMD protocol [36,40].

2.8. Measurement of Range of Motion and Elbow Flexion Soreness

Maximal elbow extension and flexion of the non-dominant arm were measured using a 12-cm goniometer (True Angle Goniometer, Bodycare, Warwickshire, UK), and the total ROM was calculated. With the participant standing, the goniometer axis was placed on the pre-marked lateral epicondyle of the elbow. One arm of the goniometer was lined up with the humerus and acromion process at a pre-marked point. The other was placed in line with the distal radius, parallel to the forearm. To maintain consistency between measures, these points were marked with permanent ink at the beginning of study.

Elbow flexor soreness was rated on a 100-mm visual analogue scale (VAS) that was anchored with "no pain" at the left end of the scale (score of 0) and "unbearable pain" at the right end (score of 100). Participants were asked to complete two bicep curls with their non-dominant arm using a 1-kg dumbell, and then indicate on the scale the intensity of soreness experienced.

2.9. Statistical Analysis

Descriptive characteristics at baseline were compared across treatment groups using one-way analysis of variance (ANOVA) tests or Kruskall–Wallis tests as appropriate. The effect of the intervention on MIVC, ROM, CK, and DOMS was assessed using mixed model ANOVAs (time as the within-group factor and treatment as the between group factor). Post-exercise data points for MIVC and ROM data were converted to the percentage of their respective baseline values before data analysis. When data failed Mauchley's test of sphericity, the degrees of freedom where corrected using Greenhouse–Geisser estimates. Statistically significant main effects of time were explored using Bonferroni post-hoc tests. The MIVC, ROM, CK, and DOMS data were analyzed with and without the participants that were classified as hyper-responders based on a post-exercise CK value >2000 U·L^{-1} All the analyses were conducted on IBM SPSS statistics version 24.0 (SPSS Inc., Chicago, IL, USA), and $p < 0.05$ was set as the critical level of significance.

3. Results

3.1. Analysis of POM and TC

Each serving of POM contained nearly three times more total phenolics than a serving of TC (878.9 ± 92.7 mg versus 294.7 ± 14.9 mg) and over six times more monomeric anthocyanins (49.4 ± 2.0 mg versus 7.7 ± 0.3 mg).

3.2. Participants

The characteristics of participants in each diet group are displayed in Table 1. There were no statistically significant differences between groups for age ($p = 0.97$), BMI ($p = 0.82$), lean muscle mass ($p = 0.86$), baseline MIVC strength ($p = 0.79$), or baseline CK level ($p = 0.63$). All 36 participants completed the study. Of these, six participants (two TC, two POM, and two PLA) were classified as hyper-responders based on a post-exercise CK > 2000 U·L^{-1}. Data is missing for one participant at 72 h (PLA) owing to illness. One CK value at 48 h (POM) and two MIVC values at 72 h (TC and POM) are missing because of technical difficulties.

Table 1. Participant characteristics ($n = 36$).

	Tart Cherry	**Pomegranate**	**Placebo**
N	12	12	12
Age (years)	24.0 (IQR 22.0, 33.0)	24.0 (IQR 21.0, 32.5)	24.0 (IQR 22.5, 32.0)
BMI (kg·m^{-2})	26.2 (± 3.9)	25.5 (± 3.9)	25.2 (± 4.5)
LMM (kg)	36.0 (± 5.2)	35.0 (± 6.1)	34.8 (± 6.2)
Baseline strength (N·cm^{-2})	41.8 (IQR 34.3, 53.2)	42.7 (IQR 33.6, 52.2)	44.1 (IQR 37.2, 51.5)
Baseline CK level (U·L^{-1})	148.9 (± 60.1)	150.3 (± 44.6)	130.9 (± 57.7)

Values are mean (± standard deviation (SD)) or median and interquartile range (IQR); BMI = body mass index; LMM = lean muscle mass; CK = creatine kinase.

3.3. Compliance with Protocol

Participants reported consuming all their test drinks, but two participants reported that on one test day, they consumed both drinks in the evening after forgetting to consume a drink in the morning. There were no statistically significant differences in the portions of polyphenol-rich foods consumed by each diet group during the study (TC = 2.81 ± 1.19; POM =2.93 ± 2.23; PLA = 3. 97 ± 2.67, $p = 0.44$).

3.4. MIVC

The muscle damage (MD) protocol induced a statistically significant reduction in MIVC ($F_{(2.9, 85.5)}$ = 23.30, $p < 0.001$), which reached a nadir at 24 h post-exercise (max mean decrement 26.8%, Figure 2). Post-hoc Bonferroni analysis revealed statistically significant differences in strength from pre-exercise to: immediately post, 24 h post, 48 h post, and 72 h post-exercise ($p < 0.05$), but not 96 h post-exercise ($p = 0.08$). The reduction in MIVC over the complete post-exercise period was fairly similar in all groups (($F_{(2, 30)} = 0.27$, $p = 0.77$; mean decrement TC = 14.6%, POM = 19.4%, and PLA = 17.3%), and the interaction between diet-group and time was not statistically significant ($F_{(5.7, 85.5)} = 0.33$, $p = 0.91$).

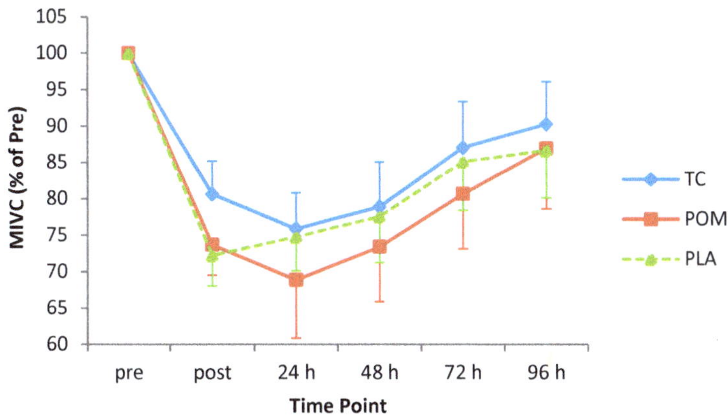

Figure 2. Change in maximal isometric elbow flexion strength following eccentric exercise (expressed as a percentage of pre-exercise strength). Values are an average of five maximal contractions and are displayed as mean (± SEM). Time by treatment, $p = 0.91$. MIVC = maximal isometric voluntary contraction; TC = tart cherry; POM = pomegranate; PLA = placebo, $n = 11$ for each group.

3.5. Elbow Flexor Soreness

There was a statistically significant increase in reported muscle soreness, which peaked 48 h (mean increase 26.9 mm) after the MD protocol ($F_{(3.2, 102.2)} = 19.22$, $p < 0.001$) (Figure 3). Post-hoc analysis showed that soreness scores were significantly different from baseline at all time-points after the MD protocol ($p < 0.05$). However, there was no statistically significant main effect of diet group ($F_{(2, 32)} = 1.19$, $p = 0.32$; mean increase in soreness: TC = 17.2 mm; POM = 17.4 mm; PLA = 11.7 mm) or diet group by time interaction ($F_{(6.4, 102.2)} = 1.81$, $p = 0.10$).

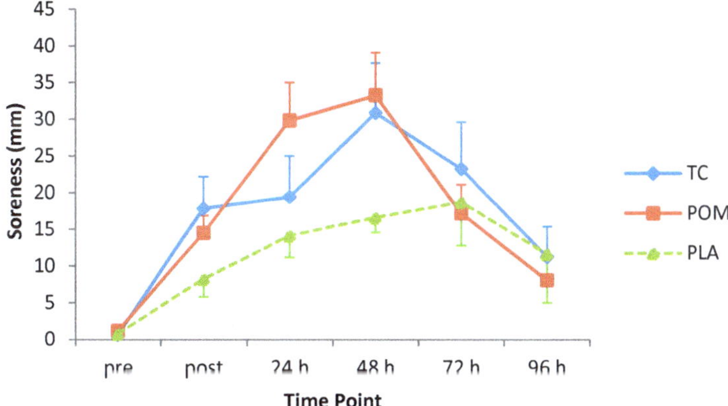

Figure 3. Change in soreness following eccentric exercise (reported on a scale of 0–100 mm). Values are displayed as mean (± SEM). Time by treatment, $p = 0.10$. TC = tart cherry; POM = pomegranate; PLA = placebo, $n = 12$ for TC, and POM and $n = 11$ for PLA.

3.6. ROM

There was a significant decrease in the ROM of the elbow flexors after the MD protocol (main effect of time: $F_{(1,6)} = 11.60$, $p < 0.001$; max mean decrement 6.8% at 48 h) (Figure 4). Post-hoc analysis showed that ROM was significantly lower at all post-exercise time-points in comparison to pre-exercise ($p < 0.05$). ROM did not differ significantly between diet groups ($F_{(1,3)} = 1.23$, $p = 0.31$; mean reduction in ROM: TC = 4.6%; POM = 6.1%; PLA = 3.5%), and there was no statistically significant interaction between diet group and time ($F_{(1,6;1,3)}$ = 0.62, p = 0.71).

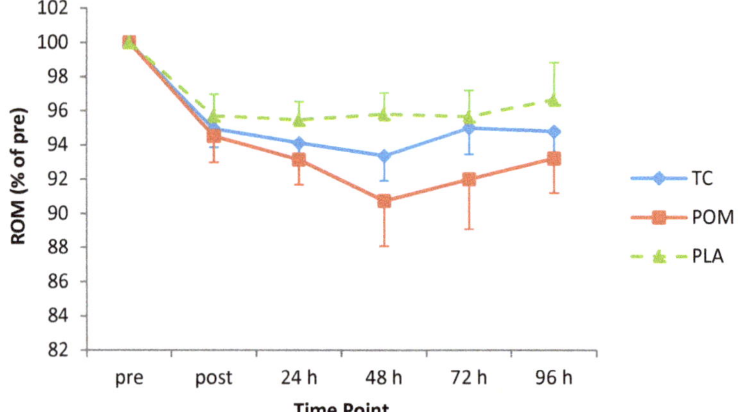

Figure 4. Range of motion (ROM) following eccentric exercise (expressed as a percentage of pre-exercise ROM). Values are displayed as mean (± SEM). Time by treatment, p = 0.71. TC = tart cherry; POM = pomegranate; PLA = placebo, n = 12 for TC and POM; and n = 11 for PLA.

3.7. Creatine Kinase Concentration

Whole-blood CK increased after the MD protocol with peak values recorded 96 h post-exercise (1383 $U \cdot L^{-1}$; main effect of time: $F_{(1.1, 34.1)}$ = 5.22; p = 0.03; Figure 5). There was no statistically significant main effect of diet group ($F_{(2, 31)}$ = 0.05, p = 0.995; mean increase TC = 478.3 $U \cdot L^{-1}$; POM = 502.9 $U \cdot L^{-1}$; PLA = 471.6 $U \cdot L^{-1}$) or statistically significant interaction of diet group by time ($F_{(2.2, 34.1)}$ = 0.09, p = 0.93).

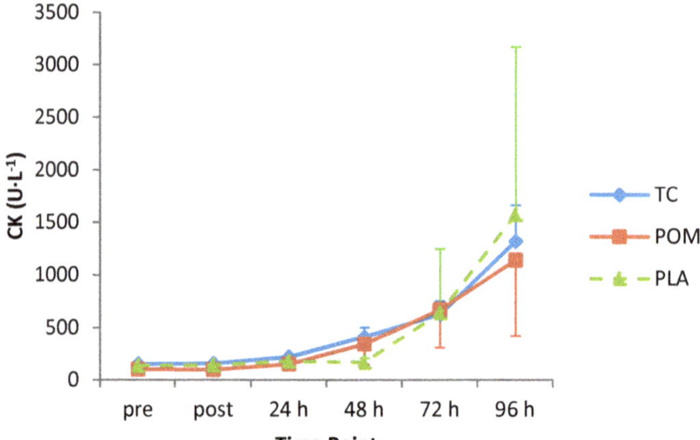

Figure 5. Creatine kinase (CK) following eccentric exercise. Values are displayed as mean (± SEM). Time by treatment, p = 0.93. TC = tart cherry; POM = pomegranate; PLA = placebo; n = 12 for TC, and n = 11 for POM and PLA.

3.8. Exploratory Analysis with Hyper-Responders Removed

The dataset contained two participants in each group that were identified as hyper-responders on the basis that the concentration of CK in their blood rose above 2000 $U \cdot L^{-1}$ in the post-exercise period.

An exploratory analysis omitting the data of these six hyper-responders did not reveal any statistically significant main effects of diet or diet × time interactions for any of the outcome markers (Table 2).

Table 2. Analysis of outcome markers with hyper-responders removed.

Outcome Marker	Main Effect of Diet Group	Main Effect of Time	Diet × Time Interaction
MIVC	$p = 0.78$	$p < 0.001$	$p = 0.97$
DOMS	$p = 0.28$	$p < 0.001$	$p = 0.26$
ROM	$p = 0.71$	$p < 0.001$	$p = 0.90$
CK	$p = 0.87$	$p = 0.01$	$p = 0.35$

MIVC, maximal isometric voluntary contraction; DOMS, delayed onset of muscle soreness; ROM, range of motion; CK, creatine kinase.

4. Discussion

This is the first study to directly compare the effects of TC and POM on recovery from EIMD. We hypothesized that both juices would accelerate recovery relative to an energy-matched placebo drink. We also speculated that the potency of the juices might vary because of differences in their content of potentially anti-inflammatory and anti-oxidative phenolic compounds. However, we found that neither juice accelerated recovery as determined by changes in MIVC, DOMS, ROM, and CK, relative to an energy-matched placebo drink. Also, despite a compositional analysis revealing that a serving of POM contained almost three times more total phenolics and six times more monomeric anthocyanins than a serving of TC, these differences did not translate to any statistically significant differences between the effects of the two juices.

The lack of effect of TC and POM on recovery of muscle strength was unexpected, because both drinks have consistently been reported to improve the recovery of MIVC [15,20,21,23,25,30,31]. To our knowledge, only one previous study has failed to report a beneficial effect of TC on recovery of strength when using a similar mode of resistance exercise [29]. However, in that study, the exercise protocol failed to invoke a substantial reduction in MIVC during the post-exercise period, thus making it impossible to detect a faster recovery of strength. In contrast, our MD protocol caused a statistically significant reduction in MIVC of a magnitude and time course that is consistent with studies that have reported beneficial effects of TC [20,21,23,25] or POM [15,30,31]. The lack of effect of TC and POM on MIVC was mirrored by a lack of effect on DOMS. Previous studies reporting the effect of TC and POM on DOMS have been inconsistent with approximately half of studies reporting a benefit [16,23,26,27,30,32,41], and half finding no treatment effects [15,20–22,24,29,31]. However, no study has reported a reduction in DOMS in the absence of a faster recovery of MIVC, so it is unsurprising that we found no effect of TC and POM on DOMS.

Our MD protocol induced a significant elevation in blood CK, but there was no significant diet group or diet group by time interactions. The lack of effect of TC and POM on CK is consistent with existing literature [20,22,24,26,29]. CK levels started to rise substantially at 72 h post-exercise, and were at their highest at the 96-h time-point at the end of the study. This pattern of response is similar to other studies using a 50-repetition eccentric elbow flexion protocol in non-resistance trained men [15,42,43]. Consistent with the literature [44], we observed large inter-individual variations in the CK response to the MD protocol. This highly variable response between individuals and the fact that the magnitude of CK response does not necessarily correlate with the extent of muscle damage means that measuring CK responses may be of limited utility as a quantitative marker of muscle damage [45].

There was an approximate 7% decrease in relaxed elbow ROM in response to the MD protocol, which did not completely return to normal at 96 h. However, there was no difference in response between diet groups. To our knowledge, relaxed elbow ROM has only been measured in two studies investigating the effects of supplementation with TC [16,29], and no studies investigating the effect of supplementation with POM. In agreement with our study, TC did not alter the change in elbow ROM in previous studies [16,29].

A recent review on fruit-derived polyphenols and exercise reported that studies demonstrating favorable effects of TC on the recovery of muscle strength/function after a MD protocol typically supplied ≥1200 mg of total phenolics per day, and that two studies that failed to report a benefit may have done so because they supplied insufficient doses of TC-derived phenolics, which were 480 mg [27] and 733 mg [29] respectively [1]. The TC used in the current study contained approximately 300 mg of total phenolics per serving, resulting in a daily intake of approximately 600 mg. This might partly explain the lack of effect that we observed. The ability of TC to accelerate recovery and exert beneficial health effects is commonly attributed to its content of anthocyanins [46]. In some publications, the brand of TC used in the current study has been reported to contain 273 mg of total anthocyanins per 30-mL serving of concentrate [21,22]. However, using the pH differential method [38] to measure total (non-degraded) anthocyanins, we could only detect 7 mg of anthocyanins per 30-mL serving of concentrate. If intact anthocyanins make a substantial contribution to the protective effects of TC, then the low levels present in the TC could have contributed to its lack of effect. Anthocyanins are not particularly stable during processing and storage [47,48], so it is possible that the TC contained substantial quantities of anthocyanin derivatives that were not detected by our assay [38]. However, a low content of total phenolics and/or anthocyanins does not explain why the POM exerted no beneficial effects. The POM used in the current study provided slightly greater amounts of total phenolics (approximately 900 mg per serving) than reported in previous studies [15,30,31] and substantial amounts of anthocyanins (approximately 50 mg per serving).

Previous studies that have investigated the effect of TC and POM to accelerate recovery from eccentric resistance exercise have most commonly used a crossover study design assigning contralateral limbs to each treatment. We used a parallel design, because we had three treatments. A parallel design has the advantage of removing the contralateral repeated bout effect that may influence the extent of damage in the limb tested second in a crossover study [49–51]. However, a disadvantage is that inter-individual variation in a parallel design dictates that large samples sizes are required to detect small and moderate effects. It is possible that we may have detected some treatment effects if we had used a larger sample size. However, others have reported significant effects of POM [31] and TC [20,24] on muscle strength or function using a parallel design with similar group sizes of between 10–15 participants, although in the case of TC with different types of damage protocols i.e., marathon running [20] and repeated sprints [24]. Moreover, in our study, there were no weak trends toward any treatment effects. Interrogation of our dataset revealed that two participants in each diet group exhibited a hyper-response to the MD protocol classified on the basis of a post-exercise CK > 2000 U·L^{-1}. We conducted an exploratory analysis of the dataset with these 'hyper-responders' removed. This failed to reveal any statistically significant treatment effects and for most measures shifted the treatment outcomes further toward a null effect.

This study has several limitations. It has been proposed that TC and POM may accelerate recovery from EIMD because of their antioxidant and anti-inflammatory effects [1]. In the present study, we did not measure any markers of oxidative stress or inflammation. It may have been informative to determine whether TC or POM reduced markers of oxidative stress such as F2-isoprostanes and protein carbonyls, with the caveat that changes measured in blood do not necessarily correlate with changes within muscles [52]. It would have been useful to document a reduction in exercise-induced inflammation after supplementation with TC and POM. However, common serum markers of inflammation such as c-reactive protein and interleukin-6 are typically not markedly elevated by muscle damage protocols based on single limbs [1]. We only measured one blood marker of muscle damage: creatine kinase. It is possible that measuring a larger number of markers may have revealed a treatment effect. It may have been illuminating to measure the appearance of polyphenolic metabolites in our participants after they consumed POM or TC. However, this would have required the repeated collection of blood and/or urine samples over a 24-h period. This was not feasible within the context of the current study design. Despite our instructions to avoid foods rich in polyphenols participants consumed some foods rich in polyphenols during the intervention. Although intake was similar in all three diet groups, it is possible that we may have detected an effect of POM or TC if participants had avoided all the foods containing polyphenols. However, this is almost impossible to achieve in a free-living study, because of the ubiquitous nature of foods and beverages rich in polyphenols. We compared serving sizes of TC and POM that have previously been reported to accelerate recovery from EIMD [21,25,30]. Our in-house analyses revealed that a typical serving of POM provided approximately three times greater amounts of total phenolics and approximately six times greater amounts of total anthocyanins than a typical serving of tart cherry concentrate. Matching the two drinks for total phenolic content may have provided a more equitable comparison, but it would have required participants to consume either six servings of TC concentrate per day or only a third of the volume of POM that has previously been reported to accelerate recovery [30].

5. Conclusions

The MD protocol used in the present study induced a substantial reduction in MIVC, a modest reduction in relaxed elbow ROM, a substantial elevation in blood CK, and a modest increase in reported muscle soreness. The TC and POM provided markedly different quantities of total phenolics and intact monomeric anthocyanins. However, there were no statistically significant differences in the effects of the two juices, and both juices failed to exert any statistically significant effect on the measured indices. The current results indicate that in non-resistance trained men neither TC nor POM enhance recovery from high force eccentric contractions of the elbow flexors.

Author Contributions: Conceptualization: A.L. & M.K.R.; Methodology: A.L., M.K.R., K.L.L. & E.J.; investigation: K.L.L., E.J., J.D. & F.D.; Data analysis: K.L.L. & A.L.; Writing original draft preparation: K.L.L., A.L., E.J. and M.K.R.; Writing, review and editing: K.L.L., A.L., M.K.R., J.D. & F.D. Project administration: A.L. & M.K.R.

Funding: This study was funded by Sheffield Hallam University and the University of Sheffield.

Acknowledgments: We would like to thank our participants for taking part in the study.

Conflicts of Interest: A.L. is a recipient of a grant from the Cherry Marketing Institute to study the effects of tart cherry on gout. K.L.L. is currently working on the gout study. The other authors declare no conflicts of interest.

References

1. Bowtell, J.; Kelly, V. Fruit-Derived polyphenol supplementation for athlete recovery and performance. *Sports Med.* **2019**, *49*, 3–23. [CrossRef] [PubMed]
2. Howatson, G.; van Someren, K.A. The prevention and treatment of exercise-induced muscle damage. *Sports Med.* **2008**, *38*, 483–503. [CrossRef] [PubMed]
3. Owens, D.J.; Twist, C.; Cobley, J.N.; Howatson, G.; Close, G.L. Exercise-induced muscle damage: What is it, what causes it and what are the nutritional solutions? *Eur. J. Sport Sci.* **2019**, *19*, 71–85. [CrossRef] [PubMed]
4. Proske, U.; Morgan, D.L. Muscle damage from eccentric exercise: Mechanism, mechanical signs, adaptation and clinical applications. *J. Physiol.* **2001**, *537*, 333–345. [CrossRef] [PubMed]
5. Hyldahl, R.D.; Hubal, M.J. Lengthening our perspective: Morphological, cellular, and molecular responses to eccentric exercise. *Muscle Nerve* **2014**, *49*, 155–170. [CrossRef] [PubMed]
6. Gissel, H.; Clausen, T. Excitation-induced Ca2+ influx and skeletal muscle cell damage. *Acta Physiol. Scand.* **2001**, *171*, 327–334. [CrossRef] [PubMed]
7. Butterfield, T.A.; Best, T.M.; Merrick, M.A. The dual roles of neutrophils and macrophages in inflammation: A critical balance between tissue damage and repair. *J. Athl. Train.* **2006**, *41*, 457–465. [PubMed]
8. Nguyen, H.X.; Tidball, J.G. Interactions between neutrophils and macrophages promote macrophage killing of rat muscle cells in vitro. *J. Physiol.* **2003**, *547*, 125–132. [CrossRef]
9. Bazzucchi, I.; Patrizio, F.; Ceci, R.; Duranti, G.; Sgrò, P.; Sabatini, S.; Di Luigi, L.; Sacchetti, M.; Felici, F. The effects of quercetin supplementation on eccentric exercise-induced muscle damage. *Nutrients* **2019**, *11*, 205. [CrossRef]
10. Clifford, T.; Bell, O.; West, D.J.; Howatson, G.; Stevenson, E.J.; Clifford, T. The effects of beetroot juice supplementation on indices of muscle damage following eccentric exercise. *Eur. J. Appl. Physiol.* **2016**, *116*, 353–362. [CrossRef]

11. Lynn, A.; Garner, S.; Nelson, N.; Simper, T.N.; Hall, A.C.; Ranchordas, M.K. Effect of bilberry juice on indices of muscle damage and inflammation in runners completing a half-marathon: A randomised, placebo-controlled trial. *J. Int. Soc. Sports Nutr.* **2018**, *15*, 22. [CrossRef] [PubMed]

12. McLeay, Y.; Barnes, M.J.; Mundel, T.; Hurst, S.M.; Hurst, R.D.; Stannard, S.R. Effect of New Zealand blueberry consumption on recovery from eccentric exercise-induced muscle damage. *J. Int. Soc. Sports Nutr.* **2012**, *9*, 19. [CrossRef] [PubMed]

13. Peschek, K.; Pritchett, R.; Bergman, E.; Pritchett, K. The effects of acute post exercise consumption of two cocoa-based beverages with varying flavanol content on indices of muscle recovery following downhill treadmill running. *Nutrients* **2013**, *6*, 50–62. [CrossRef] [PubMed]

14. Da Silva, W.; Machado, Á.S.; Souza, M.A.; Mello-Carpes, P.B.; Carpes, F.P. Effect of green tea extract supplementation on exercise-induced delayed onset muscle soreness and muscular damage. *Physiol. Behav.* **2018**, *194*, 77–82. [CrossRef] [PubMed]

15. Trombold, J.R.; Barnes, J.N.; Critchley, L.; Coyle, E.F. Ellagitannin consumption improves strength recovery 2-3 d after eccentric exercise. *Med. Sci. Sports Exerc.* **2010**, *42*, 493–498. [CrossRef]

16. Connolly, D.A.J.; McHugh, M.P.; Padilla-Zakour, O.I.; Carlson, L.; Sayers, S.P. Efficacy of a tart cherry juice blend in preventing the symptoms of muscle damage. *Br. J. Sports Med.* **2006**, *40*, 679–683. [CrossRef] [PubMed]

17. Seeram, N.P.; Aviram, M.; Zhang, Y.; Henning, S.M.; Feng, L.; Dreher, M.; Heber, D. Comparison of antioxidant potency of commonly consumed polyphenol-rich beverages in the United States. *J. Agric. Food Chem.* **2008**, *56*, 1415–1422. [CrossRef]

18. Bassiri-Jahromi, S.; Doostkam, A. Comparative evaluation of bioactive compounds of various cultivars of pomegranate (Punica granatum) in different world regions. *AIMS Agric. Food* **2018**, *4*, 41–55. [CrossRef]

19. Vallejo, F.; García-Viguera, C.; Tomás-Barberán, F.A. Changes in broccoli (Brassica oleracea L. Var. italica) health-promoting compounds with inflorescence development. *J. Agric. Food Chem.* **2003**, *51*, 3776–3782. [CrossRef]

20. Howatson, G.; McHugh, M.P.; Hill, J.A.; Brouner, J.; Jewell, A.P.; Van Someren, K.A.; Shave, R.E.; Howatson, S.A. Influence of tart cherry juice on indices of recovery following marathon running. *Scand. J. Med. Sci. Sport* **2009**, *20*, 843–852. [CrossRef]

21. Bowtell, J.L.; Sumners, D.P.; Dyer, A.; Fox, P.; Mileva, K.N. Montmorency cherry juice reduces muscle damage caused by intensive strength exercise. *Med. Sci. Sports Exerc.* **2011**, *43*, 1544–1551. [CrossRef] [PubMed]

22. Bell, P.G.; Walshe, I.H.; Davison, G.W.; Stevenson, E.J.; Howatson, G. Recovery facilitation with Montmorency cherries following high-intensity, metabolically challenging exercise. *Appl. Physiol. Nutr. Metab.* **2014**, *40*, 414–423. [CrossRef] [PubMed]

23. Bell, P.G.; Stevenson, E.; Davison, G.W.; Howatson, G. The effects of montmorency tart cherry concentrate supplementation on recovery following prolonged, intermittent exercise. *Nutrients* **2016**, *8*, 441. [CrossRef] [PubMed]

24. Brown, M.A.; Stevenson, E.J.; Howatson, G. Montmorency tart cherry (Prunus cerasus L.) supplementation accelerates recovery from exercise-induced muscle damage in females. *Eur. J. Sport Sci.* **2019**, *19*, 95–102. [CrossRef] [PubMed]

25. Bell, P.G.; Walshe, I.H.; Davison, G.W.; Stevenson, E.; Howatson, G. Montmorency cherries reduce the oxidative stress and inflammatory responses to repeated days high-intensity stochastic cycling. *Nutrients* **2014**, *6*, 829–843. [CrossRef] [PubMed]

26. Levers, K.; Dalton, R.; Galvan, E.; O'Connor, A.; Goodenough, C.; Simbo, S.; Mertens-Talcott, S.U.; Rasmussen, C.; Greenwood, M.; Riechman, S.; et al. Effects of powdered Montmorency tart cherry supplementation on acute endurance exercise performance in aerobically trained individuals. *J. Int. Soc. Sports Nutr.* **2016**, *13*, 22. [CrossRef] [PubMed]

27. Levers, K.; Dalton, R.; Galvan, E.; Goodenough, C.; O 'connor, A.; Simbo, S.; Barringer, N.; Mertens-Talcott, S.U.; Rasmussen, C.; Greenwood, M.; et al. Effects of powdered Montmorency tart cherry supplementation on an acute bout of intense lower body strength exercise in resistance trained males. *J. Int. Soc. Sports Nutr.* **2015**, *12*, 41. [CrossRef] [PubMed]

28. McCormick, R.; Peeling, P.; Binnie, M.; Dawson, B.; Sim, M. Effect of tart cherry juice on recovery and next day performance in well-trained water polo players. *J. Int. Soc. Sports Nutr.* **2016**, *13*, 4–11. [CrossRef]

29. Beals, K.; Allison, K.F.; Darnell, M.; Lovalekar, M.; Baker, R.; Nieman, D.C.; Vodovotz, Y.; Lephart, S.M. The effects of a tart cherry beverage on reducing exercise-induced muscle soreness. *Isokinet. Exerc. Sci.* **2017**, *25*, 53–63. [CrossRef]

30. Trombold, J.R.; Reinfeld, A.S.; Casler, J.R.; Coyle, E.F. The effect of pomegranate juice supplementation on strength and soreness after eccentric exercise. *J. Strength Cond. Res.* **2011**, *25*, 1782–1788. [CrossRef]

31. Machin, D.; Christmas, K.; Chou, T.; Hill, S.; Van Pelt, D.; Trombold, J.; Coyle, E. Effects of differing dosages of pomegranate juice supplementation after eccentric exercise. *Physiol. J.* **2014**, *2014*, 1–7. [CrossRef]

32. Ammar, A.; Turki, M.; Chtourou, H.; Hammouda, O.; Trabelsi, K.; Kallel, C.; Abdelkarim, O.; Hoekelmann, A.; Bouaziz, M.; Ayadi, F.; et al. Pomegranate supplementation accelerates recovery of muscle damage and soreness and inflammatory markers after a weightlifting training session. *PLoS ONE* **2016**, *11*, 0160305. [CrossRef] [PubMed]

33. Kirakosyan, A.; Seymour, E.M.; Kaufman, P.B.; Bolling, S.F. Tart cherry fruits: Implications for human health. In *Bioactive Food as Dietary Interventions for Arthritis and Related Inflammatory Diseases: Bioactive Food in Chronic*; Watson, R.R., Preedy, V.R., Eds.; Academic Press: Cambridge, MA, USA, 2013; pp. 473–484.

34. Borges, G.; Crozier, A. HPLC–PDA–MS fingerprinting to assess the authenticity of pomegranate beverages. *Food Chem.* **2012**, *135*, 1863–1867. [CrossRef] [PubMed]

35. Piccolella, S.; Fiorentino, A.; Pacifico, S.; D'Abrosca, B.; Uzzo, P.; Monaco, P. Antioxidant properties of sour cherries (Prunus cerasus L.): Role of colorless phytochemicals from the methanolic extract of ripe fruits. *J. Agric. Food Chem.* **2008**, *56*, 1928–1935. [CrossRef] [PubMed]

36. Chen, T.C. Variability in muscle damage after eccentric exercise and the repeated bout effect. *Res. Q. Exerc. Sport* **2013**, *77*, 362–371. [CrossRef] [PubMed]

37. Singleton, V.L.; Rossi, J.A. Colorimetry of total phenolics with phosphomolybdic-phosphotungstic acid reagents. *Am. J. Enol. Vitic.* **1965**, *16*, 144–158.

38. Lee, J.; Durst, R.W.; Wrolstad, R.E. Determination of total monomeric anthocyanin pigment content of fruit juices, beverages, natural colorants, and wines by the pH differential method: Collaborative study. *J. AOAC Int.* **2005**, *88*, 1269–1278.

39. Biodex Medical Systems Inc. *Biodex System 3 Pro: Application/Operation Manual*; Biodex Medical Systems Inc.: Shirley, NY, USA. Available online: https://www.biodex.com/sites/default/files/835000man_06159.pdf (accessed on 13 July 2019).

40. Clarkson, P.M.; Nosaka, K.; Braun, B. Muscle function after exercise-induced muscle damage and rapid adaptation. *Med. Sci. Sports Exerc.* **1992**, *24*, 512–520. [CrossRef]

41. Kuehl, K.; Perrier, E. Efficacy of Tart Cherry Juice in Reducing Muscle Pain During Running. *J. Int. Soc. Sports Nutr.* **2010**, *7*, 1–6. [CrossRef]

42. Rawson, E.S.; Gunn, B.; Clarkson, P.M. The Effects of Creatine Supplementation on Exercise-Induced Muscle Damage. *J. Strength Cond. Res.* **2001**, *15*, 178–184.

43. O'Fallon, K.S.; Kaushik, D.; Michniak-Kohn, B.; Dunne, C.P.; Zambraski, E.J.; Clarkson, P.M. Effects of quercetin supplementation on markers of muscle damage and inflammation after eccentric exercise. *Int. J. Sport Nutr. Exerc. Metab.* **2012**, *22*, 430–437. [CrossRef] [PubMed]

44. Koch, A.J.; Pereira, R.; Machado, M. The creatine kinase response to resistance exercise. *J. Musculoskelet. Neuronal Interact.* **2014**, *14*, 68–77. [PubMed]

45. Warren, G.L.; Lowe, D.A.; Armstrong, R.B. Measurement tools used in the study of eccentric contraction–induced injury. *Sports Med.* **1999**, *27*, 43–59. [CrossRef] [PubMed]

46. Cook, M.D.; Elisabeth, M.; Willems, T. Dietary Anthocyanin: A review of the exercise performance effects and related physiological responses. *Int. J. Sport Nutr. Exerc. Metab.* **2019**, *29*, 322–330. [CrossRef]

47. Sanchez, V.; Baeza, R.; Chirife, J. Comparison of monomeric anthocyanins and colour stability of fresh, concentrate and freeze-dried encapsulated cherry juice stored at 38ᵒC. *J. Berry Res.* **2015**, *5*, 243–251. [CrossRef]

48. Piljac-Žegarac, J.; Valek, L.; Martinez, S.; Belščak, A. Fluctuations in the phenolic content and antioxidant capacity of dark fruit juices in refrigerated storage. *Food Chem.* **2009**, *113*, 394–400. [CrossRef]

49. Howatson, G.; van Someren, K.A. Evidence of a contralateral repeated bout effect after maximal eccentric contractions. *Eur. J. Appl. Physiol.* **2007**, *101*, 207–214. [CrossRef] [PubMed]

50. Xin, L.; Hyldahl, R.D.; Chipkin, S.R.; Clarkson, P.M. A contralateral repeated bout effect attenuates induction of NF-κB DNA binding following eccentric exercise. *J. Appl. Physiol.* **2013**, *116*, 1473–1480. [CrossRef] [PubMed]
51. Starbuck, C.; Eston, R.G.; Kraemer, W.J. Exercise-induced muscle damage and the repeated bout effect: Evidence for cross transfer. *Eur. J. Appl. Physiol.* **2012**, *112*, 1005–1013. [CrossRef]
52. Powers, S.K.; Smuder, A.J.; Kavazis, A.N.; Hudson, M.B. Experimental guidelines for studies designed to investigate the impact of antioxidant supplementation on exercise performance. *Int. J. Sport Nutr. Exerc. Metab.* **2010**, *20*, 2–14. [CrossRef] [PubMed]

© 2019 by the authors. Licensee MDPI, Basel, Switzerland. This article is an open access article distributed under the terms and conditions of the Creative Commons Attribution (CC BY) license (http://creativecommons.org/licenses/by/4.0/).

 *nutrients* 

Article

Does Beef Protein Supplementation Improve Body Composition and Exercise Performance? A Systematic Review and Meta-Analysis of Randomized Controlled Trials

Pedro L. Valenzuela [1], Fernando Mata [2], Javier S. Morales [3], Adrián Castillo-García [4] and Alejandro Lucia [3,*]

1   Department of Systems Biology, University of Alcalá, 28805 Madrid, Spain; pedrol.valenzuela@edu.uah.es
2   NutriScience, Nutrition & Health Sciences, 14002 Córdoba, Spain; fernando.mata@nutriscience.pt
3   Faculty of Sport Sciences, Universidad Europea de Madrid and Research Institute 'i+12'
    (Hospital 12 de Octubre), 28670 Madrid, Spain; javiersalvador.morales@universidadeuropea.es
4   Fissac-Physiology, health and physical activity, 28015 Madrid, Spain; castillogarcia.adrian@gmail.com
*   Correspondence: alejandro.lucia@universidadeuropea.es; Tel.: +34-916-647800 (ext. 3010)

Received: 29 May 2019; Accepted: 21 June 2019; Published: 25 June 2019

Abstract: Protein supplementation might improve body composition and exercise performance. Supplements containing whey protein (WP) have received the most attention, but other protein sources such as beef protein (BP) are gaining popularity. We conducted a systematic review and meta-analysis of randomized controlled trials that compared the effects of exercise training combined with BP, WP or no protein supplementation (NP), on body composition or exercise performance. Secondary endpoints included intervention effects on total protein intake and hematological parameters. Seven studies (n = 270 participants) were included. No differences were found between BP and WP for total protein intake (standardized mean difference (SMD) = 0.04, p = 0.892), lean body mass (LBM) (SMD = −0.01, p = 0.970) or fat mass (SMD = 0.07, p = 0.760). BP significantly increased total daily protein intake (SMD = 0.68, p < 0.001), LBM (SMD = 0.34, p = 0.049) and lower-limb muscle strength (SMD = 0.40, p = 0.014) compared to NP, but no significant differences were found between both conditions for fat mass (SMD = 0.15, p = 0.256), upper-limb muscle strength (SMD = 0.16, p = 0.536) or total iron intake (SMD = 0.29, p = 0.089). In summary, BP provides similar effects to WP on protein intake and body composition and, compared to NP, might be an effective intervention to increase total daily protein intake, LBM and lower-limb muscle strength.

Keywords: resistance training; exercise; nutrition; muscle mass; hypertrophy

---

1. Introduction

Protein supplementation might increase muscle anabolism and, if combined with exercise training, could also improve physical performance [1–3]. The anabolic response to protein intake can be affected not only by factors such as individual nutritional state, digestive/absorption capacity, or sensitivity of muscle anabolic pathways, but also by the source of protein [4]. In this regard, numerous protein sources have been investigated for their effect on muscle anabolism, including milk [5], eggs [6], soy [7], rice [8] or bovine colostrum [9]. Although supplements containing whey protein (WP) have received the most attention due to their high content in leucine and potential anabolic effect [10,11], beef protein (BP) has gained popularity in recent years [12].

Acute ingestion of BP can increase muscle protein synthesis in both young and older individuals [13–15], and these benefits seem to be maximized when combined with exercise training (particularly resistance training) [16]. There is thus biological rationale to support BP supplementation as a potentially effective strategy for improving lean body mass (LBM) and physical performance (notably, muscle strength). However, the evidence on the effectiveness of BP supplementation for increasing muscle mass or performance is mixed, with some studies reporting benefits compared to no protein supplementation [17,18] but others finding no such benefits [19,20]. Moreover, the effectiveness of BP for improving body composition or performance compared to WP remains unclear [18,19,21,22].

On the other hand, owing to its high content in heme iron, BP supplementation could also increase total iron intake and thus theoretically benefit hematological parameters, and indeed some benefits have been reported [20,22]. However, to our knowledge, there is not yet meta-analytical evidence on the potential benefits of BP supplementation.

Given the purported benefits of BP supplementation on body composition and performance, and the lack of consensus on its actual effectiveness, the main aim of this systematic review and meta-analysis was to compare the effects of BP, WP or no protein supplementation (NP) combined with exercise training on body composition and exercise performance. Changes in nutritional intake and hematological parameters were also analyzed as secondary endpoints.

2. Materials and Methods

The conduct and reporting of the current systematic review and meta-analysis conform to the Preferred Reporting Items for Systematic Reviews and Meta-Analyses (PRISMA) [23].

2.1. Systematic Search

Relevant articles were identified by title and abstract in the electronic databases MEDLINE, Web of Science and SportDiscus (since inception to 5 May 2019) using the following search strategy: (beef OR meat) AND (supplement*) AND (athlet* OR train* OR "physical activity" OR exercise). The electronic search was supplemented by a manual review of reference lists from relevant publications and reviews to find additional publications on the subject. Gray literature (abstracts, conference proceedings or editorials) and reviews were excluded.

2.2. Study Selection and Data Extraction

Studies were eligible for inclusion if they met each of the following criteria: (a) analyzed the effect of BP supplementation on at least one endpoint related to body composition or exercise performance; (b) duration of the exercise training + supplementation intervention ≥4 weeks; (c) used a randomized controlled trial design that included one group performing exercise training in combination with BP supplementation and 1+ groups performing the same exercise program as those receiving BP, but with WP or NP. Case studies were excluded, as well as those in which BP was combined with WP or other protein/'anabolic' supplements (e.g., creatine).

Two reviewers (P.L.V. and F.M.) independently extracted the following data from each study: number of participants within each group, participants' characteristics, exercise and nutritional intervention's characteristics, endpoints, measurement methods and results. Disagreements were resolved through discussion with a third reviewer (J.S.M.).

Endpoints data were registered as mean and standard deviation (SD). When the standard error of the mean (SEM) but not the SD was available, the latter was calculated from the square root of the sample size multiplied by the SEM. If data were provided as a figure, we used a specific software (WebPlotDigitalizer 4.2, San Francisco, CA, USA) for data extraction. We contacted the authors when necessary to request additional information. In this regard, the authors of two studies [17,19] provided us with the required specific data upon request. Fat-free mass was taken as a surrogate of LBM if the latter was not reported.

2.3. Quality Assessment

Study quality was evaluated using the 'risk-of-bias' assessment tool following the recommendations by the Cochrane Handbook for Systematic Reviews of Interventions [24]. Two authors (A.C.G. and

J.S.M.) analyzed the following criteria: random sequence generation and allocation concealment (selection bias), blinding of participants and research staff to group allocation (performance bias), blinding of outcome assessor (detection bias), incomplete outcome data (attrition bias), and selective reporting (reporting bias). Quality assessment by both reviewers was compared and disagreements in scores were resolved through discussion with a third reviewer (P.L.V.).

2.4. Statistical Analysis

We performed a meta-analysis when a minimum of three studies compared the effects (pre- and post-intervention data) of BP vs. WP or NP for a given endpoint. The pooled standardized mean difference (SMD) between interventions (post- *minus* pre-intervention data) and 95% confidence interval (CI) were computed using a random effects model. A conservative Pearson' r value of 0.7 was used for the computation of the within-group SD [25], and a sensitivity analysis ($r = 0.2$, 0.5 or 0.9) was then performed when a significant result was found. Begg's test was used to determine the presence of publication bias. We considered that the analysis of an outcome was at risk of bias when Begg's test was significant (i.e., $p < 0.05$). The Q and I^2 statistics were used to assess heterogeneity between studies. I^2 values of 25, 50, and 75% were considered reflective of low, moderate, and large heterogeneity, respectively.

All statistical analyses were performed using the statistical software package Comprehensive Meta-analyis 2.0 (Biostat; Englewood, NJ, USA) setting the level of significance at 0.05.

3. Results

3.1. Included Studies

From the retrieved articles, seven (including 270 participants in total) [17–22,26] met all inclusion criteria and were included in the systematic review (Figure 1, Table 1).

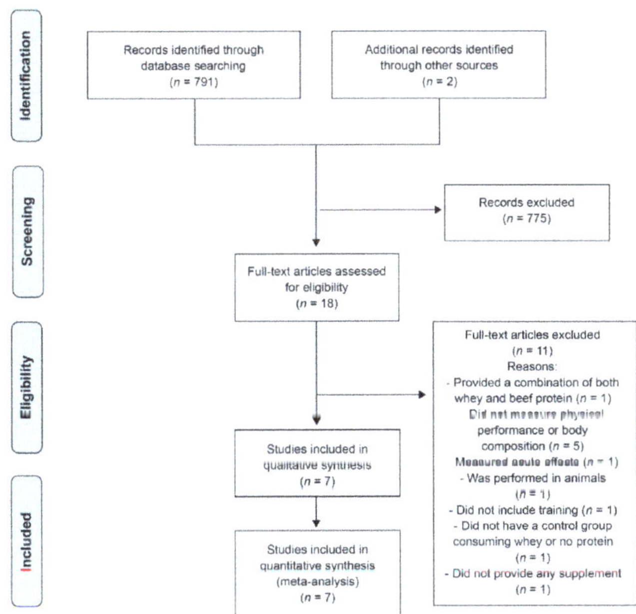

Figure 1. Flow chart of literature search.

Table 1. Main characteristics of the included studies.

Study	Participants and Group Assignment	Duration	Exercise Training Protocol (Common to all Study Groups)	Baseline Protein Intake	BP Group	WP Group	NP Group	Additional Group	Measurements	Main Outcomes
Burke et al. [20]	28 runners (18–24 years, 14 female) stratified by iron status, use of iron supplements, and gender, and randomized into a BP ($n = 14$) or NP group ($n = 14$)	8 weeks	Maintained their typical exercise regime	~1.7 g/kg/day	255 g of lean beef supplement per week + multivitamin daily	–	Multivitamin daily	–	-Dietary intake -Blood analysis -Body composition (plethysmography) -VO_{2max}	-No differences in body composition -↑heme iron intake in females -↑hematocrit in females -No differences in VO_{2max}
Daly et al. [17]	100 females (60–90 years) randomized into a BP ($n = 53$) or NP group ($n = 47$)	4 months	RT 2 times per week	~1.1–1.3 g/kg/day	220 g lean red meat (45 g protein) 6 days per week + 1 vitamin D3 capsule (1000-IU) daily	–	1 serving pasta or rice daily (25–35 g CHO) + 1 vitamin D3 capsule (1000-IU) daily	–	-Dietary intake -Physical activity -Body composition (DXA) -Muscle and fat CSA and muscle density (peripheral quantitative computed tomography) -Muscle function (TUG, FSST, and 30-s STS) -Strength (1RM estimated from 3 RM) -Blood analysis	-↑Protein intake -↑LBM -↑Leg LBM -No differences in BMD -↑Muscle strength -No differences in muscle function. -↑IGF-I -↓IL-6 -No differences in blood lipids or blood pressure.
Naclerio et al. [21]	24 active males (~26–29 years) randomized into a BP, WP or NP group ($n=8$ each)	8 weeks	RT 3 times per week	~1.5 g/kg/day	20 g of beef supplement (16.4 g protein) + 250 mL of orange juice per day	20 g of WP + 250 mL of orange juice per day	20 g of CHO + 250 mL of orange juice per day	–	-Body composition (plethysmography) -Limb circumference -Strength (1 RM) -Muscle thickness (ultrasound)	-↑*biceps brachialis* thickness -No differences in limb circumference or body composition. -No differences in strength.
Naclerio et al. [22]	24 male master triathletes (35–60 years) randomized into a BP, WP or NP group ($n = 8$ each)	10 weeks	ET 4–6 times per week	~1.3–1.5 g/kg/day	20 g of beef supplement (16.4 g protein) per day	20 g of WP per day	20 g of CHO per day	–	-Body composition (plethysmography) -VO_{2max} -Muscle thickness (ultrasound) -Blood analysis	-No differences in body composition. -↓BM -↑heme iron intake. -↑ferritin concentrations -↑muscle thickness -No differences in VO_{2max}

Table 1. *Cont.*

Study	Participants and Group Assignment	Duration	Exercise Training Protocol (Common to all Study Groups)	Baseline Protein Intake	BP Group	WP Group	NP Group	Additional Group	Measurements	Main Outcomes
Naclerio et al. [19]	27 active males and females (~24–28 years) randomized into a BP, WP or NP group (n = 8 each)	8 weeks	RT 3 times per week	~1.1–1.5 g/kg/day	20 g of beef supplement (16.4 g protein) + 250 mL of orange juice per day	20 g of WP + 250 mL of orange juice per day	20 g of CHO + 250 mL of orange juice per day	–	-Body composition (plethysmography) -Strength (total weight lifted) -Muscle thickness (ultrasound) -Blood analysis -Saliva analysis	-No differences in body composition nor muscle thickness. -↓HNP1-3 and saliva flow rate. -No differences in strength.
Negro et al. [26]	26 male and female healthy subjects (~24 years) randomized into a BP (n = 12) or NP group (n = 14)	8 weeks	RT 3 times per week	~1.0 g/kg/day	135 g (20 g protein) of tinned beef per day	–	No supplement provided	–	-Strength (1 RM) -Body composition (bioimpedance)	-↑FFM and ↓FM -No differences in LBM -No differences in strength.
Sharp et al. [18]	41 male and female trained subjects (18–30 years) randomized into a WP (n = 10, 5 male) BP (n = 10, 5 male), chicken protein (n = 11, 5 male) or NP group (n = 10, 4 male)	8 weeks	RT 3 times per week and HIIT 2 times per week.	~2.0–2.2 g/kg/day	46 g of isolated BP per day	46 g of WP per day	46 g of CHO per day	46 g of chicken protein per day	-Dietary intake -Body composition (DXA). -Anaerobic peak power (10-second sprint) -Strength (1RM) -Gastrointestinal symptoms	-↑LBM and ↓FM compared to CHO. -No differences in body composition compared to other protein sources. -No differences in strength. -↓improvement in peak power compared to WP. -No differences in gastrointestinal symptoms.

Abbreviations: BM, body mass; BMD, bone mineral density; BP, beef protein; CHO, carbohydrates; CSA, cross-sectional area; DXA, dual-energy X-ray absorptiometry; ET, endurance training; FFM, fat-free mass; FM, fat mass; FSST, 4-square step test; HIIT, high intensity interval training; HNP1-3, human neutrophil peptides; IGF-1, insulin-like growth factor 1; IL-6, interleukin-6; LBM, lean body mass; NP, no protein supplementation; RM, repetition maximum; RT, resistance training; STS, sit-to-stand test; TUG, timed-up-and-go test; VO$_{2max}$, maximal oxygen uptake; WP, whey protein. ↑, increase; ↓, reduce.

3.2. Quality Assessment and Publication Bias

The risk of bias in the included studies was overall low (Figure 2). Three of seven studies [18,20,26] failed to perform/report an appropriate random sequence generation or allocation concealment. Three studies (where BP was provided as lean beef) were at a high risk of performance bias because participants were not blinded to group allocation [17,20,26].

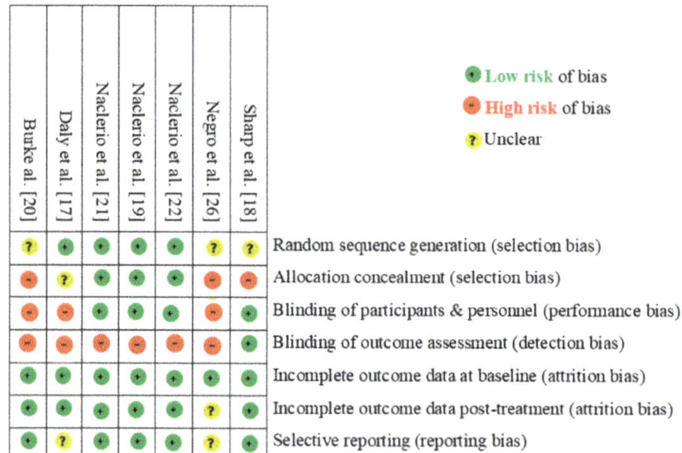

Figure 2. Quality assessment of the included studies.

3.3. Participants and Intervention Characteristics

Five studies [18–21,26] analyzed young male and female active/trained subjects whose age ranged between ~18 and 30 years, one [22] analyzed master male athletes (35 to 60 years), and another one [17] assessed women aged 60 to 90 years.

The duration of the interventions ranged from 8 to 16 weeks. BP was supplied as isolated BP (powder form) in four studies [18,19,21,22] or as a lean beef supplement in a remainder of studies [17,20,26]. The amount of protein was specified in all but one study [20], and ranged from 16.4 to 46 g/day. BP was compared with NP in all studies, and with WP in four of them [18,19,21,22]. The nutritional intervention was combined with a resistance ('strength') exercise training program (2 to 3 sessions/ week) in five studies [17–19,21,26] or with aerobic exercise training (four to six sessions/week) in one study [22]. One study [20] only reported that participants were cross-country runners who maintained their typical usual exercise regime during the intervention.

3.4. Body Composition

Four studies [18,19,21,22] compared the effects of BP and WP on body composition, with none reporting significant differences. Thus, the meta-analysis showed no differences between both conditions for LBM (SMD = −0.01, p = 0.970, Figure 3A), with no signs of heterogeneity (I^2 = 0%, Q = 0.152) or publication bias (p = 0.202). Similarly, no differences were found between BP and WP for fat mass (SMD = 0.07, p = 0.760, Figure 3B), with no signs of heterogeneity (I^2=0%, Q = 0.012) or publication bias (p = 0.054).

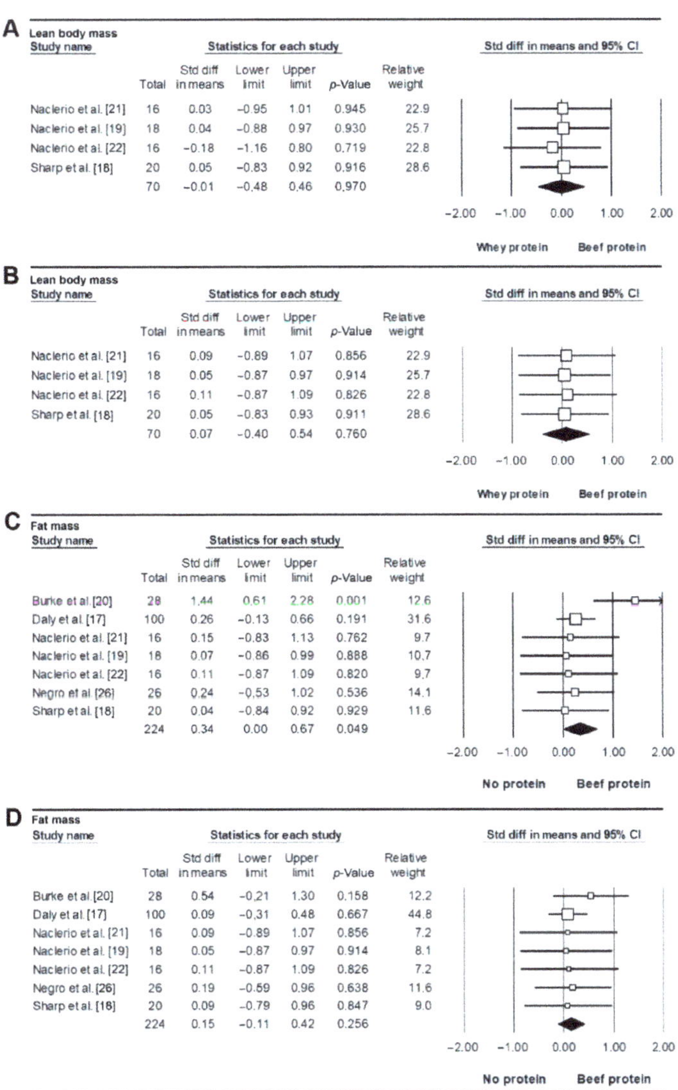

Figure 3. Effects on lean body (**A**) and fat mass (**B**) of beef vs. whey protein supplementation and effects on lean body (panel **C**) and fat mass (panel **D**) of beef vs. no protein supplementation. Each forest plot displays the pooled standardized difference in means and 95% confidence interval.

All included studies compared the effects of BP vs. NP on body composition. The meta-analysis showed a beneficial effect of BP over NP for LBM (SMD = 0.34, p = 0.049, Figure 3C), with no signs of heterogeneity (I^2 = 4.326%, Q = 6.271) or publication bias (p = 0.500). This effect remained significant for r = 0.9 (SMD = 0.52, p = 0.032) or almost significant for 0.5 (SMD = 0.26, p = 0.051), but differences did not reach statistical significance for r=0.2 (SMD = 0.21, p = 0.112). No differences were found between BP and NP for fat mass (SMD = 0.15, p = 0.256, Figure 3D), with no signs of heterogeneity (I^2 = 0%, Q = 1.238) or publication bias (p = 0.500).

Three studies [19,21,22] analyzed changes in muscle thickness using ultrasonography, but we could not meta-analyze their results because these studies analyzed different muscles. One study [19] found no benefits on muscle thickness with BP vs. WP or NP. However, the other two reported an improvement in the thickness of the *biceps brachialis* [21] or *vastus medialis* muscle [22] with BP compared to WP or NP. One study [22] also analyzed changes in thigh and arm circumference and found larger benefits with BP vs. WP or NP. Finally, one study [17] that analyzed thigh muscle cross-sectional area as well bone mineral density (femoral neck, total hip and lumbar spine) using dual-energy X-ray absorptiometry found no differences between BP and NP.

3.5. Exercise Performance

All retrieved studies analyzed the effects of BP on 1+ markers of exercise performance. Four studies [17,18,21,26] analyzed the effects of BP vs. NP on lower-limb maximal strength (deadlift, leg press, knee extension). Of these, three [18,21,26] found similar increases in strength with both conditions, and one study [17] found an 18% greater increase in leg extension strength with BP than with NP. The meta-analysis showed a beneficial effect of BP over NP (SMD = 0.40, p = 0.014, Figure 4A), with no signs of heterogeneity (I^2 = 0%, Q = 2.954) or publication bias (p = 0.367). This effect remained significant for r=0.5 (SMD = 0.33, p = 0.040), but not for r = 0.2 (SMD = 0.26, p = 0.098) or 0.9 (SMD = 0.42, p = 0.150). Three studies [18,21,26] that assessed changes in upper-limb maximal muscle strength (bench press, 1-repetition maximum [1RM]) found no differences between BP and NP, and the meta-analysis showed no differences between both conditions for this outcome (SMD = 0.16, p = 0.536, I^2 = 0%, Q = 0.058, Begg's p-value = 0.148, Figure 4B).

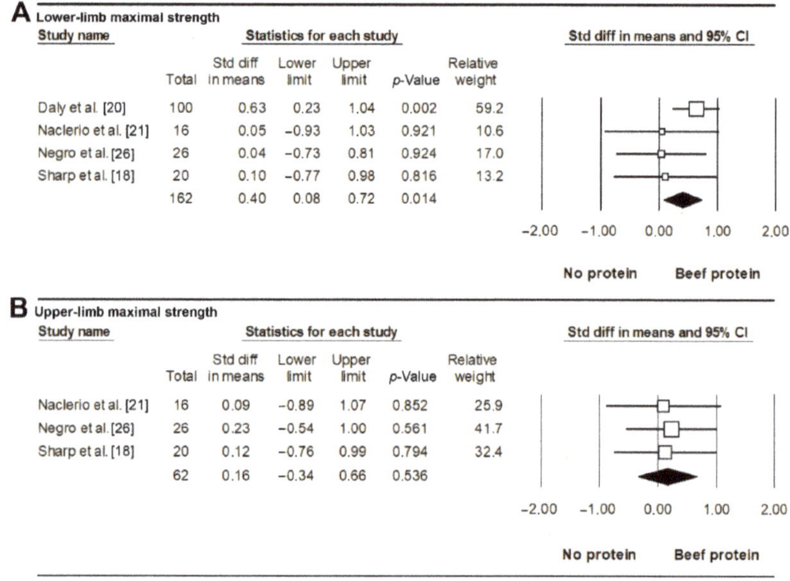

Figure 4. Effects on lower- (**A**) and upper-limb maximal strength (**B**) of beef vs. no protein supplementation. Each forest plot displays the pooled standardized difference in means and 95% confidence interval.

Two studies [18,21] also compared the gains in 1RM with BP over WP and neither found additional benefits. Finally, no benefits were reported with BP vs. NP for maximal oxygen uptake [20,22], anaerobic peak power [18] or total weight lifted during a resistance training session [19].

3.6. Nutritional Intake

Except for one study [26], where the baseline protein intake of the participants was described as ~1.0 g/kg/day, the mean total protein intake (in g/kg/day) during the intervention was consistently reported and ranged from 1.3 to 2.2 g/kg/day (BP), 1.7 to 2.2 g/kg/day (WP), and 1.1 to 2.0 g/kg/day (NP). The total protein intake was higher in BP than NP (SMD = 0.68, $p < 0.001$, Figure 5A; $I^2 = 0\%$, $Q = 1.226$, Begg's $p = 0.403$). This effect remained for $r = 0.2$ (SMD = 0.42, $p = 0.006$), 0.5 (SMD = 0.53, $p = 0.001$) or 0.9 (SMD = 1.17, $p < 0.001$). By contrast, no differences in protein intake were found for BP vs. WP (SMD = 0.04, $p = 0.892$, Figure 5B), with no heterogeneity between studies ($I^2 = 0\%$, $Q = 0.016$) and no signs of publication bias ($p = 0.500$).

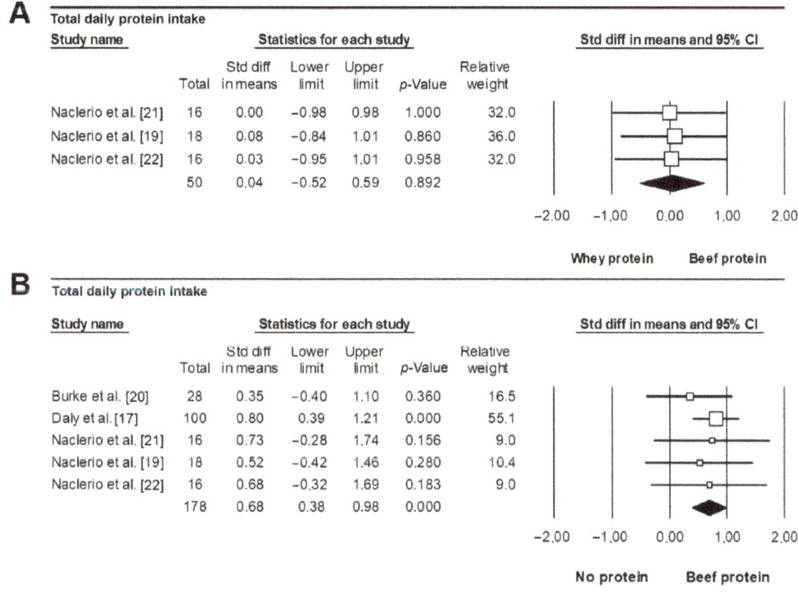

Figure 5. Effects on total daily protein intake of beef vs. whey (**A**) or no protein supplementation (**B**). Each forest plot displays the pooled standardized difference in means and 95% confidence interval.

Five studies [17,19–22] analyzed changes in fat intake with BP vs. NP, with no significant differences between conditions (SMD = −0.11, $p = 0.587$) and with no heterogeneity between studies ($I^2 = 3.8\%$, $Q = 4.158$) and no publication bias ($p = 0.110$). Three of these studies [19,21,22] also analyzed the changes in fat intake with BP vs. WP, and the meta-analysis showed a higher fat intake with the former (SMD = 0.71, $p = 0.015$), with no heterogeneity ($I^2 = 0\%$, $Q = 0.195$) or bias between studies ($p = 0.148$). This effect remained significant for $r = 0.9$ (SMD = 1.22, $p < 0.001$), but not for 0.5 (SMD = 0.55, $p = 0.056$) or 0.2 (SMD = 0.44, $p = 0.128$).

Three studies [17,20,22] analyzed changes in iron intake with BP vs. NP. The meta-analysis showed a non-significant trend towards a higher iron intake with BP (SMD – 0.29, $p = 0.089$), which was significant for $r = 0.9$ (SMD = 0.46, $p = 0.006$), but not for $r = 0.5$ (SMD = 0.22, $p = 0.180$) or 0.2 (SMD = 0.18, $p = 0.285$).

3.7. Hematological Parameters

Two studies [20,22] analyzed the effects of BP vs. NP on hematological parameters, with one study reporting an increased ferritin concentration [22] and the other study finding an increased hematocrit (although significance was only reached for the female participants) [20].

Three studies [17–19] analyzed the effects of BP vs. NP on lipid profile, and two of them [18,19] also compared BP and WP. None of them found differences between interventions for any of the analyzed variables (i.e., total cholesterol, LDL and HDL-cholesterol, or triglycerides). However, no meta-analysis could be performed because the mean and SD values for baseline and post-intervention were not available for all the studies.

4. Discussion

Protein supplementation might be a potentially beneficial strategy for maximizing exercise training-related gains in body composition and performance [2], with WP receiving the most attention to date. This systematic review and meta-analysis showed that, when combined with exercise training, BP supplementation provides benefits on protein intake and LBM that are similar to those elicited by WP. Our results also show that BP supplementation might be an effective means of increasing total daily protein intake compared to NP, and suggest it might be also useful for improving LBM and lower-limb muscle strength, although these results should be corroborated in further studies.

The popularity of WP compared to other protein sources is mainly based on a higher digestibility of the former together with a greater content of essential amino acids such as leucine [27]. Indeed, WP has proven to stimulate muscle protein synthesis to a greater extent than other popular protein sources such as casein or soy [28,29]. The ingestion of BP has also demonstrated promising anabolic effects, with the ingestion of 30 g (equivalent to 113 g of lean beef) increasing muscle protein synthesis by ~50% compared with fasting conditions [13–15]. Moreover, the combination of BP and exercise training might provide additional benefits, as the increase in muscle protein synthesis observed after BP intake followed by a resistance training session is higher than that observed with the former alone [15,16]. However, WP has been reported to have a higher digestibility (as reflected by a higher Protein Digestibility Corrected Amino Acid Score, 1.00 vs. 0.92) as well as a greater content of essential amino acids than BP (52% vs. 44% of total protein, respectively), including more leucine (13.6% vs. 8.8%) and lysine (10.6% vs. 8.9%) and a similar methionine content (2.5% vs. 2.3%) [27]. Of note, in human muscle, essential amino acids make up to 45% of the total protein content, and 9.4%, 8.7% and 2.2% of the protein comes from leucine, lysine and methionine, respectively [27]. Burd et al. [30] compared the mean muscle protein synthesis response to the ingestion of 30 g of protein coming from skimmed milk or beef after a resistance training session. Skimmed milk (from which WP is obtained) was more rapidly digested and absorbed than beef, thereby resulting in a greater leucine availability, and a higher stimulation of muscle protein synthesis in the early phase after exercise (0–2 h), but not for the whole recovery period (0–5 h post-exercise) [30]. Thus, these results suggest that milk/WP might be more effective than BP for the stimulation of muscle anabolism during the early postprandial stage, but that no overall differences are found between these protein sources. In line with these findings, our results support similar mid/long-term (≥4 weeks) benefits of WP or BP supplementation on LBM.

On the other hand, the present study shows that BP might be effective compared to NP for increasing LBM and lower-limb muscle strength, although the magnitude of the benefit was small (SMD <0.50 in both cases) and this result should be confirmed in further studies as sensitivity analyses raised concern about statistical significance. Furthermore, no consistent benefits were observed on physical performance, as reflected by the lack of differences for upper-limb muscle strength. Given the abovementioned anabolic potential of BP, which also resulted in a higher total daily protein intake than NP, we expected greater benefits of BP on LBM and physical performance compared to NP. One of the factors that might explain the small benefits obtained in these endpoints is that participants already consumed an 'optimal' quantity of protein in their diet (1.1 to 2.0 g/kg/day for the NP group, above the current international recommendations for the general adult population of 0.8 g/kg/day [31]). Of note, a protein intake higher than 0.8 g/kg/day is recommended to promote skeletal muscle anabolism in physically active individuals (i.e., 1.0, 1.3, and 1.6 g/kg/day for individuals with minimal, moderate, and intense physical activity levels, respectively; [32]) as well as in older people (>1.2 g/kg/day, [33]) such as those included in the study of Daly et al. [17]. However, a recent meta-analysis found no additional benefits on LBM when protein intake was greater than 1.6 g/kg/day [2]. Thus, the daily protein intake of the participants in the studies analyzed here could be considered within the 'optimal' range even without protein supplementation. Greater benefits might perhaps have been observed in individuals with a baseline protein intake below the recommended levels.

Another purported benefit of supplementation with BP is the potential increase in iron intake, which might theoretically improve hematological parameters. In this regard, our results show a non-significant trend towards an increased iron intake with BP compared to NP. Two studies [20,22] analyzed the specific changes in heme iron intake and both found increased values with BP—although only in the female participants in one study [20]—compared to WP or NP. In line with these results, although we could not perform a meta-analysis of hematological parameters, preliminary evidence suggests that BP could provide benefits at hematological level, such as increased hematocrit levels in collegiate distance runners [20] or enhanced iron (ferritin) deposits in master-age triathletes [22]. Future research should confirm if BP might serve as an effective intervention to counteract hematological conditions such as anemia in individuals with deficient iron intake.

An interesting finding was that, when compared to WP, BP supplementation seemed to increase the total daily intake of fat. However, in sensitivity analysis, this effect did not reach statistical significance and no differences in total daily fat intake were observed between BP and NP. Thus, there is not yet clear evidence to support that BP *per se* significantly increases fat intake. Moreover, no differences have been reported in lipid profile (e.g., cholesterol levels) or in fat mass with BP compared to WP or NP. However, more research is needed in the field.

Some limitations must be noted, notably the small number of studies included, the small sample size of most studies or the lack of participants' blinding to group allocation in some of them. The included studies provided BP in different forms (powder or lean beef), which might have affected digestibility/absorption rates and potentially, anabolic responses [34]. However, although minced beef is more rapidly digested/absorbed than beef steak resulting in a greater amino acid availability, no differences have been reported in post-prandial muscle protein synthesis [35]. Future research should confirm if providing BP in isolate form (powder) results in a greater anabolic response than lean beef. The type of exercise training intervention did also differ between studies, with some including resistance training and others using endurance training. In this regard, resistance training is more effective for promoting muscle protein synthesis and muscle mass accretion than endurance exercise [36], and this might have acted as a confounding variable in our findings. However, protein supplementation can also provide benefits in individuals performing endurance exercise training, notably by avoiding muscle catabolism—especially during periods of negative energy intake—and through the enhancement of recovery and the promotion of exercise-induced adaptations [37]. On the other hand, we combined studies with participants of wide age ranges, including individuals aged 60 to 90 years. In this regard, aging is associated with an impaired anabolic response to both aminoacidemia and acute physical exercise [38], which might have acted as a confounding factor in our results. Interestingly, the study that included older adults was the one reporting the greatest increases in muscle strength [17]. Thus, BP supplementation might be particularly useful for this population segment. However, statistical analyses showed no heterogeneity between studies for the outcomes that we analyzed.

On the other hand, the benefits of supplementing with BP—and of every diet in general—should be viewed under the context of its sustainability, that is, it should have a low environmental impact that contributes to food and nutrition security and to a healthy lifestyle for present and future generations [39]. Diets that primarily contain animal-derived food sources are associated with higher greenhouse gas emissions, and indeed beef production/consumption is the food source that results in greater gas emissions (>70 kg CO_2/kg) [40]. By contrast, the production of dairy products (such as milk or WP), and especially of plant-based protein foods, is associated with lower moderate greenhouse gas emissions [41]. Thus, a balance between environmental sustainability and optimal dietary intake (particularly regarding protein consumption) should be sought, and WP might be a more recommendable option in this regard.

5. Conclusions

This systematic review and meta-analysis of randomized controlled trials showed no mid/long-term ($\geq$4 weeks) differences between BP and WP on total daily protein intake or body composition (LBM, fat mass), and thus both protein sources might be similarly effective to promote increases in LBM. However, BP resulted in a greater total daily protein intake than NP, and our results suggest that BP might represent an effective strategy for increasing LBM and lower-limb muscle strength. However, these and other potential benefits (including preliminary data on heme iron intake and ferritin levels) should be confirmed in future studies.

Author Contributions: Conceptualization, P.L.V. and F.M.; Data curation, P.L.V. and F.M.; Formal analysis, P.L.V.; Funding acquisition, A.L.; Investigation, P.L.V.; Methodology, P.L.V., J.S.M. and A.C.-G.; Project administration, P.L.V.; Resources, P.L.V.; Software, P.L.V.; Supervision, P.L.V. and A.L.; Validation, A.L.; Visualization, A.C.-G.; Writing—original draft, P.L.V. and A.L.; Writing—review and editing, P.L.V., F.M., J.S.M., A.C.-G. and A.L.

Funding: Pedro L. Valenzuela is supported by the University of Alcalá (FPI2016). Javier S. Morales is supported by the Ministry of Education, Culture and Sport (FPU14/03435). Research by Alejandro Lucia is supported by grants from the Spanish Ministry of Science, Innovation and Universities and Fondos FEDER (Fondo de Investigaciones Sanitarias [FIS], Grant No. PI15/00558 and PI18/00139).

Conflicts of Interest: The authors declare no conflict of interest.

References

1. Cermak, N.M.; Res, P.T.; de Groot, L.C.; Saris, W.H.M.; Loon, L.J.C. Van Protein supplementation augments the adaptive response of skeletal muscle to resistance-type exercise training: A meta-analysis. *Am. J. Clin. Nutr.* **2012**, *96*, 1454–1464. [CrossRef]

2. Morton, R.W.; Murphy, K.T.; McKellar, S.R.; Schoenfeld, B.J.; Henselmans, M.; Helms, E.; Aragon, A.A.; Devries, M.C.; Banfield, L.; Krieger, J.W.; et al. A systematic review, meta-analysis and meta-regression of the effect of protein supplementation on resistance training-induced gains in muscle mass and strength in healthy adults. *Br. J. Sports Med.* **2018**, *52*, 376–384. [CrossRef]

3. Pasiakos, S.M.; McLellan, T.M.; Lieberman, H.R. The Effects of Protein Supplements on Muscle Mass, Strength, and Aerobic and Anaerobic Power in Healthy Adults: A Systematic Review. *Sports Med.* **2014**, *45*, 111–131. [CrossRef]

4. Stokes, T.; Hector, A.J.; Morton, R.W.; McGlory, C.; Phillips, S.M. Recent perspectives regarding the role of dietary protein for the promotion of muscle hypertrophy with resistance exercise training. *Nutrients* **2018**, *10*, 180. [CrossRef]

5. Elliot, T.A.; Cree, M.G.; Sanford, A.P.; Wolfe, R.R.; Tipton, K.D. Milk ingestion stimulates net muscle protein synthesis following resistance exercise. *Med. Sci. Sports Exerc.* **2006**, *38*, 667–674. [CrossRef]

6. Moore, D.; Robinson, M.; Fry, J.; Tang, J.; Glover, E.; Wilkinson, S.; Prior, T.; Tarnopolsky, M.; Philips, S. Ingested protein dose response of muscle and albumin protein synthesis after resistance exercise in young men. *Am. J. Clin. Nutr.* **2009**, *89*, 161–168. [CrossRef]

7. Yang, Y.; Churchward-Venne, T.A.; Burd, N.A.; Breen, L.; Tarnopolsky, M.A.; Phillips, S.M. Myofibrillar protein synthesis following ingestion of soy protein isolate at rest and after resistance exercise in elderly men. *Nutr. Metab.* **2012**, *9*, 1. [CrossRef]

8. Joy, J.M.; Lowery, R.P.; Wilson, J.M.; Purpura, M.; De Souza, E.O.; Wilson, S.M.; Kalman, D.S.; Dudeck, J.E.; Jäger, R. The effects of 8 weeks of whey or rice protein supplementation on body composition and exercise performance. *Nutr. J.* **2013**, *12*, 1–7. [CrossRef]

9. Duff, W.; Chilibeck, P.; Rooke, J.; Kaviani, M.; Krentz, J.; Haines, D. The Effect of Bovine Colostrum Supplementation in Older Adults during Resistance Training. *Int. J. Sport Nutr. Exerc. Metab.* **2014**, *24*, 276–285. [CrossRef]

10. Cribb, P.J.; Williams, A.D.; Hayes, A.; Carey, M.F. The Effect of Whey Isolate and Resistance Training on Strength, Body Composition and Plasma Glutamine. *Int. J. Sport Nutr. Exerc. Metab.* **2006**, *16*, 494–509. [CrossRef]

11. Volek, J.S.; Volk, B.M.; Gómez, A.L.; Kunces, L.J.; Kupchak, B.R.; Freidenreich, D.J.; Aristizabal, J.C.; Saenz, C.; Dunn-Lewis, C.; Ballard, K.D.; et al. Whey Protein Supplementation During Resistance Training Augments Lean Body Mass. *J. Am. Coll. Nutr.* **2013**, *32*, 122–135. [CrossRef]

12. Gorissen, S.H.M.; Rémond, D.; van Loon, L.J.C. The muscle protein synthetic response to food ingestion. *Meat Sci.* **2015**, *109*, 96–100. [CrossRef]

13. Symons, T.; Schutzler, S.E.; Cocke, T.L.; Chinkes, D.L.; Wolfe, R.R.; Paddon-Jones, D. Aging does not impair the anabolic response to a protein-rich meal. *Am. J. Clin. Nutr.* **2007**, *86*, 451–456. [CrossRef]

14. Symons, T.; Sheffield-Moore, M.; Wolfe, R.R.; Paddon-Jones, D. A Moderate Serving of High-Quality Protein Maximally Stimulates Skeletal Muscle Protein Synthesis in Young and Elderly Subjects. *J. Am. Diet. Assoc.* **2009**, *109*, 1582–1586. [CrossRef]

15. Robinson, M.J.; Burd, N.A.; Breen, L.; Rerecich, T.; Yang, Y.; Hector, A.J.; Baker, S.K.; Phillips, S.M. Dose-dependent responses of myofibrillar protein synthesis with beef ingestion are enhanced with resistance exercise in middle-aged men. *Appl. Physiol. Nutr. Metab.* **2013**, *38*, 120–125. [CrossRef]

16. Symons, T.; Sheffield-Moore, M.; Mamerow, M.M.; Wolfe, R.R.; Paddon-Jones, D. The anabolic response to resistance exercise and a protein-rich meal is not diminished by age. *J. Nutr. Health Aging* **2011**, *15*, 376–381. [CrossRef]

17. Daly, R.M.; O'Connell, S.L.; Mundell, N.L.; Grimes, C.A.; Dunstan, D.W.; Nowson, C.A. Protein-enriched diet, with the use of lean red meat, combined with progressive resistance training enhances lean tissue mass and muscle strength and reduces circulating IL-6 concentrations in elderly women: A cluster randomized controlled trial. *Am. J. Clin. Nutr.* **2014**, *99*, 899–910. [CrossRef]

18. Sharp, M.H.; Lowery, R.P.; Shields, K.A.; Lane, J.R.; Gray, J.L.; Partl, J.M.; Hayes, D.W.; Wilson, G.J.; Hollmer, C.A.; Minivich, J.R.; et al. The Effects of Beef, Chicken, or Whey Protein Post-Workout on Body Composition and Muscle Performance. *J. Strength Cond. Res.* **2018**, *32*, 2233–2242. [CrossRef]

19. Naclerio, F.; Larumbe-Zabala, E.; Ashrafi, N.; Seijo, M.; Nielsen, B.; Allgrove, J.; Earnest, C.P. Effects of protein–carbohydrate supplementation on immunity and resistance training outcomes: A double-blind, randomized, controlled clinical trial. *Eur. J. Appl. Physiol.* **2017**, *117*, 267–277. [CrossRef]

20. Burke, D.E.; Johnson, J.V.; Vukovich, M.D.; Kattelmann, K.K. Effects of Lean Beef Supplementation on Iron Status, Body Composition and Performance of Collegiate Distance Runners. *Food Nutr. Sci.* **2012**, *03*, 810–821. [CrossRef]

21. Naclerio, F.; Seijo-Bujia, M.; Larumbe-Zabala, E.; Earnes, C. Carbohydrates Alone or Mixing With Beef or Whey Protein Promote Similar Training Outcomes in Resistance Training Males: A Double Blind, Randomized Controlled Clinical Trial. *Int. J. Sport Nutr. Exerc. Metab.* **2017**, *27*, 408–420. [CrossRef] [PubMed]

22. Naclerio, F.; Seijo, M.; Larumbe-Zabala, E.; Ashrafi, N.; Christides, T.; Karsten, B.; Nielsen, B.V. Effects of Supplementation with Beef or Whey Protein Versus Carbohydrate in Master Triathletes. *J. Am. Coll. Nutr.* **2017**, *36*, 593–601. [CrossRef] [PubMed]

23. Moher, D.; Liberati, A.; Tetzlaff, J.; Altman, D.G.; Altman, D.; Antes, G.; Atkins, D.; Barbour, V.; Barrowman, N.; Berlin, J.A.; et al. Preferred reporting items for systematic reviews and meta-analyses: The PRISMA statement. *PLoS Med.* **2009**, *6*, e1000097. [CrossRef] [PubMed]

24. Higgins, J.P.T.; Altman, D.G.; Gøtzsche, P.C.; Jüni, P.; Moher, D.; Oxman, A.D.; Savović, J.; Schulz, K.F.; Weeks, L.; Sterne, J.A.C. The Cochrane Collaboration's tool for assessing risk of bias in randomised trials. *BMJ* **2011**, *343*, 1–9. [CrossRef] [PubMed]

25. Rosenthal, R. *Meta-Analytic Procedures for Social Research*; Sage: Newbury Park, CA, USA, 1991.

26. Negro, M.; Vandoni, M.; Ottobrini, S.; Codrons, E.; Correale, L.; Buonocore, D.; Marzatico, F. Protein supplementation with low fat meat after resistance training: Effects on body composition and strength. *Nutrients* **2014**, *6*, 3040–3049. [CrossRef] [PubMed]

27. Van Vliet, S.; Burd, N.A.; van Loon, L.J.C. The Skeletal Muscle Anabolic Response to Plant- versus Animal-Based Protein Consumption. *J. Nutr.* **2015**, *145*, 1981–1991. [CrossRef] [PubMed]

28. Tang, J.E.; Moore, D.R.; Kujbida, G.W.; Tarnopolsky, M.A.; Phillips, S.M. Ingestion of whey hydrolysate, casein, or soy protein isolate: Effects on mixed muscle protein synthesis at rest and following resistance exercise in young men. *J. Appl. Physiol.* **2009**, *107*, 987–992. [CrossRef] [PubMed]

29. Burd, N.A.; Yang, Y.; Moore, D.R.; Tang, J.E.; Tarnopolsky, M.A.; Phillips, S.M. Greater stimulation of myofibrillar protein synthesis with ingestion of whey protein isolate v. micellar casein at rest and after resistance exercise in elderly men. *Br. J. Nutr.* **2012**, *108*, 958–962. [CrossRef]

30. Burd, N.A.; Gorissen, S.H.; Van Vliet, S.; Snijders, T.; Van Loon, L.J.C. Differences in postprandial protein handling after beef compared with milk ingestion during postexercise recovery: A randomized controlled trial. *Am. J. Clin. Nutr.* **2015**, *102*, 828–836. [CrossRef]

31. Trumbo, P.; Schlicker, S.; Yates, A.; Poos, M. Dietary reference intakes for energy, carbohydrate, fiber, fat, fatty acids, cholesterol, protein and amino acids. *J. Am. Diet. Assoc.* **2002**, *102*, 1621–1630. [CrossRef]

32. Wu, G. Dietary protein intake and human health. *Food Funct.* **2016**, *7*, 1251–1265. [CrossRef] [PubMed]

33. Traylor, D.A.; Gorissen, S.H.M.; Phillips, S.M. Perspective: Protein requirements and optimal intakes in aging: Arewe ready to recommend more than the recommended daily allowance? *Adv. Nutr.* **2018**, *9*, 171–182. [CrossRef] [PubMed]

34. Trommelen, J.; Betz, M.W.; van Loon, L.J.C. The Muscle Protein Synthetic Response to Meal Ingestion Following Resistance-Type Exercise. *Sports Med.* **2019**, *49*, 185–197. [CrossRef] [PubMed]

35. Pennings, B.; Groen, B.B.L.; Van Dijk, J.W.; De Lange, A.; Kiskini, A.; Kuklinski, M.; Senden, J.M.G.; Van Loon, L.J.C. Minced beef is more rapidly digested and absorbed than beef steak, resulting in greater postprandial protein retention in older men. *Am. J. Clin. Nutr.* **2013**, *98*, 121–128. [CrossRef] [PubMed]

36. Egan, B.; Zierath, J.R. Exercise metabolism and the molecular regulation of skeletal muscle adaptation. *Cell Metab.* **2013**, *17*, 162–184. [CrossRef] [PubMed]

37. Moore, D.R.; Camera, D.M.; Areta, J.L.; Hawley, J. A Beyond muscle hypertrophy: Why dietary protein is important for endurance athletes. *Appl. Physiol. Nutr. Metab.* **2014**, *11*, 1–11. [CrossRef] [PubMed]

38. Breen, L.; Phillips, S.M. Skeletal muscle protein metabolism in the elderly: Interventions to counteract the "anabolic resistance" of ageing. *Nutr. Metab.* **2011**, *8*, 68. [CrossRef] [PubMed]

39. MacDiarmid, J.I. Is a healthy diet an environmentally sustainable diet? *Proc. Nutr. Soc.* **2013**, *72*, 13–20. [CrossRef]

40. Scarborough, P.; Appleby, P.N.; Mizdrak, A.; Briggs, A.D.M.; Travis, R.C.; Bradbury, K.E.; Key, T.J. Dietary greenhouse gas emissions of meat-eaters, fish-eaters, vegetarians and vegans in the UK. *Clim. Chang.* **2014**, *125*, 179–192. [CrossRef]

41. Gorissen, S.H.M.; Witard, O.C. Characterising the muscle anabolic potential of dairy, meat and plant-based protein sources in older adults. *Proc. Nutr. Soc.* **2018**, *77*, 20–31. [CrossRef]

© 2019 by the authors. Licensee MDPI, Basel, Switzerland. This article is an open access article distributed under the terms and conditions of the Creative Commons Attribution (CC BY) license (http://creativecommons.org/licenses/by/4.0/).

 *nutrients*

Article

Nutritional Intake, Sports Nutrition Knowledge and Energy Availability in Female Australian Rules Football Players

Dominique Condo [1,2,3,*], **Rachel Lohman** [1], **Monica Kelly** [3] **and Amelia Carr** [1]

[1]   Centre for Sport Research, School of Exercise and Nutrition Sciences, Faculty of Health, Deakin University, 221 Burwood Highway, Burwood, Melbourne, VIC 3125, Australia; rlohma@deakin.edu.au (R.L.); amelia.carr@deakin.edu.au (A.C.)

[2]   Institute for Physical Activity and Nutrition, School of Exercise and Nutrition Sciences, Faculty of Health, Deakin University, 221 Burwood Highway, Burwood, Melbourne, VIC 3125, Australia

[3]   Geelong Football Club, GMHBA Stadium, Geelong, VIC 3125, Australia; mkelly@geelongcats.com.au

*   Correspondence: dominique.condo@deakin.edu

Received: 21 February 2019; Accepted: 19 April 2019; Published: 28 April 2019

Abstract: This study aimed to assess nutritional intake, sports nutrition knowledge and risk of Low Energy Availability (LEA) in female Australian rules football players. Victorian Football League Women's competition (VFLW) players (n = 30) aged 18–35 (weight: 64.5 kg ± 8.0; height: 168.2 cm ± 7.6) were recruited from Victoria, Australia. Nutritional intake was quantified on training days using the Automated 24 h Dietary Assessment Tool (ASA24-Australia), and sports nutrition knowledge was measured by the 88-item Sports Nutrition Knowledge Questionnaire (SNKQ). The risk of LEA was assessed using the Low Energy Availability in Females Questionnaire (LEAF-Q). Daily mean carbohydrate intake in the current investigation was 3 $g \cdot kg^{-1} \cdot d^{-1}$, therefore, below the minimum carbohydrate recommendation for moderate exercise of approximately one hour per day (5–7 $g \cdot kg^{-1} \cdot d^{-1}$) and for moderate to intense exercise for 1–3 h per day (6–10 $g \cdot kg^{-1} \cdot d^{-1}$) for 96.3% and 100% of players, respectively. Daily mean protein intake was 1.5 $g \cdot kg^{-1} \cdot d^{-1}$, therefore, consistent with recommendations (1.2–2.0 $g \cdot kg^{-1} \cdot d^{-1}$) for 77.8% of players. Daily mean calcium intake was 924.8 $mg \cdot d^{-1}$, therefore, below recommendations (1000 $mg \cdot d^{-1}$) for 65.5% of players, while mean iron intake was 12.2 $mg \cdot d^{-1}$, also below recommendations (18 $mg \cdot d^{-1}$) for 100% of players. Players answered 54.5% of SNKQ questions correctly, with the lowest scores observed in the section on supplements. Risk of LEA was evident in 30% of players, with no differences in carbohydrate ($p = 0.238$), protein ($p = 0.296$), fat ($p = 0.490$) or energy ($p = 0.971$) intakes between players at risk of LEA and those not at risk. The results suggest that female Australian rules football players have an inadequate intake of carbohydrate and calcium and low sports nutrition knowledge. Further investigation to assess the risk of LEA using direct measures is required.

Keywords: carbohydrate intake; team sports; female athletes; nutritional recommendations; nutrition knowledge

---

1. Introduction

In 2017, the Australian Football League Women's (AFLW) competition commenced. In addition to the elite AFL competition, Australian rules football is comprised of numerous sub-elite competitions including the Victorian Football League (VFL) and regional leagues [1,2]. Prior to the initiation of the AFLW, there was no elite-level competition available to female Australian rules football players, and, therefore, many current players have transferred from other sports, including basketball and soccer [3]. Although AFLW in Australia is rapidly evolving, many AFLW and Victorian Football

League Women's (VFLW) players have experienced minimal elite sport experience within a structured training program [3]. It has been established that differences exist between the male elite AFL and sub-elite VFL competition, in terms of increased match intensity, greater running distance and overall game speeds for AFL players [2,4]. However, these differences may not exist between AFLW and VFLW players because the competitions are similar, both characterised by shorter quarters (up to 20 min), no limit on the number of rotations, and players' part-time status [3].

The nutritional requirements of Australian rules football players vary throughout the season [5], which may impact on macronutrient consumption, including appropriate carbohydrate, protein and total energy intake. Current daily carbohydrate recommendations for moderate- to high-intensity exercise (1–4 h duration) range between 5 and 12 $g \cdot k^{-1}$ body mass (BM) [6]. In the context of Australian rules football, moderate- to high-intensity exercise between 1 and 3 h is performed during pre-season training, in-season main training sessions and during a match. Carbohydrate requirements are reduced during low-intensity exercise of up to one hour (3–5 $g \cdot k^{-1} \cdot d^{-1}$ BM) [6]. Contrary to recommendations, previous studies have reported relatively low carbohydrate intakes in male AFL players (<5 $g \cdot kg^{-1} \cdot d^{-1}$) [5,7,8], and similar intakes have been identified in Australian female athletes from a range of individual and team sports [9]. While the recommended carbohydrate intake for high-intensity exercise is well established, the consistently reported low carbohydrate intake in team sport athletes may suggest that the physical demands of these sports require less carbohydrate than current recommendations. In contrast, athletes tend to exceed the recommended protein intake [9]. Protein requirements for athletes are higher than those for the general population (0.8–1.0 $g \cdot kg^{-1} \cdot d^{-1}$ [10]) to assist with muscle protein synthesis and recovery, with the current recommendations specifying that athletes consume 1.2–2 $g \cdot kg^{-1} \cdot d^{-1}$, depending on training program goals and overall energy budgets [6]. The average reported protein intake for Australian female athletes is 1.6 $g \cdot kg^{-1} \cdot d^{-1}$ [9]. At present, recommendations for micronutrient intake in athletes are consistent with general population health guidelines [10]. However, it is likely that certain athletes may have an increased requirement for specific micronutrients, including vitamins B for athletes who restrict their energy intake, which is common in female athletes [9], iron for endurance athletes and calcium and vitamin D for amenorrheic females [6,9]. The current literature assessing micronutrient intake in female athletes is limited, with one Australian study reporting that mean micronutrient intake met the relevant Australian Nutrient Reference Values [10]. Despite this, a significant proportion of individual athletes in this study failed to meet the general population health recommendations, in particular for folate (44%), calcium (22%), iron (19%) and magnesium (15%) [9].

Nutrition education is one strategy used to assist athletes to consume an adequate diet [11]. Athletes with higher nutrition knowledge are more likely to consume more fruit, vegetables and carbohydrate-rich foods than those with low nutritional understanding [12], which suggests that sports nutrition knowledge may be associated with appropriate dietary intake. Women have been reported to have a higher level of nutrition knowledge than men, with a recent study reporting that Australian female athletes had a higher total knowledge score than other participants [13]. In male Australian rules footballers, three studies have reported poor sports nutrition knowledge, with only approximately half of the questions being answered correctly [7,8,14]. Currently, sports nutrition knowledge in female football players who have not been training within an elite program or had regular exposure to nutrition support is unknown.

Energy availability (EA) is defined as the ingested energy remaining for all other metabolic processes after the energy cost of training has been subtracted [15]. Low EA (LEA) occurs when the amount of energy available for basic physiological functions becomes insufficient, which, if continued, may lead to a significant health risk for female athletes [16]. LEA may be present with or without a clinical eating disorder, with the causes often being unintentional and due to increased energy expenditure without an increase in energy intake [17]. Directly measuring EA can be challenging as it requires the quantification of 24 h energy intake and expenditure over a period of time, which is often impractical and burdensome for athletes. Furthermore, there is currently no gold standard approach for the measurement of exercise energy expenditure, as many tools used in an applied context, including accelerometers, underestimate energy expenditure at high exercise intensities [18]. Because of the challenges associated with directly measuring EA, the validated Low Energy Availability in Females Questionnaire (LEAF-Q) has been used to assess the risk of LEA in athletes, with the prevalence reported to range between 33 and 46% [19,20]. However, most of the existing literature has focused on endurance athletes, and little is known about team sport athletes. Furthermore, female Australian rules football players may be at an increased risk of low EA due to an increase in energy expenditure after joining a professional program. Thus, screening for the risk of LEA in this group of athletes may have important long-term implications for health and performance and may determine the need for further education to ensure appropriate dietary intake to meet the demands of the sport.

To date, no previous studies have assessed dietary intake, nutritional knowledge or risk of low energy availability in female Australian rules football players. Further, data assessing nutritional adequacy in Australian female athletes, in particular team sport athletes, are limited. As the AFLW league is new and rapidly growing, this information will be beneficial for coaches, fitness trainers, medical staff, nutritionists and dietitians working with these female athletes. Thus, the aims of the current investigation were to: (a) quantify energy, macronutrient and micronutrient intake in female Australian rules football players and compare these to current recommendations, (b) quantify sports nutrition knowledge in female Australian rules football players, and (c) quantify the risk of low EA in female Australian rules football players and compare nutrient intake between players who are at risk of low EA and those not at risk.

2. Materials and Methods

2.1. Subjects and Study Design

This study followed a cross-sectional design, involving VFLW players (n = 30) aged 18–35 years. Players were recruited in Victoria from one elite-level professional AFL club. Data collection was performed during the 2017 VFLW preseason (March–April). The study was conducted in accordance with the Declaration of Helsinki, and the Deakin University Human Research Ethics Committee approved the study (Project Approval Code: EC00213). All players provided their written informed consent prior to their involvement in the study.

2.2. Data Collection

2.2.1. Anthropometry

Height (cm) was measured using a portable stadiometer (Seca, 213, Hamburg, Germany), and body mass (kg) was measured using electronic scales (A&D Australasia, HW-PW200, Adelaide, South Australia) by a qualified research assistant.

2.2.2. Nutritional Intake

The Automated Self-Administered 24 Hour (ASA24) Dietary Assessment Tool [21], based on an Australian-based food and beverage database, was used to assess the previous 24 h (midnight to midnight) nutritional intake on three non-consecutive training days within a seven-day period. The data entered were checked via a follow-up interview with a qualified nutritionist. The mean daily intake of energy, macronutrients and micronutrients over the three 24 h recalls was calculated. The mean daily protein and carbohydrate intake was compared with the values reported in the current American College of Sports Medicine (ACSM) sports nutrition guidelines [6,22] and were expressed in grams (g), relative to BM (g·kg·BM^{-1}). Carbohydrate intake was compared to relevant recommended values (5–7 g·kg·BM^{-1} and 6–10 g·kg·BM^{-1}) [6], given the physical demands and duration of pre-season training. Daily fat intake, expressed as a percentage of total daily energy intake, and micronutrient intake, expressed as milligrams (mg) or micrograms (µg), was compared with those established by the Australian general population health recommendations [10] as there are currently no guidelines specific to athletes. The current type of supplement use was captured via an online software program (Qualtrics, Provo, UT, USA).

2.2.3. Sports Nutrition Knowledge

Players completed the 88-item Sports Nutrition Knowledge Questionnaire (SNKQ) [23] via an online software program (Qualtrics, Provo, UT, USA). The SNKQ has been assessed for validity (content and construct) and reliability (test–retest), with findings indicating a high construct validity and good test–retest concordance and therefore suitability to be used to determine sports nutrition knowledge. The SNKQ consists of five sub-sections (general nutrition concepts, fluid, recovery, weight control and supplements). One point was awarded for each correct answer, and an 'unsure' or incorrect response received zero points. The overall score (out of 88) was expressed as a percentage of correctly answered questions.

2.2.4. Energy Availability

The 25-item LEAF-Q was used to assess the risk of LEA considering three factors: gastrointestinal function, injuries and menstrual function. The LEAF-Q has been validated in female athletes aged 18–39 training ≥5 times/week, with findings producing an acceptable sensitivity (78%) and specificity (90%) to classify current energy availability [16]. Consistent with the original validation study [16], players completed a paper version of the LEAF-Q to ensure validity and reliability were maintained. Scoring was based on the original validation study, with those who scored ≤7 being classified as 'not at risk' of LEA, and those who scored ≥8 being classified as 'at risk' of LEA [16].

2.3. Statistical Analysis

All statistical analyses were conducted using the Statistical Package for Social Sciences (SPSS 24.0, Chicago, IL, USA). Results were reported as mean ± standard deviation (SD) for normally distributed variables and median and interquartile range (IQR) for non-normally distributed variables. Independent samples *t*-tests were used to compare energy and macronutrient intake between players who were deemed to be at risk of low EA (≥8) and those who were shown to be not at risk of low EA (≤7). Statistical significance was set to a level of $p \leq 0.05$.

3. Results

The demographic characteristics of players (n = 30) are provided in Table 1. The completion rate of the ASA-24 dietary recalls, SNKQ and LEAF-Q was 29 (97%), 30 (100%) and 27 (90%), respectively, for 30 participants.

Nutritional intake is presented in Table 2. Mean carbohydrate intake was below recommendations for training days (moderate- to high-intensity activity), while protein and fat intake met current recommendations (Table 2). In comparison with current sports nutrition guidelines, carbohydrate intake was less than the recommended value for moderate exercise of approximately one hour (5–7 g·kg^{-1}) in 96.3% of players and for moderate to intense exercise (6–10 g·kg^{-1}) for 1–3 h [6] in 100% of players.

Micronutrient intake was consistent with current general population guidelines, with the exception of iron and calcium, while sodium intake exceeded the current recommendations (Table 3). In comparison to the general population guidelines, 65.5% of players' calcium intake was below recommendations, while none of the players met the minimum requirement for iron. Despite the average intake recommendations, 38% of players' magnesium and 20.7% of players' zinc was below the general population guidelines.

The median score for total sports nutrition knowledge was 48 correct answers out of possible 88 (54.5%) [23]. The lowest median scores were observed in the sub-section on supplements (18% correctly answered) (Table 4).

Table 1. Demographic characteristic of VFLW players (n = 30).

	(n = 30)
(Mean ± SD)	
Age (years)	24.15 (±4.1)
Mass (kg)	64.5 (±8.0)
Height (cm)	168.2 (±7.6)
Birth Place [N, (%)]	
Australia	30 (100)
Relationship Status [N, (%)]	
Married	3 (10.0)
De facto	0 (0.0)
Committed dating/Engaged	13 (43.3)
Never married/Single	14 (46.7)
Level of Education [N, (%)]	
Year 12 or equivalent	13 (43.3)
Trade/apprenticeship	1 (3.3)
Certificate/diploma	4 (13.3)
University degree	9 (30.0)
Higher University degree	3 (10.0)

VFLW: Victorian Football League Women's; SD: standard deviation; N: number.

Table 2. Macronutrient intake of VFLW players (mean ± SD) and current recommendations.

Nutrient Intake	VFLW (n = 29)	Recommendation	Percent Meeting Minimum Recommendation
Energy, kJ	7826 ± 2411.6	NA	NA
Energy kJ·kg^{-1}·d^{-1}	199.5 ± 37.4		
Protein, g	98 ± 32.1	NA	NA
Protein, g·kg^{-1}·d^{-1}	1.5 ± 0.5	1.2–2.0 [a]	77.8%
Carbohydrate, g	192.4 ± 51.8	NA	NA
Carbohydrate, g·kg^{-1}·d^{-1}	3.0 ± 0.8	5–7 [b]	3.7%
		6–10 [c]	0%
Sugar, g	86.2 ± 33.1	NA	NA
Sugar, % of E	18.6 ± 4.4	NA	NA
Fibre, g	25.5 ± 8	25g [d]	43.8%
Total fat, g	72.2 ± 33.4	NA	NA
Total fat, % of E	33.2 ± 6.5	20–35 [d]	96.4%
Saturated fat, g	25.7 ± 14.6	NA	NA
Saturated fat, % of E	11.6 ± 3.2	<10 [d]	34.4%
Monounsaturated fat, g	29 ± 14.1	NA	NA
Polyunsaturated fat, g	11.4 ± 4.8	NA	NA

% of E: percentage of total energy intake; NA: Not Applicable. [a] source: reference [6], [b] source: reference [6], up to 1 h of moderate exercise/day [c] source: reference [6], 1–3 h of moderate- to high-intensity exercise/day, [d] source: reference [10].

Table 3. Micronutrient intake of VFLW players (mean ± SD) and current recommendations.

Nutrient Intake	VFLW (n = 29)	Recommended Dietary Intake (RDI) or Adequate Intake (AI) [a]	Upper Limit	Percent Meeting Minimum Recommendation
Calcium, mg	924.8 ± 544.7	1000	2500	34.5%
Iron, mg	12.2 ± 3.2	18	45	0%
Magnesium, mg	367.5 ± 137.8	310	350 (as a supplement)	82.8%
Phosphorus, mg	1569.3 ± 549.4	1000	4000	96.6%
Potassium, mg	3109 ± 1173	2800	NA	58.6%
Sodium, mg	2063.3 ± 957	460–920	Not Determined	100%
Zinc, mg	11.7 ± 4	8	40	79.3%
Selenium, µg	98.1 ± 64.7	60	400	79.5%
Vitamin C, mg	106.8 ± 115.3	45	NA	82.8%
Thiamine, mg	1.9 ± 1.9	1.1	NA	69.0%
Riboflavin, mg	2.8 ± 2.2	1.1	NA	96.6%
Niacin, mg	25.5 ± 8.9	14	35	96.7%
Folate, µg	484.6 ± 149.8	400	1000	69.0%
Vitamin B12, µg	13.7 ± 46.8	2.4	NA	93.1%

[a] source: reference [10].

Table 4. Nutrition knowledge scores in VFLW players (median (IQR), percent (%) of correct answers).

Sections (No. Questions)	VFLW (n = 30) Median (IQR), %
General Nutrition Concepts (46)	28 (7), 60.8%
Fluid (9)	6 (7), 66.7%
Recovery (7)	4 (3), 57.1%
Weight Control (15)	7 (3), 46.7%
Supplements (11)	2 (3), 18.2%
Total Nutrition Knowledge (88)	48 (12), 54.5%

IQR: Interquartile range.

The median score on the LEAF-Q was 7, with 8 out of 27 players (30%) scoring above 8, indicating a risk of LEA [16]. Energy and macronutrient intake did not differ ($p > 0.05$) between players who were classified as at risk of LEA and players who were classified as not at risk of LEA (Table 5). Table 6 highlights the key LEAF-Q responses. Many players reported injuries in the previous year, the most common being ankle- or foot-related. Nine of the 27 players (33%) reported exercise-induced changes to menstrual function. Gastrointestinal symptoms not related to menstruation were less frequent.

The most common forms of supplementation used by players were protein supplementation (65%) and vitamins and minerals supplementation (70%) (Figure 1). Performance-enhancing supplements such as creatine, B-alanine, caffeine and nitrates were not used by any player (Figure 1).

Table 5. Energy and macronutrient intake (mean ± SD) in VFLW players at risk of LEA and in players not at risk of LEA (n = 27).

	Not at Risk of LEA (LEAF-Q ≤ 7) (n = 19)	At Risk of LEA (LEAF-Q ≤ 8) (n = 8)	*p*-Value
Energy (kJ)	7734.7 ± 2192.4	7699.3 ± 2552.6	0.971
Protein (g·kg^{-1}·d^{-1})	1.5 ± 0.4	1.7 ± 0.8	0.296
Carbohydrate (g·kg^{-1}·d^{-1})	2.9 ± 0.8	3.4 ± 1.1	0.248
Fat (g)	73.1 ± 30.1	63.9 ± 31.7	0.490

LEA: Low Energy Availability; LEAF-Q: LEA in females Questionnaire.

Table 6. Responses to key components of the LEAF-Q (n = 27).

Leaf Questionnaire Component	Frequency (N)	Percent (%)
1. Injury history		
Number of days lost from participation due to injury in past year:		
0	11	40.7
1–7	6	22.2
8–14	4	14.8
15–21	3	11.1
≥22	3	11.1
Most common injuries reported (respondents could choose more than one)		
Back injury	3	11.1
Knee injury	2	7.4
Head injury/concussion	4	14.8
Groin injury	2	7.4
Shin splints/calf tightness	4	14.8
Ligament/tendon injury of the thumb	4	14.8
Rolled ankle/Achilles soreness/broken foot	6	22.2
Illness	2	7.4
Hamstring strain	1	3.7
Hip injury	1	3.7
Shoulder injury strained ac joint	2	7.4
2. Menstrual function exercise-related menstrual changes:		
Bleed less	3	11.1
Menstruation stops	4	14.8
Increased bleeding	2	7.4
Number of periods in the last year (n = 21) (if still menstruating)		
≥9	19	90.5
5–8	0	0
0–2	2	9.5
3. Gastrointestinal disturbances abdominal bloated/gaseous when not having periods:		
Daily-weekly	2	7.4
Seldom	12	44.4
Rarely or never	13	48.1
Cramps/stomach ache not related to your menstruation:		
Daily-weekly	1	3.6
Seldom	8	28.6
Rarely or never	19	67.9

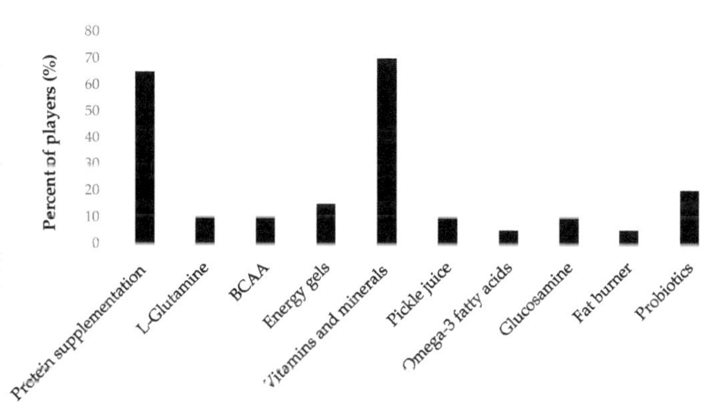

Figure 1. Current supplement use in VFLW players (n = 30). BCAA: branched-chain amino acids

4. Discussion

This is the first study to assess nutrient intake, sports nutrition knowledge and risk of LEA within female Australian rules football players. The key findings from this study were that carbohydrate intake on training days was below recommendations for players performing moderate training of approximately one hour per day (5–7 $g·kg^{-1}·d^{-1}$) and moderate- to high-intensity training of 1–3 h per day (6–10 $g·kg^{-1}·d^{-1}$) [6]. Protein intake was consistent with current recommendations (1.2–2 $g·kg^{-1}·d^{-1}$) [6] and consistent with previous findings [5,7,8]. The assessment of micronutrient intake showed that the average intake of calcium and iron was below the general population recommendations for females (1000 $mg·d^{-1}$ and 18 $mg·d^{-1}$, respectively), while other micronutrients were in line with recommendations. Sodium intake exceeded the recommended daily intake for the general population (460–920 $mg·d^{-1}$). The overall sports nutrition knowledge scores were low, with the lowest level of knowledge related to supplementation (18% correctly answered). Furthermore, 30% of players were deemed to be at risk of LEA, which is similar to previously reported rates in female athletes [19,20].

Mean energy intake in this group of female Australian rules football players was 7826 ± 2411 $kJ·kg^{-1}$, which is slightly below that reported in previous studies in Australian female athletes, with mean energy intakes ranging from 8674 kJ to 10,511 $kJ·kg^{-1}$ [9,24,25]. The lower energy intake in the current investigation may be due to the nature of Australian rules football. Endurance athletes (such as cyclists and triathletes) have been reported to have higher energy intakes (11,850 kJ per day) than team sport athletes such as softball players (8959 kJ per day) and power sport athletes within individual sports such as track and field (8951 kJ per day) [9]. Furthermore, the method of estimating energy intake may also contribute to the differences in results, with previous studies using food frequency questionnaires (FFQ), which can overestimate energy intake [9], or 24 h recalls, which can underestimate energy intake [24]. In the current study, the repeated online 24 h recalls may have underestimated energy intake, since the ASA-24 is a self-administered tool. The estimation of energy requirements in football players is challenging, as requirements can increase or decrease depending on age, athlete caliber, overall daily activity levels and body composition [26]. Objective measures of energy expenditure such as those obtained by using accelerometers can provide a more accurate representation of energy requirements; however, there are limitations to their use in high-intensity sport as well as when performing resistance training [18,27]. To overcome differences in exercise modes, the use of weighed food diaries has been suggested as the gold standard for assessing dietary intake in athletes [9]. Therefore, further research is required to estimate energy expenditure and intake using the most accurate tools currently available, in order to more comprehensively determine whether energy intake within this population is sufficient.

Carbohydrate intake assessed on training days (involving a minimum of one hour of moderate exercise) in this group of VFLW players was below the minimum recommended 5 $g·kg^{-1}·d^{-1}$ [6]. This finding is consistent with previous findings in both Australian female athletes [9,24] and male AFL players [5,7,8]. Furthermore, the majority of players (96.3%) did not meet the recommendations for athletes performing moderate exercise for approximately 1 h per day (5–7 $g·kg^{-1}$), while no players (0%) met the recommendation for moderate- to high-intensity exercise for 1–3 h per day (6–10 $g·kg^{-1}$) [6], consistently with previous reports of carbohydrate intake during pre-season training [5,28]. Female athletes from a range of different sports have previously been reported to have a similar daily intake of carbohydrate to that of the current investigation (4.5 $g·kg^{-1}·d^{-1}$), with the lowest intakes in softball players (3.3 $g·kg^{-1}·d^{-1}$) [9]. Furthermore, it was reported that female endurance athletes' carbohydrate intake was 6.7 $g·kg^{-1}·d^{-1}$, higher than that for other female athletes [9], a result which may have been due to either a higher training demand or a greater knowledge of the importance of carbohydrate for performance in this sport [9]. Low-carbohydrate diets have become popular amongst the general public as well as female athletes, as they are often perceived to be a tool for weight or fat loss [29]. In athletes, a sustained low-carbohydrate intake may be insufficient to support the training demands, possibly impacting on performance and long-term health [29]. The carbohydrate intakes observed in the current investigation, however, are consistent with those previously reported for both male and female team sport athletes [30]. Furthermore, in the current study, there was no significant difference in carbohydrate intake between players who were at risk of LEA and those not at risk, suggesting that the carbohydrate intake in this group may be sufficient to meet the training demands. The current carbohydrate intake recommendations are based on the outcomes of studies that include athletes from varied populations, with a large proportion from endurance sport [6], and therefore it should be considered whether they are applicable to team-based sports, which often involve intermittent and skill-based training [31]. Thus, further research in female team sport athletes' habitual carbohydrate intake, glycogen depletion rates and optimal carbohydrate periodization interventions [32,33] is required to provide evidence for guidelines specific to a female team sport athlete population.

The mean intake of most micronutrients in the current investigation was above the Australian RDI for the general population. Only one other Australian study has measured micronutrient intake in female athletes, reporting similar findings [9], the key differences being that within the current study, calcium (924.8 mg·d^{-1}) and iron (12.2 mg·d^{-1}) intakes were below the RDI, and a higher proportion of athletes did not meet the RDI for calcium (65%) and iron (100%). Adequate intakes of calcium, which plays a vital role in bone health and injury risk [34,35], and iron, essential to carry oxygen and maintaining energy release [36], are important for females, in particular when performing high-intensity contact sport. The main sources of calcium in the diet are dairy foods, including milk, cheese and yoghurt [10], while the main sources of iron include meat products, in particular red meat [10]. In the current investigation, the mean intake of dairy food per day was 1.2 serves, which is below the recommended 2.5 serves for the age group of the participants [10]. Furthermore, 17% of players did not consume any dairy food, and only 53% of players reported drinking milk, while only 50% consumed red meat 2–4 times per week. A low intake of dairy [37] and meat [36] foods is commonly reported in young Australian females and may be a result of the perception that these foods can contribute to weight gain [38,39]. Furthermore, intolerances to dairy food and lactose products have become more common [40], and veganism has become more visible and accepted in the sports industry [41], possibly contributing to the low intakes of these minerals. Nonetheless, there is a need for further education focused on the importance of calcium intake and the appropriate dietary sources for female football players.

Sodium intake in the current investigation was substantially higher than the recommended intake for the Australian general population. In the current study, 20% of players reported consumption of sports drinks, which are typically high in sodium, on a daily basis, which may therefore have contributed to the overall sodium intake [42]. There is limited available evidence on the required sodium intake in an athletic population, as there is a high degree of inter-variability between athletes, often dependent on individual sweat sodium concentrations and sweat rates [43]. Furthermore, the current recommendations are aimed at reducing the risk of chronic disease [10], and a recent study found that the general health recommendations of a low-sodium diet may not be appropriate for athletes, especially in warm conditions [44]. Overall, there is limited understanding of the micronutrient requirements of athletes and whether their training demands can be adequately met by intakes that match the general population guidelines. For female athletes, menstruation may affect micronutrient status, especially that of iron, vitamin B12 and zinc, which is another consideration that warrants further investigation.

The results of the current investigation suggest that female Australian football players do not have a high level of sports nutrition knowledge. This is the first study to investigate sports nutrition knowledge within the female population of Australian rules football players; however, the median score for overall sports nutrition knowledge (54.4%) was similar to that previously reported in male Australian football players [7,14]. The aforementioned similarity in sports nutrition knowledge was found despite the previous studies' use of a different sports nutrition knowledge questionnaire [45]. In a recent study from our group using the same sports nutrition knowledge questionnaire as that used in the current investigation [46], we found scores to be 51% in both elite and sub elite male

Australian football players [8]. These observations collectively suggest that sports nutrition knowledge levels may be similar between male and female Australian rules football players. Improving upon low sports nutrition knowledge could be beneficial to male and female Australian rules football players, as it has recently been suggested that for team sport athletes, an increased level of sports nutrition knowledge might facilitate improvements in performance [47]. Furthermore, in the current investigation, knowledge specific to supplements elicited in the lowest median scores (18%), which is consistent with the result reported for male sub-elite players in our recent investigation [8], suggesting that deficits in specific areas of sports nutrition knowledge may affect both males and female Australian rules football players similarly.

In the current study, 30% of players were deemed to be at risk of LEA. This is the first investigation to explore the risk of low energy availability for female Australian rules football players. However, recent investigations have reported an incidence of LEA of 33.3% in a similar participant population and stage of the training year (competitive female soccer players in either their pre-season or competitive season) [48,49]. In the current study, energy and macronutrient intake did not differ between players who were classified as at risk of LEA and players who were classified as not at risk of LEA. This observation contrasts with the result of a previous investigation [45] which reported significantly lower daily energy intake in females with LEA. The aforementioned study, however, directly measured energy availability, whereas only the risk of LEA was quantified in the current investigation, highlighting the need for the further investigation and direct measurement of EA in female Australian football players. It has been conclusively determined that LEA can induce adverse health effects such as increased injury rates and menstrual dysfunction [50], and indeed, in the current investigation, players who were identified as "at-risk" of LEA reported injuries in the past year. Furthermore, exercise-induced changes to menstrual function were reported in one-third of players in the current investigation. There is at least some positive association between knowledge of appropriate nutritional practices and nutritional intake [51]. Therefore, the implementation of nutritional education programs that incorporate specific information about the potentially deleterious health effects of long-term LEA may assist to improve health outcomes for female Australian rules football players.

Several limitations need to be addressed in the current study. Firstly, the players in this study were recruited from one professional sporting club, and thus the sample size was limited by the number of players in the VFLW team. As a result, additional analysis investigating the correlation between nutritional intake and knowledge could not be included within the current investigation. This data would have provided insight into the potential factors influencing nutritional intake, forming the basis of targeted education programs. However, the aim of this study was to gather detailed dietary information in an emerging athlete population, which future studies can build upon. Furthermore, the findings may be specific to players from this club. However, given that this study was conducted during the early phase of the pre-season, in the inaugural year of the VFLW competition, it is unlikely that the club's philosophies, training or education programs would have impacted on the players. Although dietary intake data were captured on training days, training logs were not kept, which limited the detail of records in terms of training session duration and intensity. The dietary assessment tool also has limitations, as, although it is validated [21], the online ASA-24 is self-administered and, therefore, subject to memory bias and underreporting. Finally, the LEAF-Q used in this study can only assess the risk of LEA. To objectively and accurately investigate energy availability, gold standard methods to assess energy availability include doubly labelled water, portable metabolic carts and weighed food diaries over numerous days. However, these methods are costly, burdensome on athletes and often impractical in field-based studies. Other methods to assess energy expenditure such as the use of accelerometers and weighed food diaries (used for a limited number of days) may be more realistic to implement for athletes and would provide further information on the status of energy availability.

5. Conclusions

The results of the current investigation suggest that the female population of Australian rules football players exhibits inadequate carbohydrate, iron and calcium intake with respect to the current recommendations for athletes. Almost one-third of players may be at risk of LEA, and therefore the prevalence of LEA within this population may be similar to that reported for other female athletes. Protein intake was in line with athlete recommendations, while sodium intake within this athlete population appeared to exceed the recommended intake for general populations. Overall, sports nutrition knowledge was found to be low for female Australian football players, especially regarding supplementation. Female football players may benefit from education programs that provide details of the deleterious effects of prolonged LEA.

Author Contributions: All authors conceived and designed the experiments; R.L. and D.C. analysed the data; All authors wrote the paper.

Funding: This research received no external funding.

Acknowledgments: The authors acknowledge the participants for their involvement in the study.

Conflicts of Interest: The authors declare no conflict of interest.

References

1. Gray, A.; Jenkins, D. Match analysis and the physiological demands of Australian Football. *Sports Med.* **2010**, *40*, 347–360. [CrossRef] [PubMed]
2. Johnston, R.D.; Black, G.M.; Harrison, P.W.; Murray, N.B.; Austin, D.J. Applied sport science of Australian football: A systematic review. *Sports Med.* **2018**, *48*, 1673–1694. [PubMed]
3. Clarke, A.C.; Ryan, S.; Couvalias, G.; Dascombe, B.J.; Coutts, A.J.; Kempton, T. Physical demands and technical performance in Australian Football League Women's (AFLW) competition match-play. *J. Sci. Med. Sport* **2018**, *21*, 748–752. [CrossRef] [PubMed]
4. Aughey, R.J. Widening margin in activity profile between elite and sub-elite Australian football: A case study. *J. Sci. Med. Sport.* **2013**, *16*, 382–386. [CrossRef]
5. Bilsborough, J.C.; Greenway, K.; Livingston, S.; Cordy, J.; Coutts, A.J. Changes in anthropometry, upper-body strength and nutrient intake in professional Australian Football players during a season. *Int. J. Sports Physiol. Perform.* **2016**, *11*, 290–300. [CrossRef] [PubMed]
6. Thomas, T.; Erdman, K.; Burke, L. American college of sports medicine joint position statement nutrition and athletic performance. *Med. Sci. Sports Exerc.* **2016**, *48*, 543–568. [PubMed]
7. Devlin, B.; Leveritt, M.; Kingsley, M.; Belski, R. Dietary intake, body composition, and nutrition knowledge of Australian Football and soccer players: Implications for sports nutrition professionals in practice. *Int. J. Sport Nutr. Exerc. Metab.* **2017**, *27*, 130–138.
8. Lohman, R.; Carr, A.; Condo, D. Nutritional intake in Australian football players: Sports nutrition knowledge, macronutrient and micronutrient intake. *Int. J. Sport Nutr. Exerc. Metab.* **2018**, in press. [CrossRef] [PubMed]
9. Heaney, S.; O'Connor, H.; Gifford, J.; Naughton, G. Comparison of strategies for assessing nutritional adequacy in elite female athletes' dietary intake. *Int. J. Sport Nutr. Exerc. Metab.* **2010**, *20*, 245–256. [CrossRef]
10. National Health and Medical Research Council (NHMRC). *Nutrient Reference Values for Australian and New Zealand*; NHMRC: Canberra, Australia, 2017.
11. Birkenhead, K.; Slater, G. A review of factors influencing athletes' food choices. *Sports Med.* **2015**, *45*, 1511–1522. [CrossRef]
12. Alaunyte, I.; Perry, J.L.; Aubrey, T. Nutritional knowledge and eating habits of professional rugby league players: Does knowledge translate into practice? *J. Int. Soc. Sports Nutr.* **2015**, *12*, 18. [CrossRef]
13. Spendlove, J.K.; Heaney, S.E.; Gifford, J.A.; Prvan, T.; Denyer, G.S.; O'Connor, H.T. Evaluation of general nutrition knowledge in elite Australian athletes. *Br. J. Nutr.* **2012**, *107*, 1871–1880. [CrossRef]
14. Devlin, B.; Belski, R. Exploring general and sports nutrition and food knowledge in elite male Australian athletes. *Int. J. Sport Nutr. Exerc. Metab.* **2015**, *25*, 225–232. [CrossRef]
15. Marquez, S.; Molinero, O. Energy availability, menstrual dysfunction and bone health in sports; an overview of the female athlete triad. *Nutr. Hosp.* **2013**, *28*, 1010–1017.

16. Melin, A.; Tornberg, A.B.; Skouby, S.; Faber, J.; Ritz, C.; Sjodin, A.; Sundgot-Borgen, J. The LEAF questionnaire: A screening tool for the identification of female athletes at risk for the female athlete triad. *Br. J. Sports Med.* **2014**, *48*, 540–545. [CrossRef] [PubMed]

17. Logue, D.; Madigan, S.M.; Delahunt, E.; Heinen, M.; Mc Donnell, S.J.; Corish, C.A. Low energy availability in athletes: A review of prevalence, dietary patterns, physiological health, and sports performance. *Sports Med.* **2018**, *48*, 73–96. [CrossRef]

18. Crouter, S.E.; Churilla, J.R.; Bassett, D.R., Jr. Estimating energy expenditure using accelerometers. *Eur. J. Appl. Physiol.* **2006**, *98*, 601–612. [CrossRef] [PubMed]

19. Folscher, L.L.; Grant, C.C.; Fletcher, L.; Janse van Rensberg, D.C. Ultra-marathon athletes at risk for the female athlete triad. *Sports Med. Open* **2015**, *1*, 29. [CrossRef] [PubMed]

20. Slater, J.; McLay-Cooke, R.; Brown, R.; Black, K. Female recreational exercisers at risk for low energy availability. *Int. J. Sport Nutr. Exerc. Metab.* **2016**, *26*, 421–427. [CrossRef]

21. Subar, A.; Kirkpatrick, S.; Mittl, B.; Zimmermanm, T.; Thompson, F.; Bingley, C.; Willis, G.; Islam, N.; Baranowski, T.; McNutt, S.; et al. The automated self-administered 24-hour dietary recall (ASA24): A resource for researchers, clinicians, and educators from the National Cancer Institute. *J. Acad. Nutr. Diet.* **2012**, *112*, 1134–1137. [CrossRef]

22. Campbell, B.; Kreider, R.; Ziegenfuss, T.; Bounty, P.; Roberts, M.; Burke, D.; Landis, J.; Lopez, H.; Antonio, J. International society of sports nutrition position stand: Protein and exercise. *J. Int. Soc. Sports Nutr.* **2007**, *4*, 8. [CrossRef] [PubMed]

23. Zinn, C.; Schofield, G.; Wall, C. Development of a psychometrically valid and reliable sports nutrition knowledge questionnaire. *J. Sci. Med. Sport* **2005**, *8*, 346–351. [CrossRef]

24. Baker, L.; Heaton, L.; Nuccio, R.; Stein, K. Dietitian-observed macronutrient intakes of young skill and team-sport athletes: Adequacy of pre, during and postexercise nutrition. *Int. J. Sport Nutr. Exerc. Metab.* **2014**, *24*, 166–176. [CrossRef] [PubMed]

25. Burkhart, S.J.; Pelly, F.E. Dietary intake of athletes seeking nutrition advice at a major international competition. *Nutrients* **2016**, *8*, 638. [CrossRef]

26. Mielgo-Ayuso, J.; Maroto-Sanchez, B.; Luzardo-Socorro, R.; Palacios, G.; Palacios Gil-Antunano, N.; Gonzalez-Gross, M. Evaluation of nutritional status and energy expenditure in athletes. *Nutr. Hosp.* **2015**, *31*, 227–236.

27. Rawson, E.S.; Walsh, T.M. Estimation of resistance exercise energy expenditure using accelerometry. *Med. Sci. Sports Exerc.* **2010**, *42*, 622–628. [CrossRef]

28. Walker, E.; McAinch, A.; Sweeting, A.; Aughey, R. Inertial sensors to estimate the energy expenditure of team-sport athletes. *J. Sci. Med. Sport* **2016**, *19*, 177–181. [CrossRef]

29. Burke, L.M.; Hawley, J.A.; Wong, S.H.; Jeukendrup, A.E. Carbohydrates for training and competition. *J. Sports Sci.* **2011**, *29*, S17–S27. [CrossRef]

30. Holway, F.; Spriet, L. Sport-specific nutrition: Practical strategies for team sports. *J. Sports Sci. Med.* **2011**, *29*, S115–S125. [CrossRef]

31. Bishop, D.; Spencer, M. Determinants of repeated-sprint ability in well-trained team-sport athletes and endurance-trained athletes. *J. Sports Med. Phys. Fit.* **2004**, *44*, 1–7.

32. Krustrup, P.; Mohr, M.; Steensberg, A.; Bencke, J.; Kjær, M.; Bangsbo, K. Muscle and blood metabolites during a soccer game: Implications for sprint performance. *Med. Sci. Sports Exerc.* **2006**, *38*, 1165–1174. [CrossRef] [PubMed]

33. Balsom, P.; Wood, K.; Olsson, P.; Ekblom, B. Carbohydrate intake and multiple sprint sports: With special reference to football (soccer). *J. Sci. Med. Sport* **1999**, *20*, 48–52. [CrossRef]

34. Warden, S.J.; Davis, I.S.; Fredericson, M. Management and prevention of bone stress injuries in long-distance runners. *J. Orthop. Sports Phys. Ther.* **2014**, *44*, 749–765. [CrossRef] [PubMed]

35. Goolsby, M.A.; Boniquit, N. Bone Health in Athletes. *Sports Health* **2017**, *9*, 108–117. [CrossRef] [PubMed]

36. Hinton, P.S. Iron and the endurance athlete. *Appl. Physiol. Nutr. Metab.* **2014**, *39*, 1012–1018. [CrossRef] [PubMed]

37. Parker, C.E.; Vivian, W.J.; Oddy, W.H.; Beilin, L.J.; Mori, T.A.; O'Sullivan, T.A. Changes in dairy food and nutrient intakes in Australian adolescents. *Nutrients* **2012**, *4*, 1794–1811. [CrossRef] [PubMed]

38. Barr, S.I. Dieting attitudes and behavior in urban high school students: Implications for calcium intake. *J. Adolesc. Health* **1995**, *16*, 458–464. [CrossRef]

39.	Askovic, B.; Kirchengast, S. Gender differences in nutritional behavior and weight status during early and late adolescence. *Anthropol. Anz.* **2012**, *69*, 289–304. [CrossRef] [PubMed]

40.	Lukito, W.; Malik, S.G.; Surono, I.S.; Wahlqvist, M.L. From 'lactose intolerance' to 'lactose nutrition'. *Asia Pac. J. Clin. Nutr.* **2015**, *24*, S1–S8.

41.	Rogerson, D. Vegan diets: Practical advice for athletes and exercisers. *J. Int. Soc. Sports Nutr.* **2017**, *14*, 36. [CrossRef]

42.	Valentine, V. The importance of salt in the athlete's diet. *Curr. Sports Med. Rep.* **2007**, *6*, 237–240. [CrossRef]

43.	Baker, L.B. Sweating rate and sweat sodium concentration in athletes: A review of methodology and intra/interindividual variability. *Sports Med.* **2017**, *47*, 111–128. [CrossRef] [PubMed]

44.	Holmes, N.; Miller, V.; Bates, G.; Zheo, Y. The effect of exercise intensity on sweat rate and sweat sodium loss in well trained athletes. *J. Sci. Med. Sport* **2011**, *14*, e112. [CrossRef]

45.	Shifflett, B.; Timm, C.; Kananov, L. Understanding athletes' nutritional needs among athletes, coaches and athletic trainers. *Res. Quart. Exerc. Sport.* **2002**, *73*, 357–362. [CrossRef] [PubMed]

46.	National Cancer Institute. ASA24-Australia: National Cancer Institute 2018. Available online: https://epi.grants.cancer.gov/asa24/respondent/australia.html (accessed on 10 February 2017).

47.	Nikolaidis, P.T.; Theodoropoulou, E. Relationship between nutrition knowledge and physical fitness in semiprofessional soccer players. *Scientifica* **2014**, *2014*, 180353. [CrossRef] [PubMed]

48.	Reed, J.; De Souza, M.; Williams, N. Changes in energy availability across the season in Division I female soccer players. *J. Sports Sci. Med.* **2013**, *31*, 314–324. [CrossRef] [PubMed]

49.	Reed, J.; De Souza, M.; Kindler, J.; Williams, N. Nutritional practices associated with low energy availability in Division I female soccer players. *J. Sports Sci. Med.* **2014**, *32*, 1499–1509. [CrossRef] [PubMed]

50.	Mountjoy, M.; Sundgot-Borgen, J.; Burke, L.; Ackerman, K.; Blauwet, C.; Constantini, N.; Lebrun, C.; Lundy, B.; Melin, A.; Meyer, N. IOC consensus statement on relative energy deficiency in sport (RED-S): 2018 update. *Br. J. Sports Med.* **2018**, *52*, 687–697. [CrossRef]

51.	Spronk, I.; Heaney, S.E.; Prvan, T.; O'Connor, H.T. Relationship between general nutrition knowledge and dietary quality in elite athletes. *Int. J. Sport Nutr. Exerc. Metab.* **2015**, *25*, 243–251. [CrossRef]

© 2019 by the authors. Licensee MDPI, Basel, Switzerland. This article is an open access article distributed under the terms and conditions of the Creative Commons Attribution (CC BY) license (http://creativecommons.org/licenses/by/4.0/).

Article

Anaerobic Performance after a Low-Carbohydrate Diet (LCD) Followed by 7 Days of Carbohydrate Loading in Male Basketball Players

Małgorzata Magdalena Michalczyk [1], Jakub Chycki [2,*], Adam Zając [2], Adam Maszczyk [3], Grzegorz Zydek [1] and Józef Langfort [1]

[1] Department of Sports Nutrition, The Jerzy Kukuczka Academy of Physical Education in Katowice, Mikolowska 72a, 40-065 Katowice, Poland; m.michalczyk@awf.katowice.pl (M.M.M.); g.zydek@awf.katowice.pl (G.Z.); j.langfort@awf.katowice.pl (J.L.)

[2] Department of Sports Training, The Jerzy Kukuczka Academy of Physical Education in Katowice, Mikolowska 72a, 40-065 Katowice, Poland; a.zajac@awf.katowice.pl

[3] Department of Methodology and Statistics, The Jerzy Kukuczka Academy of Physical Education in Katowice, Mikolowska 72a, 40-065 Katowice, Poland; a.maszczyk@awf.katowice.pl

* Correspondence: j.chycki@awf.katowice.pl

Received: 31 January 2019; Accepted: 1 April 2019; Published: 4 April 2019

Abstract: Despite increasing interest among athletes and scientists on the influence of different dietary interventions on sport performance, the association between a low-carbohydrate, high-fat diet and anaerobic capacity has not been studied extensively. The aim of this study was to evaluate the effects of a low-carbohydrate diet (LCD) followed by seven days of carbohydrate loading (Carbo-L) on anaerobic performance in male basketball players. Fifteen competitive basketball players took part in the experiment. They performed the Wingate test on three occasions: after the conventional diet (CD), following 4 weeks of the LCD, and after the weekly Carbo-L, to evaluate changes in peak power (PP), total work (TW), time to peak power (TTP), blood lactate concentration (LA), blood pH, and bicarbonate (HCO_3^-). Additionally, the concentrations of testosterone, growth hormone, cortisol, and insulin were measured after each dietary intervention. The low-carbohydrate diet procedure significantly decreased total work, resting values of pH, and blood lactate concentration. After the low-carbohydrate diet, testosterone and growth hormone concentrations increased, while the level of insulin decreased. After the Carbo-L, total work, resting values of pH, bicarbonate, and lactate increased significantly compared with the results obtained after the low-carbohydrate diet. Significant differences after the low-carbohydrate diet and Carbo-L procedures, in values of blood lactate concentration, pH, and bicarbonate, between baseline and post exercise values were also observed. Four weeks of the low-carbohydrate diet decreased total work capacity, which returned to baseline values after the carbohydrate loading procedure. Moreover, neither the low-carbohydrate feeding nor carbohydrate loading affected peak power.

Keywords: athletes; dietary intervention; anaerobic power; work capacity

---

1. Introduction

Dietary manipulations are an integral part of an athletes training process, related significantly to optimal performance. The mechanisms responsible for improved exercise performance are best recognized for high-carbohydrate diets (HCHO-D), and are attributed to maximizing muscle glycogen content and thereby its availability and utilization during exercise [1,2]. It has been evidenced that a HCHO-D improves performance in both prolonged and low-to-moderate intensity as well as short, high-intensity exercises [3–5]. Interestingly, the effect of exercise on markers of mitochondrial biogenesis, expressed by an increase of citrate synthase suscinate dehydrogenase and β-hydroxyacyl-CoA dehydrogenase activities as well as cytochrome c oxidase IV total protein content, a crucial factor for improving aerobic performance, has been observed when carbohydrate restricted diets were utilized [6]. The molecular mechanism that leads to mitochondrial biogenesis is attributed to activation of peroxisome proliferator-activated receptor-γ coactivator-1 (PGC-1α), recognized as the master regulator of mitochondrial gene expression [7]. Thus, dietary modifications such as carbohydrate restriction have emerged as an alternative strategy for athletes to help improve their performance.

Among carbohydrate-restricted diets, high-fat diets (HFD) have been considered as a possible nourishment manipulation to support progress in athlete's endurance since Phinney et al. [8] showed that duration time of moderate intensity exercise (60% VO_{2max}) was maintained in highly trained cyclists being fed a high-fat ketogenic diet (KD). Interestingly, individual data from this study indicated that some athletes consuming the KD were able to extend the time to exhaustion up to 155%, which suggested the occurrence of individuals more or less vulnerable to the HFD. Considering the moderate intensity of exercise applied, the ergogenic effect of the dietary intervention may be explained by higher rates of muscle ketone bodies/fat utilization in relation to carbohydrate metabolism during the aerobic exercise protocol [9–11]. Recently Zajac et al. [12] reported a significant increase in VO_{2max} and improved lactate threshold in off-road cyclists after a HFD as a result of favorable changes in body mass and body composition. There are also some studies showing improvements or no detrimental effects of an HFD on some markers of aerobic and anaerobic exercise performance [13–16]. However, despite several promising results in favor of improved performance during exercise of moderate intensity, there is a lack of ambiguous evidence to support the benefits of HFD in anaerobic endurance in athletes [14,17]. There are also studies that revealed a negative impact of an HFD on exercise performance in healthy sedentary adults [18,19]. Thus, HFD potentially compromises physical performance during high-intensity exercises and thereby may fail implementation of a training program based on such exercises.

Data on the effects of HFD on high-intensity exercise performance, including strength and speed endurance exercise capacity, are less recognized. A very recent study by Paoli et al. [20] showed a reduction in fat mass, with no changes in both muscle mass and strength in gymnasts who followed a HFD for 30 days. One of the first studies exploring this issue revealed that a 3-day HFD decreased the mean power output without changing peak power values in sedentary young men [21]. In contrast to the proposed benefits of fat metabolism for physical performance of moderate exercise, another study showed increased fat oxidation but impaired high-intensity work output [19,22,23]. In case of high intensity exercise (>75% VO_{2max}) fat oxidation is gradually suppressed, and replaced by accelerated glycolysis via the pyruvate dehydrogenase (PDH) reaction [24,25]. This phenomenon has been described as the crossover concept [26]. Moreover, during high-intensity exercise, fat metabolism shifts into higher utilization of its intramuscular deposits with a profound reduction of plasma non-esterified fatty acid metabolism [5,27].

In the last few decades or so, athletes and scientists have experimented extensively with different dietary procedures to improve body composition, muscular strength and power, as well as overall work capacity [28,29]. Recent studies have focused on a novel concept called "train low" to prepare athletes for competition [30]. This strategy is based on alternative application of low glycogen stores followed by high CHO availability in the training process. This paradigm, which is more frequently applied as the "sleep low" method, assumes beneficial effects of low glycogen availability on mitochondrial biogenesis and aerobic substrate metabolism [31,32], whereas CHO loading allows for high intensity exercise to be performed.

The organization of the training process is a complex task which is normally based on periodization approaches [33]. In competitive sports, the use of short training periods (up to 4 weeks) focused on improving a particular motor ability, while maintaining the level of other abilities, has been named block periodization [34]. A poorly recognized strategy in the training process includes the application of dietary manipulations implemented into consecutive micro-cycles. Considering the above issue, we attempted to verify if a 4-week low-carbohydrate diet (LCD) diet followed by 7 days of HCHO-D affected anaerobic performance measured by means of an all-out lower limb Wingate test. The LCD was implemented into four consecutive micro-cycles in the precompetitive period, followed by a weekly tapering period, in which sport-specific training was combined with a HCHO-D. For the purpose of our study we chose competitive basketball players because there is significant empirical evidence supporting the view that basketball is a sports discipline that requires the performance of multiple dynamic activities relying on both aerobic and anaerobic metabolism. Such an approach is geared at improving tolerance to high intensity workloads in the competitive period of the annual training cycle.

2. Material and Methods

2.1. Participants

Fifteen apparently well-trained male basketball players were enrolled in the study. Their basic somatic characteristics expressed as mean values $\pm$ SD were as follows: age 23.5 $\pm$ 2.2 years; height 194.3 $\pm$ 6.4 cm; body mass 92.18 $\pm$ 5.1 kg; body mass index (BMI) 24.98 $\pm$ 1.86 kg/m^2. All study participants had at least five years of training experience and competed at the division I level of the Polish Basketball League. Basic somatic characteristics of the study participants are presented in Table 1. During the five weeks of the experiment the athletes were fed a low-carbohydrate diet (LCD) for 4 weeks, followed by 7 days (1 week) of carbohydrate loading (Carbo-L) (Figure 1). There was no washout period between the two feeding procedures. One month before the experiment began all participants consumed a standard conventional diet (CD) (Table 2). Participants were informed of the nature of the investigation and written informed consent was obtained prior to study commencement. All participants were free from any diseases and were not taking medications nor dietary supplements during the study. None of the athletes had previous experience with the LCD and carbohydrate-loading procedures. The experimental protocol was approved by the Ethics Committee of the Jerzy Kukuczka Academy of Physical Education in Katowice, Poland (ethics reference KB-5/2015) and conformed to the principles of the Declaration of Helsinki.

Table 1. Body mass and body composition results after CD, LCD, and Carbo-L diets.

Variables	After CD	After LCD	After Carbo-L
	Mean $\pm$ SD	Mean $\pm$ SD	Mean $\pm$ SD
BM (kg)	92.18 $\pm$ 5.17	90.38 $\pm$ 3.12 *	91.82 $\pm$ 4.32
BMI (kg/m^2)	24.48 $\pm$ 0.18	23.93 $\pm$ 0.17	24.19 $\pm$ 0.21
FFM (kg)	79.62 $\pm$ 4.88	78.20 $\pm$ 3.65	79.92 $\pm$ 3.84 #
FM (%)	12.42 $\pm$ 2.25	11.1 $\pm$ 1.25 *	11.8 $\pm$ 1.23

Note: BM—body mass, BMI—body mass index, FFM—fat-free mass, FM—fat mass, LCD—low-carbohydrate diet, CD—conventional diet, Carbo-L—carbohydrate loading. * $p < 0.05$ significant difference to the after CD ($p < 0.05$), # statistically significant difference ($p < 0.05$) Carbo-L vs. LCD.

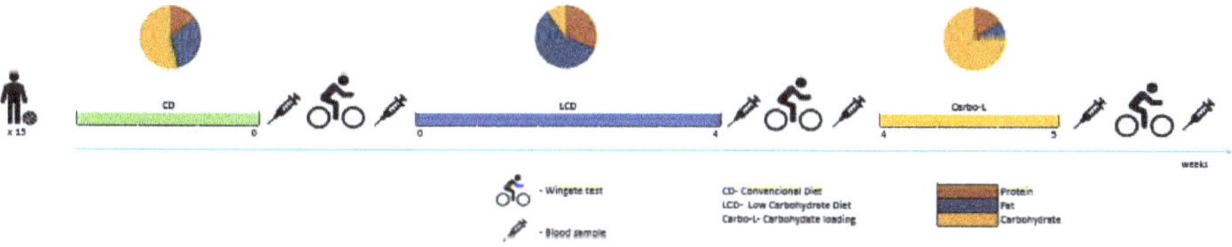

Figure 1. Scheme of the experimental protocol.

Table 2. Average macronutrients and total energy intake during the CD, LCD, and Carbo–L dietary procedures.

Nutrients	CD	LCD	Carbo-L
	Mean ± SD	Mean ± SD	Mean ± SD
Carbohydrates (%)	54 ± 6.1	10 ± 0.5	75 ± 3
Protein (%)	15 ± 6.3	31 ± 2.3	16 ± 3
Fat (%)	31 ± 4.3	59 ± 3.6	9 ± 1.6
SFAs (g)	48 ± 6.1	30 ± 4.2	11 ± 2.4
MUFAs (g)	61 ± 5.2	128 ± 12.3	13 ± 1.6
PUFAs (g)	20 ± 2	68 ± 4.5	10 ± 1.7
n-3 (g)	3.2 ± 0.2	24.4 ± 0.6	1.8 ± 0.3
n-6 (g)	16.1 ± 6	47.7 ± 2.7	7.5 ± 1.2
n-6/*n*-3	5 ± 1	2 ± 1	4 ± 1
TEI (kcal)	3740 ± 53	3758 ± 42	3752 ± 15
TEI (kJ)	15,658.63 ± 221	15,733.99 ± 175	15,708.87 ± 62

Note: SFAs—saturated fatty acids, MUFAs—monounsaturated fatty acids, PUFAs—polyunsaturated fatty acids, *n*-3—omega 3, *n*-6—omega 6, TEI—total energy intake.

2.2. Experimental Design

2.2.1. Dietary Guidelines—Monitoring of Nutritional Intake

The dietary intervention lasted 5 weeks. Before constructing individual isocaloric LCD, and Carbo-L diets, the resting metabolic rate (RMR) and the total daily energy expenditure (TDEE) associated with training were estimated. The TDEE was calculated according to the commonly accepted model (TDEE = AF (Activity Factor) × RMR (Resting Metabolic Rate) [35]. The RMR was measured at the beginning of the experiment, before the 4-week LCD, as well as before the 7-day Carb-L procedure, by means of an ergo spirometer MetaLyzer 3B (Cortex, Leipzig, Germany). AF was determined based on available indicators for athletes 2.0 (high activity) [35]. Also, before the experiment, the subjects were asked to complete the 72-h food diary (two weekdays and one weekend day). The dietary records were estimated by a nutritionist to assess previous feeding habits and daily calorie consumption. The composition of particular diets is presented in Table 2. Before the experiment all participants consumed an isocaloric conventional diet (CD) (Table 2). The CD was composed of 55% carbohydrates, 15% protein, and 30% fat. During the 5 weeks of the experiment the study participants lived in the dormitory and were fed at the academy cafeteria. The meals were prepared in the form of 24-h menus for seven days of the week. All meals were planed and supervised by a nutritionist. The quality and quantity of the food products was strictly controlled, maintaining proper proportions between the major macronutrients. During the five weeks of the experiment the athletes were fed a low-carbohydrate diet (LCD) for 4 weeks, followed by 7 days (1 week) of carbohydrate loading (Carbo-L) (Figure 1). There was no washout period between the two feeding procedures. One month before the experiment began all participants consumed a standard conventional diet (CD). The LCD was composed in such way that from all consumed fats, unsaturated fatty acids (mono and polyunsaturated) constituted 80% of the daily calorie intake. The LCD consisted of 10% carbohydrates, 31% proteins, and 59% fat. In the LCD, the subjects consumed healthy fats, mainly monounsaturated fatty acids from olive oil, dairy products, and nuts, which accounted for more than 50% of all fatty acids consumed. The LCD also contained polyunsaturated fatty acids *n*-6 and *n*-3, in a ratio not exceeding 4–5:1. The diet included the consumption of fish, like mackerel and sardines, which are rich in *n*-3 fatty acids. Additionally, the LCD included high-quality protein products such as fish, meat, eggs and dairy products. While on the LCD the athletes consumed: poultry, fish, beef, veal and lamb, dried beef, chopped meat tartare, carpaccio and cured ham; olive oil, butter, green vegetables without restriction (raw and cooked), boiled eggs, and seasoned cheese (e.g., mozzarella, halloumi). Warm drinks were restricted to tea and coffee without sugar and herbal extracts. The foods and drinks that athletes avoided included alcohol and any sweets like sugar or honey. They also did not consume white bread, pasta, white rice, sweet milk, fruit yogurt, sweets, soluble tea, and barley coffee. During the 4 weeks of the LCD the athletes consumed four main meals and one snack.

The Carbo-L consisted of 75% carbohydrates, 16% proteins and 9% fat. The Carbo-L diet contained carbohydrates, mainly with a low glycaemic index like whole grain bread and pasta, graham rolls, whole grain rice, legumes, raw vegetables, poultry, beef, pork and fish. Only after training, the athletes consumed medium or high glycaemic index snacks like bananas, honey, figs, dactyls or meals with white rice, potatoes, boiled carrots, and beetroots. In the Carbo-L diet, the subjects did not eat processed carbohydrates (fast foods), sweets and carbonated drinks. In the Carbo-L protocol, the participants ate healthy carbohydrates such as cereals, rice, buckwheat, millet, and fruits. They also consumed high-quality protein and fat products, similar to the LCD. Participants consumed four main meals and two snacks. The main meals were prepared and consumed at the cafeteria, while the snacks were packed and eaten after the training sessions.

2.2.2. Training Program

During the 5 weeks of the experiment three series of laboratory analyses were performed. Baseline evaluations (after CD diet), after four weeks of the LCD, as well as after the 7 day Carbo-L procedure. The study was conducted during the precompetitive period of the annual training cycle. The basketball players performed five training units per week with a scrimmage game played on Saturday. Each training session lasted from 90 to 120 min and included specific technical and tactical drills as well as conditioning exercises. The training intensity varied significantly from low (stretching, free throws—HR $\leq$ 120 bts/min) to submaximal during transition or full court press drills (HR $\geq$ 170 bts/min). The participants refrained from exercise for 2 days before testing to minimize the effects of fatigue.

2.3. Body Mass, Body Composition

The subjects underwent medical examinations and somatic measurements. Body composition was evaluated in the morning, between 08.00 and 08.30 h. The day before, the participants had the last meal at 20.00 h. They reported to the laboratory after an overnight fast, refraining from exercise for 48 h. The measurements of body mass were performed on a medical scale with a precision of 0.1 kg. Body composition was evaluated using the electrical impedance technique (Inbody 720, Biospace Co., Japan).

2.4. Anaerobic Performance

Anaerobic performance was evaluated by the 30-s Wingate test for lower limbs. The test was preceded by a 5 min warm-up with a resistance of 100 W and cadence within 70–80 rpm. Following the warm-up, the test trial started, in which the objective was to reach the highest cadence in the shortest possible time, and to maintain it throughout the test. The lower limb Wingate protocol was performed on an Excalibur Sport ergocycle with a resistance of 0.8 Nm·kg^{-1} (Lode BV, Groningen, The Netherlands). The recorded variables included: time to peak power (TTP (s)), peak power (PP (W/kg)), and total work performed (TW (J/kg)), (Lode Ergometer Manager—LEM, software package, Groningen, The Netherlands).

2.5. Biochemical Analysis

To determine lactate concentration (LA) and acid base equilibrium, the following variables were evaluated: LA (mmol/L), blood pH, bicarbonate (mmol/L). The measurements were performed on fingertip capillary blood samples at rest and after 3 min of recovery. Determination of LA was based on an enzymatic method (Biosen C-line Clinic, EKF-diagnostic GmbH, Barleben, Germany).

The remaining variables were measured using a Blood Gas Analyzer GEM 3500 (GEM Premier 3500, Germany). β-hydroxybutyrate (β-HGB-mmol/L) was measured using Randox UK diagnostic kits (Ranbut). Testosterone (T), growth hormone (GH), and insulin (I) concentrations were measured in duplicate using EDTA plasma and immunoassay kits customized on an automated analyzer (Cobas e411, Roche Diagnostics, Mannheim, Germany). The intra assay coefficient of variation was for C 2.2%, 2.5% for T, for GH 2.3%, and 4.6% for insulin.

2.6. Statistical Analysis

Age, body mass and body composition, as well as biochemical variables were expressed as mean ± SD. Before using the parametric test, the assumption of normality was verified using the Kolmogorov–Smirnov test. A one-way ANOVA and two way ANOVA with repeated measures were used with significance set at $p < 0.05$. When appropriate, a Bonferroni post hoc test was used to compare selected data. The remaining analyses were performed using STATISTICA (StatSoft, Inc., Tulsa, OK, USA, 2018, version 12).

3. Results

Changes in basic somatic variables after particular dietary interventions in basketball players are presented in Table 1. The analysis of variance revealed statistically significant differences between LCD and CD for body mass (BM) and fat mass (FM) and between Carbo-L and LCD for fat-free mass (FFM) values. There was no significant differences between CD and Carbo-L.

The differences between the CD, the LCD and the Carbo-L diets are shown in Table 2. When comparing the LCD with the CD and the Carbo–L, the LCD contained a substantially higher amount of monounsaturated fatty acids (MUFAs) and polyunsaturated fatty acids (PUFAs) as well as significantly greater amounts of protein. Also, the LCD contained smaller amounts of CHO.

The results of ANOVA for anaerobic variables of the Wingate test are presented in Table 3. The analysis of variance revealed statistically significant differences between CD versus LCD ($p = 0.002$) and between Carbo-L and LCD ($p = 0.021$) for TW values.

Table 3. Anaerobic variables of the Wingate test after particular dietary interventions in basketball players.

Variables	CD	LCD	Carbo-L
	Mean ± SD	Mean ± SD	Mean ± SD
TPP (s)	2.65 ± 0.61	2.73 ± 0.57	2.58 ± 0.39
PP (W/kg)	20.35 ± 3.44	19.94 ± 3.42	20.87 ± 0.39
TW/kg (J/kg)	301.17 ± 12.42	**266.69 ± 6.46 ***	**302.46 ± 8.50 #**

Note: TPP—time to peak power; PP—peak power; TW—total work; * statistically significant difference LCD vs. CD; # statistically significant difference Carbo-L vs. LCD.

The analysis of variance for blood acid–base equilibrium after the LCD revealed statistically significant differences between CD versus LCD and between Carbo-L versus LCD for LA and pH values. The same analysis showed statistically significant differences in values of LA, pH, and HCO_3^- between baseline and post exercise values (Table 4). The same analysis revealed statistically significant differences between rest and post exercise for LA, pH, and HCO_3^- in CD, LCD, and Carbo-L.

Table 4. The differences in the concentration of blood plasma lactate and acid-base variables, as well as the resting concentration of β-HB after particular dietary interventions in basketball players.

Variables		CD	LCD	Carbo-L
		Mean ± SD	Mean ± SD	Mean ± SD
LA (mmol/L)	Rest	1.65 ± 0.06	1.26 ± 0.01 *	1.69 ± 0.04 #
	Post exercise	9.47 ± 1.04 $^{\&}$	8.36 ± 0.62 $^{\&}$	9.62 ± 0.54 $^{\&}$
pH ($-$Log[H$^+$])	Rest	7.412 ± 0.003	7.381 ± 0.001 *	7.420 ± 0.01 #
	Post exercise	7.275 ± 0.005 $^{\&}$	7.322 ± 0.03 $^{\&}$	7.261 ± 0.008 $^{\&}$
HCO$_3^-$ (mmol/L)	Rest	24.10 ± 0.07	23.72 ± 0.11	24.48 ± 0.07
	Post exercise	12.80 ± 0.09 $^{\&}$	13.12 ± 0.14 $^{\&}$	12.7 ± 0.05 $^{\&}$
β-HB (mmol/L)	Rest	0.041 ± 0.02 *	0.161 ± 0.11	0.035 ± 0.01 #

Note: LA—lactate; β-HB—β-hydroxybutyrate, HCO$_3^-$—bicarbonate * statistically significant differences with $p < 0.05$ between LCD vs. CD; # statistically significant differences with $p < 0.05$ between Carbo-L vs. LCD; $^{\&}$ statistically significant differences with $p < 0.05$ between rest vs. post exercise.

The results of ANOVA for hormone concentrations after particular dietary interventions in basketball players, revealed statistically significant differences between CD versus LCD and between Carbo-L versus LCD for GH ($p = 0.003$) and I ($p = 0.002$) values (Table 5). The same analysis showed significant differences of T concentration between LCD vs CD and between Carbo-L vs CD ($p = 0.002$).

Table 5. Differences in hormone concentrations after particular dietary interventions in basketball players.

Variables	CD	LCD	Carbo-L
	Mean ± SD	Mean ± SD	Mean ± SD
Testosterone (nmol/L)	546.67 ± 167.16	642.37 ± 194.47 *	643.14 ± 186.52 $^{\$}$
Growth hormone (ng/mL)	0.15 ± 0.07	0.21 ± 0.09 *	0.11 ± 0.08 #
Insulin (IU/mL)	5.49 ± 3.25	3.99 ± 2.61 *	7.28 ± 3.65 #
Cortisol (µg/dL)	16.38 ± 6.81	16.22 ± 6.40	16.02 ± 5.79

Note: * statistically significant differences with $p < 0.05$ between LCD vs. CD; # statistically significant differences with $p < 0.05$ between Carbo-L vs. LCD; $^{\$}$ statistically significant differences with $p < 0.05$ between Carbo-L vs. CD.

4. Discussion

Particular sport disciplines are characterized by different exercise metabolism that requires specific nutrition. Substantial research has been completed on the impact of both carbohydrate and high-fat diets on substrate utilization during exercise, both in sedentary subjects and in athletes. A common observation of these studies is a positive impact on performance during prolonged submaximal exercise [3,36]. However, these ergogenic effects on physical performance are induced by various mechanisms. Chronic ingestion of a high-fat diet results in a greater reliance on fat oxidation at rest and during exercise [37] that allows the muscle to spare glycogen during moderate exercise and therefore improve exercise capacity [38,39]. On the other hand, high-carbohydrate diets elevate muscle glycogen content. This effect can be beneficial not only for endurance performance, but may in fact improve high-intensity exercise, during which muscle glycogen is the predominant fuel source [5]. More recent studies concerning athlete performance and specific dietary interventions have focused on the combined use of high-fat, low-carbohydrate diet, followed by carbohydrate loading [40]. This paradigm assumes that these two different diet interventions can influence exercise glycogen metabolism, which results in favorable effects on performance. We assumed that the glycogen sparing effect of LCD should enhance endurance performance, while the HCHO-D strategy could benefit anaerobic performance after carbohydrate re-feeding. In the present study we utilized this protocol to investigate anaerobic performance in well-trained basketball players with the following new approaches: (1) adaptation to fat metabolism was induced by the LCD diet, (2) a prolonged state of carbohydrate loading (up to one week) was applied, (3) the dietary interventions were implemented into subsequent training micro-cycles, and (4) the 30-s all-out Wingate test was used to evaluate anaerobic power and capacity. In this study, we found that 4 weeks of LCD feeding decreased total work capacity, which returned to baseline after the carbohydrate loading procedure. Moreover, neither the low-carbohydrate feeding nor the carbohydrate loading affected peak power. In case of the LCD diet used in our study, the obtained results are similar to those observed in sedentary, healthy individuals after using a short-term (3-days) ketogenic diet [21]. This means that the training process applied by the basketball players, despite the use of explosive activities and short repeated bursts of high-intensity efforts, did not trigger adaptive changes towards the elimination of negative effects of the LCD on anaerobic capacity, and carbohydrate refeeding did not bring additional benefits related to anaerobic performance. The LCD used by our subjects contained substantially high amounts of MUFAs and PUFAs as well as significant amounts of protein. The high dietary content of the later nutrient was used to protect athletes against protein loss induced by heavy training loads, whereas larger content of PUFAs and MUFAS was consumed to protect against post exercise inflammation and to enhance recovery. Despite the above mentioned modifications, the diet used in our experiment met the criteria of LCD, what was confirmed by four times higher concentrations of β-hydroxybutyrate (β-HB) compared to baseline values.

The results of the all-out Wingate test depend almost exclusively on activation of anaerobic glycolysis [41]. One possible mechanism for reduced anaerobic capacity may be attributed to the LCD-induced keto-acidosis, which was reflected in the present study by reduction of blood pH. It is well established that anaerobic glycolysis is limited by acidosis via inhibition of the rate limiting enzyme, phosphofructokinase [42]. However, the reduced anaerobic capacity may be as well attributed to aerobic re-synthesis of ATP, which depending on the physiological state of the body provides 10–40% of the energy utilized during the 30-s Wingate test. It is known that this metabolic system is less effective during high-intensity exercise after low-carbohydrate diets [43,44]. While a beneficial impact of LCDs on athletes' performance during low- to moderate-intensity exercise has delivered conflicting results [22], most of the data regarding anaerobic performance indicated its impairment [19,43], although no differences in maximal power output were found when the effects of consuming a high-fat diet were compared with carbohydrate loading, even when both diets were accompanied by intensive training [29].

During the first few seconds of the Wingate test ATP re-synthesis is mainly provided by anaerobic, alactic metabolism, i.e., muscle ad hoc available ATP and breakdown of creatine-phosphate. The later process contributes close to 70% of anaerobic synthesis of ATP and is indirectly expressed by the measurement of peak power. During the current study, in contrast to anaerobic capacity, peak power did not change significantly during the experiment. It is known that alterations in systemic pH strongly affect anaerobic exercise performance [45]. This means that conversely to anaerobic capacity, the LCD which induced subclinical keto-acidosis did not affect this variable. It also indicates that muscle acidosis was fully compensated during the first seconds of the Wingate test by transient muscle alkalization, as a result of creatine-phosphocreatine breakdown. A similar effect, with no changes in peak power, was previously seen in sedentary individuals following a 3-day ketogenic diet [21]. It is well established that high cellular H$^+$ concentration is a crucial factor that reduces peak power [46]. These results also suggest that CHO restriction may not be detrimental to perform acute all-out explosive efforts. This result is in agreement with previous data reported by Dipla et al. [47] showing that strength, measured isotonically, isometrically, or isokinetically was maintained during short-term carbohydrate restriction. The above mentioned results suggest that during the first few seconds of an all-out effort, delivery of free energy from ATP breakdown is not impaired by LCD in skeletal muscle metabolism and contracting muscles are able to maintain metabolic stability [48,49].

During this study, macronutrient redistribution towards carbohydrate loading restored total work capacity and blood pH to baseline (pre-experimental) values, yet no benefits appeared in any of the investigated anaerobic variables after carbohydrate re-feeding. These results are consistent with the findings of Burke at al. [14] and Carey et al. [50], who demonstrated an increase in fat oxidation with short-term high-fat feeding that persisted even after restoration of CHO stores. In skeletal muscles, fat feeding promotes utilization of fatty acids and intramuscular triglycerides as a primary fuel source [28,51] by decreasing the activation of pyruvate dehydrogenase complex (PDH) due to increased pyruvate dehydrogenase kinase (PDK4) activity [52], causing fat utilization as a main substrate for ATP re-synthesis. However, this mechanism was perceived only in individuals subjected to aerobic or intermittent exercises after pre-exercise carbohydrate feeding. It is worth mentioning that this type of exercise is predominantly performed with recruitment of slow-twitch fibers. There is still very little information regarding how fast the metabolism of LCD-adapted muscles could be reversed with carbohydrate re-feeding and if metabolic regulation during exercise in fast-twitch fibers which are mostly engaged in execution of all-out efforts is similar to that which takes place in slow-twitch fibers.

The supply of substrate for muscle energy metabolism and its biological effects are regulated by hormones. Research has shown that dietary depletion of CHO can shift hormonal milieu and cellular mechanisms to increased utilization of non-esterified fatty acids (NEFAs) and to a much lesser extent amino acids [53]. The following hormones showing a widespread metabolic effect have been identified as anabolic: T, I, and GH, while C is considered to possess a more potent catabolic property. Hormonal anabolic action is primary mediated by amino acid uptake and transcriptional regulation of selected genes that leads to an increase in intramuscular protein synthesis. This means that elevated levels of these hormones combined with exercise can potentiate skeletal muscle hypertrophy, which plays a significant role in the performance of anaerobic efforts. Such an effect did not occur in our subjects despite the elevation of T levels and relatively low C concentrations, since both the BM and FFM of the tested athletes did not change substantially throughout our study. This could have been caused by the selection of study participants, which included well-trained athletes with a low body fat content and significant amounts of fat free mass due to regular resistance training. Testosterone is also a strong fat-reducing hormone, and exerts this effect by inhibiting lipid uptake and lipoprotein lipase activity in adipocytes, stimulating at the same time lipolysis through increasing the number of lipolytic beta-adrenergic receptors [54]. The latter can elevate blood NEFA levels, thus providing more substrate for ketogenesis. Our study indicates that this effect cannot be considered as a primary source for stimulation of ketogenesis in our subjects because the elevated T level was stable throughout the dietary intervention. In favor of such an assumption is the fact that despite elevated T levels, the concentration of ketone bodies (β-HB) returned to baseline values after the CHO re-feeding. Human studies have demonstrated strong positive relationships between dietary saturated fat and T, which can be considered as a predictor of plasma T level [55]. The fact the increases in T concentrations in our athletes were similar after both dietary interventions indicates that despite increased fat consumption, other exercise-induced factors may play an important role in triggering T production. In one of his studies Volek et al. [55] observed that none of the fat-enriched dietary variables were significantly correlated with C concentrations. A similar effect is confirmed by the present study which showed a lack of changes in plasma C levels after the adherence to both diets. The fact that T and C responded differently in our experimental paradigm suggests involvement of numerous mechanisms for these two hormonal responses.

As has been pointed out by other authors [53,56], the substantial impact of accelerating the rate of ketogenesis may be a consequence of insulin deficiency. This leads to metabolic changes towards ketone body formation [57]. In line with such an assumption, in the present study ketoacidosis was completely reversed after carbohydrate re-feeding, once the insulin concentration increased above baseline level. The observed ketogenesis in this study was accompanied by significant and parallel changes in insulin concentration, which indicates superior influence of this hormone in regulation of metabolism during the LCD in comparison to other investigated hormones. However, the question also arises as to whether elevated testosterone levels are able to compensate skeletal muscle amino acids uptake during the LCD with decreased insulin concentration. Changes in the GH also contribute to maintenance of metabolic rate in muscles. Its physiological function in keto-adapted individuals has been poorly investigated. It is known that for both endurance and resistance exercise greater activation of anaerobic glycolysis and lactate formation increases the amount of GH released. Similarly to T, GH seems to enhances lipolysis during exercise, while sparing protein as a metabolic source, and causes an increase in muscle mass and strength [58].

5. Conclusions

Until now, there was no evidence how chronic fat adaptation followed by carbohydrate loading compromises all-out anaerobic exercise metabolism and performance when such a dietary strategy is implemented into the training process. The recommendations for such dietary interventions rely to a great extent on findings from endurance exercise. Results of the present study suggest that supplying the body with an alternative fuel through nutritional manipulation, followed by CHO re-feeding, may not bring additional benefits to athletes involved in sport disciplines with a predominance of anaerobic metabolism. Thus, our findings call into question such dietary interventions when optimizing all-out intense anaerobic exercise performance. Several studies have focused on the potential role of KD interventions on athletic performance, which could be used by coaches in parallel with training. In various sports that rely on anaerobic performance (i.e., wrestling, martial arts, boxing, gymnastics, etc.), athletes are often required to reduce body mass in a short period of time. Rapid weight loss via reduced food intake, sweat suits, diet pills, and dehydration methods is often associated with loss of an optimal physical and mental fitness. This problem can be partially solved by implementation of the LCD into the training program. Although it is very likely that such nutritional manipulation may reduce anaerobic performance, the results of our study indicate that this detrimental side-effect can be reversed by re-feeding athletes with a HCHD. However, such nutrition interventions should be applied several weeks before competition.

Author Contributions: There were six authors who significantly contributed to the manuscript submitted to Nutrients entitled *"Anaerobic performance after a low-carbohydrate diet (LCD) followed by 7 days of carbohydrate loading in basketball players"*. A.Z., M.M.M., and J.C. as the main authors were responsible for creating the concept of the research, as it was part of a bigger project which considered the effects of high-fat, low carbohydrate diets on exercise metabolism and sports performance in different sport disciplines. A.Z. made most of the interpretations of the obtained results while preparing the discussion and conclusions. J.L. performed the literature review and prepared the introduction, while helping in most of the biochemical analysis. A.M. conducted the statistical analysis and prepared the results in the form of tables and graphs. J.C. was responsible for all of the physiological testing, and the interpretation of obtained results. A.Z. and G.Z. were responsible for preparing the research protocol, including the selection of subjects, choosing and verifying the test protocol and supervising all exercise testing and blood collection. All of the tests were conducted in the Human Performance Laboratory at the Academy of Physical Education in Katowice, Poland.

Funding: This research was funded by the Ministry of Science and Higher Education of Poland under Grant NRSA3 03953 and NRSA4 040 54.

Acknowledgments: This work was supported by the Ministry of Science and Higher Education of Poland under Grant NRSA3 03953 and NRSA4 040 54.

Conflicts of Interest: The authors declare no conflict of interest. The Ministry of Science and Higher Education of Poland had no role in the design of the study; in the collection, analyses, or interpretation of data; in the writing of the manuscript, and in the decision to publish the results.

References

1. Coyle, E.F.; Coggan, A.R.; Hemmert, M.K.; Ivy, J.L. Muscle glycogen utilization during prolonged strenuous exercise when fed carbohydrate. *J. Appl. Physiol.* **1986**, *61*, 165–172. [CrossRef] [PubMed]
2. Neufer, P.D.; Costill, D.L.; Flynn, M.G.; Kirwan, J.P.; Mitchell, J.B.; Houmard, J. Improvements in exercise performance: Effects of carbohydrate feedings and diet. *J. Appl. Physiol.* **1987**, *62*, 983–988. [CrossRef] [PubMed]

3. Hargreaves, M. Exercise, muscle, and CHO metabolism. *Scand. J. Med. Sci. Sports* **2015**, *25* (Suppl. 4), 29–33. [CrossRef]

4. Hawley, J.A.; Burke, L.M. Carbohydrate availability and training adaptation: Effect on cell metabolism. *Sport Sci. Res.* **2010**, *38*, 152–160. [CrossRef]

5. Romijn, J.A.; Coyle, E.F.; Sidossis, L.S.; Gastaldelli, A.; Horowitz, J.F.; Endert, E.; Wolfe, R.R. Regulation of endogenous fat and carbohydrate metabolism in relation to exercise intensity and duration. *Am. J. Physiol. Endocrinol. Metab.* **1993**, *265*, E380–E391. [CrossRef]

6. Margolis, L.M.; Pasiakos, S.M. Optimizing intramuscular adaptations to aerobic exercise: Effects of carbohydrate restriction and protein supplementation on mitochondrial biogenesis. *Adv. Nutr.* **2013**, *4*, 657–664. [CrossRef]

7. Lin, J.; Handschin, C.; Spiegelman, B.M. Metabolic control through the PGC-1 family of transcription coactivators. *Cell Metab.* **2005**, *1*, 361–370. [CrossRef]

8. Phinney, S.D.; Bistrian, B.R.; Evans, W.J.; Gervino, E.; Blackburn, G.L. The human metabolic response to chronic ketosis without caloric restriction: Preservation of submaximal exercise capability with reduced carbohydrate oxidation. *Metabolism* **1983**, *32*, 769–776. [CrossRef]

9. Langfort, J.; Pilis, W.; Zarzeczny, R.; Nazar, K.; Kaciuba-Uścitko, H. Effect of low-carbohydrate-ketogenic diet on metabolic and hormonal responses to graded exercise in men. *J. Physiol. Pharmacol.* **1996**, *47*, 361–371. [PubMed]

10. Phinney, S.D. Ketogenic diets and physical performance. *Nutr. Metab.* **2004**, *1*, 2. [CrossRef]

11. Jeppesen, J.; Kiens, B. Regulation and limitations to fatty acid oxidation during exercise. *J. Physiol.* **2012**, *590*, 1059–1068. [CrossRef] [PubMed]

12. Zajac, A.; Poprzecki, S.; Maszczyk, A.; Czuba, M.; Michalczyk, M.; Zydek, G. The effects of a ketogenic diet on exercise metabolism and physical performance in off-road cyclists. *Nutrients* **2014**, *6*, 2493–2508. [CrossRef] [PubMed]

13. Rhyu, H.S.; Cho, S.Y. The effect of weight loss by ketogenic diet on the body composition, performance-related physical fitness factors and cytokines of Taekwondo athletes. *J. Exerc. Rehabil.* **2014**, *10*, 326–331. [CrossRef]

14. Burke, L.M.; Ross, M.L.; Garvican-Lewis, L.A.; Welvaert, M.; Heikura, I.A.; Forbes, S.G.; Mirtschin, J.G.; Cato, L.E.; Strobel, N.; Sharma, A.P.; et al. Low carbohydrate, high fat diet impairs exercise economy and negates the performance benefit from intensified training in elite race walkers. *J. Physiol.* **2017**, *595*, 2785–2807. [CrossRef]

15. McSwiney, F.T.; Wardrop, B.; Hyde, P.N.; Lafountain, R.A.; Volek, J.S.; Doyle, L. Keto-adaptation enhances exercise performance and body composition responses to training in endurance athletes. *Metabolism* **2018**, *81*, 25–34. [CrossRef] [PubMed]

16. Gregory, R.M.; Hamdan, H.; Torisky, D.M.; Akers, J.D. A low-carbohydrate ketogenic diet combined with 6-weeks of crossfit training improves body composition and performance. *Int. J. Sports Exerc. Med.* **2017**, *3*, 1–10. [CrossRef]

17. Michalczyk, M.; Zajac, A.; Mikolajec, K.; Zydek, G.; Langfort, J. No Modification in Blood Lipoprotein Concentration but Changes in Body Composition after 4 Weeks of Low Carbohydrate Diet (LCD) Followed by 7 Days of Carbohydrate Loading in Basketball Players. *J. Hum. Kinet.* **2018**, *65*, 125–137. [CrossRef]

18. Wroble, K.A.; Trott, M.N.; Schweitzer, G.G.; Rahman, R.S.; Kelly, P.V.; Weiss, E.P. Low carbohydrate, ketogenic diet impairs anaerobic exercise performance in exercise-trained women and men: A randomized-sequence crossover trial. *J. Sport Med. Phys. Fit.* **2018**. [CrossRef]

19. Urbain, P.; Strom, L.; Morawski, L.; Wehrle, A.; Deibert, P.; Bertz, H. Impact of a 6-week non-energy-restricted ketogenic diet on physical fitness, body composition and biochemical parameters in healthy adults. *Nutr. Metab.* **2017**, *14*, 17. [CrossRef]

20. Paoli, A.; Grimaldi, K.; D'Agostino, D. Ketogenic diet does not affect strength performance in elite artistic gymnasts. *J. Int. Soc. Sport Nutr.* **2012**, *9*, 34. [CrossRef]

21. Langfort, J.; Zarzeczny, R.; Pilis, W.; Nazar, K.; Kaciuba-Uścitko, H. The effect of a low-carbohydrate diet on performance, hormonal and metabolic responses to a 30-s bout of supramaximal exercise. *Eur. J. Appl. Physiol. Occup. Physiol.* **1997**, *76*, 128–133. [CrossRef]

22. Burke, L.M. Re-examinig high-fat diets for sport performance: Did we call the "nail in the coffin" to soon? *Sports Med.* **2015**, *45* (Suppl. 1), S33–S49. [CrossRef] [PubMed]

23. Havemann, L.; West, S.J.; Goedecke, J.H.; Macdonald, I.A.; St Clair Gibson, A.; Noakes, T.D.; Lambert, E.V. Fat adaptation followed by carbohydrate loading compromises high-intensity sprint performance. *J. Appl. Physiol.* **2006**, *100*, 194–202. [CrossRef] [PubMed]

24. Yeo, W.K.; Carey, A.; Burke, L.M.; Spriet, L.L.; Hawley, J. Fat adaptation in well-trained athletes: Effect of Cell Metabolism. *Appl. Physiol. Nutr. Metab.* **2011**, *36*, 12–22. [CrossRef] [PubMed]

25. Stellingweff, T.; Spriet, L.L.; Watt, M.; Kimber, N.; Hargreaves, M.; Hawley, J.; Burkey, L.M. Decreased PDH activation and glycogenolysis during exercise following fat adaptation with carbohydrate restoration. *Am. J. Physiol. Endocrinol. Metab.* **2006**, *29*, E380–E388. [CrossRef] [PubMed]

26. Brooks, G.A.; Mercier, J. Balance of carbohydrate and lipid utilization during exercise: The "crossover concept". *J. Appl. Physiol.* **1994**, *76*, 13–20. [CrossRef] [PubMed]

27. Langfort, J.; Ploug, T.; Ihlemann, J.; Baranczuk, E.; Donsmark, M.; Górski, J.; Galbo, H. Additivity of adrenaline and contractions on hormone-sensitive lipase, but not on glycogen phosphorylase, in rat muscle. *Acta Physiol. Scand.* **2003**, *178*, 51–60. [CrossRef] [PubMed]

28. Paoli, A.; Bianco, A.; Grimaldi, K.A. The ketogenic diet and sport: A possible marriage? *Exerc. Sport Sci. Rev.* **2015**, *43*, 153–162. [CrossRef] [PubMed]

29. Noakes, T.; Volek, J.S.; Phinney, S.D. Low-carbohydrate diets for athletes: What evidence? *Br. J. Sports Med.* **2014**, *48*, 1077–1078. [CrossRef] [PubMed]

30. Cochran, A.J.; Little, J.P.; Tarnopolsky, M.A.; Gibala, M.J. Carbohydrate feeding during recovery alters the skeletal muscle metabolic response to repeated sessions of high-intensity interval exercise in humans. *J. Appl. Physiol.* **2010**, *108*, 628–636. [CrossRef] [PubMed]

31. Van Praekers, K.; Szlufcik, K.; Nielsen, H.; Ramaekers, M.; Hespel, P. Beneficial metabolic adaptation due to endurance exercise training in the fasted state. *J. Appl. Physiol.* **2011**, *110*, 236–245.

32. Yeo, W.K.; Paton, C.D.; Garnham, A.P.; Burke, L.M.; Carey, A.L.; Hawley, J.A. Skeletal muscle adaptation and performance responses to once a day versus twice every second day endurance training regimes. *J. Appl. Physiol.* **2008**, *105*, 1462–1470. [CrossRef] [PubMed]

33. Issurin, V.B. Benefits and limitations of block periodized training approaches to athletes' preparation: A review. *Sport Med.* **2016**, *46*, 329–338. [CrossRef] [PubMed]

34. Issurin, V.B. New horizons for the methodology and physiology of training periodization. *Sports Med.* **2010**, *40*, 189–206. [CrossRef]

35. Jarosz, M.; Traczyk, I.; Rychlik, E. *Nutrition Norms for the Polish Population*; Institute of Nutrition: Warsaw, Poland, 2012; pp. 18–31.

36. Ortenblad, N.; Westerblad, H.; Nielsaen, J. Muscle glycogen stores and fatigue. *J. Physiol.* **2013**, *591*, 4405–4413. [CrossRef] [PubMed]

37. Brooks, G. Importance of the "crossover" concept in exercise metabolism. *Clin. Exerc. Pharm. Physiol.* **1997**, *124*, 889–895. [CrossRef]

38. Hickson, R.C.; Rennie, M.J.; Conlee, R.K.; Winder, W.W.; Holloszy, J.O. Effects of increase plasma fatty acids on glycogen utilization and endurance. *J. Appl. Physiol.* **1997**, *43*, 829–833. [CrossRef]

39. Erlenbusch, M.; Haub, M.; Munoz, K.; Mac Connie, S.; Stillwell, B. Effect of high-fat or high-carbohyddrate diets on endurance exercise: A meta-analysis. *Int. J. Sport Nutr. Exerc. Metab.* **2005**, *15*, 1–14. [CrossRef]

40. Colombani, P.O.; Mannhart, C.; Mettler, S. Carbohydrates and exercise performance in nin-fasted athletes: A systematic review of studies mimicking real-life. *Nutr. J.* **2013**, *12*, 16. [CrossRef]

41. Medbo, J.I.; Tabata, I. Anaerobic energy release in working muscle during 30 s to 3 min of exhausting bicycling. *J. Appl. Physiol.* **1993**, *75*, 1654–1660. [CrossRef]

42. Mullins, G.; Hallam, C.L.; Broom, I. Ketosis, ketoacidosis and very-low-calorie diets: Putting the record straight. *Nutr. Bull.* **2011**, *36*, 397–402. [CrossRef]

43. Fleming, J.; Sharman, M.J.; Avery, N.G.; Love, D.M.; Gómez, A.L.; Scheett, T.P.; Kraemer, W.J.; Volek, J.S. Endurance capacity and high-intensity exercises performance responses to a high-fat diet. *Int. J. Sport Nutr. Exerc. Metab.* **2003**, *13*, 466–478. [CrossRef] [PubMed]

44. Wolfe, R.R. Metabolic interactions between glucose and fatty acids in humans. *Am. J. Clin. Nutr.* **1998**, *67*, 519S–526S. [CrossRef] [PubMed]

45. Cairns, S.P. Lactic acid and exercise performance: Culprit or friend? *Sports Med.* **2006**, *36*, 279–291. [CrossRef] [PubMed]

46. Kunth, S.T.; Dave, H.; Peters, J.R.; Fitts, R.H. Low cell pH depresses peak power in rat skeletal muscle fibres at both 30 degrees C and 15 degrees C: Implication for muscle fatigue. *J. Physiol.* **2006**, *575*, 887–899.

47. Dipla, K.; Makri, M.; Zafeiridis, A.; Soulas, D.; Tsalouhidou, S.; Mougios, V.; Kellis, S. An isoenergetic high-protein, moderate-fat diet does not compromise strength and fatigue during resistance exercise in women. *Br. J. Nutr.* **2008**, *100*, 283–286. [CrossRef]

48. Grasii, B.; Rossiter, H.B.; Zoladz, J.A. Skeletal muscle fatigue and decreased efficiency: Two sides of the same coin? *Exerc. Sport Sci. Rev.* **2015**, *43*, 75–83. [CrossRef] [PubMed]

49. Zoladz, J.A.; Korzeniewski, B.; Grassi, B. Training-induced acceleration of oxygen uptake kinetics in skeletal muscle: The underlying mechanisms. *J. Physiol. Pharmacol.* **2006**, *57* (Suppl. 10), 67–84.

50. Carey, A.L.; Staudacher, H.M.; Cummings, N.K.; Stepto, N.K.; Nikolopoulos, V.; Burke, L.M.; Hawley, J.A. Effects of fat adaptation carbohydrate restoration on prolonged endurance exercise. *J. Appl. Phyiol.* **2001**, *91*, 15–22. [CrossRef]

51. Ormsbee, M.J.; Bach, C.W.; Baur, D.A. Pre-exercise nutrition: The role of macronutrients, modified starches and supplements on metabolism and endurance performance. *Nutrients* **2014**, *6*, 1782–1808. [CrossRef] [PubMed]

52. Holness, M.J.; Kraus, A.; Harris, R.A.; Sugden, M.C. Target upregulation of pyruvate dehydrogenase kinase (PDK)-4 in slow-twitch skeletal muscle underlies the stable modification of the regulatory characteristics of PDK induced by high-fat feeding. *Diabetes* **2000**, *49*, 775–781. [CrossRef]

53. Cox, P.J.; Kirk, T.; Ashmore, T.; Willerton, K.; Evans, R.; Smith, A.; Murray, A.J.; Stubbs, B.; West, J.; McLure, S.W.; et al. Nutritional ketosis alters fuel preference and thereby endurance performance in athletes. *Cell Metab.* **2016**, *24*, 256–268. [CrossRef] [PubMed]

54. De Pergola, G. The adipose tissue metabolism: Role of testosterone and dehydroepiandrosterone. *Int. J. Obes. Relat. Metab. Disord.* **2000**, *24* (Suppl. 2), S59–S63. [CrossRef] [PubMed]

55. Volek, J.S.; Kraemer, W.J.; Bush, J.A.; Incledon, T.; Boetes, M. Testosterone and cortisol in relationship to dietary nutrient band resistance exercise. *J. Appl. Physiol.* **1997**, *82*, 49–54. [CrossRef] [PubMed]

56. Laffel, L. Ketone bodies: A review of physiology, pathophysiology and application of monitoring to diabetes. *Diabetes Metab. Res. Rev.* **1999**, *15*, 412–426. [CrossRef]

57. Westman, E.C.; Mavropoulos, J.; Yancy, W.S.; Volek, J.S. A review of low-carbohydrate ketogenic diets. *Curr. Atheroscler. Rep.* **2003**, *5*, 476–483. [CrossRef] [PubMed]

58. Berggren, A.; Ehrnborg, C.; Rosen, T.; Ellegard, L.; Bengtsson, B.A.; Caidahl, K. Short-term administration of supraphysiological recombinant human growth hormone (GH) does not increase maximum endurance capacity in healthy, active young men and women with normal GH-insulin-like growth factor I axes. *J. Clin. Endocrinol. Metab.* **2005**, *90*, 3268–3273. [CrossRef] [PubMed]

© 2019 by the authors. Licensee MDPI, Basel, Switzerland. This article is an open access article distributed under the terms and conditions of the Creative Commons Attribution (CC BY) license (http://creativecommons.org/licenses/by/4.0/).

Article

Analysis of the Effects of Dietary Pattern on the Oral Microbiome of Elite Endurance Athletes

Nida Murtaza [1], Louise M. Burke [2,3,*], Nicole Vlahovich [3,4], Bronwen Charlesson [3], Hayley M. O'Neill [4], Megan L. Ross [2,3], Katrina L. Campbell [4], Lutz Krause [1] and Mark Morrison [1,*]

[1] Faculty of Medicine, Translational Research Institute, University of Queensland Diamantina Institute, Brisbane, QLD 4102, Australia; nida.murtaza@uqconnect.edu.au (N.M.); l.krause@uq.edu.au (L.K.)

[2] Centre for Exercise & Nutrition, Mary MacKillop Institute for Health Research, Australian Catholic University, Melbourne, VIC 3000, Australia; meg.ross@ausport.gov.au

[3] Australian Institute of Sport, Canberra, ACT 2617, Australia; Nicole.Vlahovich@ausport.gov.au (N.V.); bronwen.charlesson@outlook.com (B.C.)

[4] Faculty of Health Sciences & Medicine, Bond University, Robina, QLD 4226, Australia; haoneill@bond.edu.au (H.M.O.); Katrina.Campbell@health.qld.gov.au (K.L.C.)

* Correspondence: Louise.Burke@ausport.gov.au (L.M.B.); m.morrison1@uq.edu.au (M.M.); Tel.: +61-2-621-41351 (L.M.B.); +61-7-344-36957 (M.M.)

Received: 17 January 2019; Accepted: 4 March 2019; Published: 13 March 2019

Abstract: Although the oral microbiota is known to play a crucial role in human health, there are few studies of diet x oral microbiota interactions, and none in elite athletes who may manipulate their intakes of macronutrients to achieve different metabolic adaptations in pursuit of optimal endurance performance. The aim of this study was to investigate the shifts in the oral microbiome of elite male endurance race walkers from Europe, Asia, the Americas and Australia, in response to one of three dietary patterns often used by athletes during a period of intensified training: a High Carbohydrate (HCHO; $n = 9$; with 60% energy intake from carbohydrates; ~8.5 g kg^{-1} day^{-1} carbohydrate, ~2.1 g kg^{-1} day^{-1} protein, 1.2 g kg^{-1} day^{-1} fat) diet, a Periodised Carbohydrate (PCHO; $n = 10$; same macronutrient composition as HCHO, but the intake of carbohydrates is different across the day and throughout the week to support training sessions with high or low carbohydrate availability) diet or a ketogenic Low Carbohydrate High Fat (LCHF; $n = 10$; 0.5 g kg^{-1} day^{-1} carbohydrate; 78% energy as fat; 2.1 g kg^{-1} day^{-1} protein) diet. Saliva samples were collected both before (Baseline; BL) and after the three-week period (Post treatment; PT) and the oral microbiota profiles for each athlete were produced by 16S rRNA gene amplicon sequencing. Principal coordinates analysis of the oral microbiota profiles based on the weighted UniFrac distance measure did not reveal any specific clustering with respect to diet or athlete ethnic origin, either at baseline (BL) or following the diet-training period. However, discriminant analyses of the oral microbiota profiles by Linear Discriminant Analysis (LDA) Effect Size (LEfSe) and sparse Partial Least Squares Discriminant Analysis (sPLS-DA) did reveal changes in the relative abundance of specific bacterial taxa, and, particularly, when comparing the microbiota profiles following consumption of the carbohydrate-based diets with the LCHF diet. These analyses showed that following consumption of the LCHF diet the relative abundances of *Haemophilus*, *Neisseria* and *Prevotella* spp. were decreased, and the relative abundance of *Streptococcus* spp. was increased. Such findings suggest that diet, and, in particular, the LCHF diet can induce changes in the oral microbiota of elite endurance walkers.

Keywords: oral microbiome; elite athletes; diet

---

1. Introduction

Recent technological advances have enabled a more holistic definition and characterisation of the microbes that colonise the human body, the "microbiomes". The human oral cavity serves as the habitat for a numerically large and diverse microbiome [1], which has been extensively characterised with respect to infectious and periodontal diseases, and caries, as well as for its contributions to the onset and progression of chronic conditions such as diabetes, cardiovascular disease and cancer [2]. However, the impacts of dietary pattern on the oral microbiome are not well defined, neither for the general population nor for cohorts who may follow specialised diets, such as elite athletes. In that context, recent studies have revealed a positive symbiotic association between the oral bacteria and host with respect to an enterosalivary nitrate-nitrite-nitric oxide pathway, which contributes to nitric oxide (NO) homeostasis [3,4]. Here, facultative anaerobic bacteria in the mouth reduce salivary gland concentrated nitrate to nitrite, which is then swallowed and absorbed into the bloodstream before further reduction to NO. The critical role of the oral microbiota in this effect has been demonstrated, where a seven-day period of antiseptic mouth wash treatment was shown to disrupt the oral microbiota of healthy non-athletes and, in the absence of any dietary modifications, was associated with reductions in plasma and oral nitrite levels and an increase in blood pressure [5]. These findings raise the spectre that diet may also invoke changes in the oral microbiota that manifest in alterations of this enterosalivary pathway and NO homeostasis but remains unexplored.

A recent investigation [6] of the effect of diet and training on exercise metabolism and performance in elite endurance athletes provided an opportunity for pilot work on this theme. The "Supernova 1" study investigated parameters around endurance capacity in a cohort of elite endurance race walkers who followed one of the three popular dietary approaches during a three-week period of intensified training: a ketogenic Low Carbohydrate High Fat diet (LCHF), or a diet high in carbohydrates consumed either ad libitum (HCHO) or at specific periods on a daily/weekly basis (PCHO). While the HCHO diet is focused on optimal muscle and brain carbohydrate (CHO) stores for each training session, the PCHO diet involves a strategic combination of sessions with such dietary support as well as other which are undertaken with low muscle glycogen availability to promote greater metabolic stress and cellular adaptation [7,8]. Finally, the LCHF diet involves severe CHO restriction to promote adaptations that increase muscle capacity for fat oxidation [7,8]. Details on the rationale for these radically different types of nutrition [6,8] and the actual protocols employed in this study can be found elsewhere [6,7]. In summary, the Supernova 1 study found that each group of athletes achieved a significant improvement in their aerobic capacity over the training block, which was undertaken during the base phase of the annual training plan. However, while this was associated with improved economy and real-world race performance in the two groups who trained while consuming the HCHO and PCHO diets, the LCHF group experienced an increase in the oxygen cost of exercise supported by high rates of fat oxidation, and thereby failed to improve their race performance despite the gain in aerobic capacity [6]. Based on these differential results, the overall aim of this study was to examine whether and how the oral microbiome of these athletes was affected by their diet during intensified training. This appears to be the first study that provides an in-depth investigation of diet x oral microbiome interactions in elite athletes.

2. Materials and Methods

2.1. Study Design

The group of world-class race walkers and the design of the "Supernova 1" study are described in detail by Burke et al. [6] and Mirtschin et al. [7]. In summary, these male race walkers (aged 20–35 years, BMI range 16–23 kg/m^2) were from Australia, Canada, Japan, Italy, Poland, Sweden, Chile and South Africa, and all met International Association of Athletics Federations (IAAF) standards for international competition, with more than 75% participating in the major championships during the year of the study (i.e., 2016 Rio Olympic Games and 2016 World Walking Cup). Twenty-nine study experiences were gained from 21 elite athletes who participated in either one ($n = 13$) or both ($n = 8$) of the Supernova 1 research camps conducted at the Australian Institute of Sport. Each camp involved three weeks of intensified training and rigorously supervised dietary interventions.

2.2. Allocation to Dietary Interventions

The athletes were involved over several months of planning and received education about the range of likely effects of the diets on various aspects of health and performance. Each had ample time to choose a diet(s) according to his beliefs of the performance benefits from the chosen diet. Although this type of assignment was non-random, given that all athletes choose freely to be in the study and to be fed their diet of choice, this approach both promoted adherence to the intervention and controlled for the random effects of such rigorous dietary control (e.g., feeling anxious about losing personal freedom of dietary choice). Therefore, any effects on the oral microbiome could be attributed to the diet, including any additional intrinsic biochemical, physiological or psychological overlay that belongs to the diet itself.

Three diets were compared: (i) a diet high in carbohydrate availability (HCHO; $n = 9$) comprised of 60% of energy intake from CHO (~8.5 g/kg body mass (BM)/day), 16% protein (~2.1 g/kg BM/day), 20% fat; (ii) a diet with periodised carbohydrate availability (PCHO; $n = 10$) of similar overall macronutrient composition as HCHO but consumed at different intervals across the day and throughout the week to support different training sessions with high or low CHO availability and (iii) a ketogenic low carbohydrate-high fat diet (LCHF; $n = 10$) comprised of 78% fat, 17% protein (~2.2 g/kg/day) and 0.50 g/kg/day carbohydrate (3.5% energy). All the meals were prepared taking into consideration the nutritional requirement within the allocated dietary intervention. Mirtschin et al. reported the detailed nutritional information and meal plans for the above three dietary interventions of this study [7].

2.3. Sample Collection, Genomic DNA Extraction and 16S rRNA Gene Amplicon Preparation

Saliva samples were collected from the athletes prior (baseline, BL) and after the three -week training-diet intervention using the OMNIgene saliva collection and preservative kit and according to the manufacturer's instruction (fasted collection, saliva collected by spitting into the tube). Total DNA was extracted from 0.25 mL aliquots of the preserved saliva samples using the repeated bead-beating procedure for cell lysis [9] and an automated column-based DNA purification procedure (Maxwell® 16MDx system, Promega Corporation, WI, USA) as described by Shanahan et al. [10]. Bar-coded PCR amplicon libraries of the V6-V8 hypervariable regions of 16S rRNA genes from Bacteria/Archaea were produced also following the protocols described by Shanahan et al. [10], and then sequenced via the Illumina MiSeq platform and workflows established by the Australian Centre for Ecogenomics at the University of Queensland.

2.4. Bioinformatics Analysis

The sequence data were analysed using the Quantitative Insights into Microbial Ecology (QIIME) software package on an Ubuntu Linux virtual machine. QIIME was used to demultiplex and perform quality control checking and filtering of the sequence data [11]. USEARCH 6.1 was used for the removal of candidate chimeric sequences [12]. The chimera checked filtered sequences were then clustered into Operational Taxonomic Units (OTUs) using the open reference OTU picking method. Threshold setting of 97% sequence identity was applied and Greengenes (version 13.8) database was used as the reference database [13]. Following the OTU picking step, OTUs that were not identified as Bacteria or Archaea, and/or OTUs that comprised $\leq 0.01\%$ of the total sample sequence count, were discarded from further analysis. All of the samples with less than 1000 reads were also excluded from the OTU table. Single rarefaction was done by random sampling to the minimum read count (6861 reads) to generate a subsampled OTU table.

2.5. Statistical Analysis

The rarefied OTU table was used to generate the taxonomy plots from phylum to genus levels and used to calculate alpha- and beta-diversity metrics. Alpha diversity was measured by Shannon index as an estimator for richness and evenness of microbiota communities. Weighted and unweighted UniFrac distance matrices were constructed in QIIME and used for Principle Coordinate Analysis (PCoA) of beta (between sample) diversity analysis. Redundancy Analysis (RDA) and Analysis of Similarity (ANOSIM) measures were also performed on these data within the Calypso web server to identify any clustering with respect to BL or the specific dietary patterns. Linear Discriminant Analysis (LDA) Effect Size (LEfSe) and sparse Partial Least Squares Discriminant Analysis (sPLS-DA) analyses via the Mixomics mixMC package within Calypso were then used to identify whether any individual taxa are discriminatory for the different dietary patterns [14,15]. The multiple linear regression analysis available via Calypso was used to perform pairwise comparisons between the BL and post-diet training microbiome profiles. A p-value < 0.05 was considered significant for all the statistical analysis. Tukey's test was used to compare age and BMI data for the athletes assigned to each dietary group.

2.6. Ethics Approval, Trial Registration and Consent to Participate

The study was approved by the Ethics committee of the Australian Institute of Sports (AIS, no. 20150802) and UQ-HREC 2015001965. The clinical trial is registered by the Australian New Zealand Clinical Trials Registry (ANZCTR) and has been assigned the number ACTRN12618001529235. The raw sequence paired-end files are deposited in the European Nucleotide Archive with the primary accession number PRJEB29801. All the subjects were informed and consented to do this research.

3. Results

As mentioned previously, 21 elite race walkers were recruited in the study (with eight athletes recruited in both the camps). Table 1 provides the anthropometric details of the subjects enrolled in the study. According to Tukey's multiple comparisons test, there were no significant differences in the age ($p = 0.53$ for HCHO vs. PCHO; $p = 0.28$ for HCHO vs. LCHF and $p = 0.87$ for PCHO vs. LCHF) and the BMI scores ($p = 0.36$ for HCHO vs. PCHO; $p = 0.84$ for HCHO vs. LCHF and $p = 0.67$ for PCHO vs. LCHF) when athletes were grouped according to the dietary intervention they received.

Table 1. Athlete cohort characteristics.

	High Carbohydrate (HCHO) Diet	Periodised Carbohydrate (PCHO) Diet	Low Carbohydrate High Fat (LCHF) Diet
Sample size	$n = 9$	$n = 10$	$n = 10$
Age (years)	25.4 ± 4	27.4 ± 4.6	28.3 ± 3.5
BMI (kg/m^2)	20 ± 1.6	21 ± 1.3	20.4 ± 1.8
Country of origin	Australia, Canada, Japan, South Africa	Australia, Canada, Japan, Poland, Sweden, Italy	Australia, Canada, Japan, Poland, Sweden, Chile, South Africa
Gender	Male	Male	Male

Note: Data for Age and body mass index (BMI) are shown as mean $\pm$ standard deviation.

The Shannon alpha diversity was reduced following the diet-training interventions when compared to their subject-matched BL measures; however, these reductions were not statistically significant ($p = 0.1$ for BL vs. HCHO and BL vs PCHO; $p = 0.62$ for BL vs. LCHF). The PCoA analysis of the weighted UniFrac distances are shown in Figure 1 and did not reveal any distinct clustering of the saliva microbiome profiles, either with respect to the ethnic origin of the athletes or the dietary intervention. Similarly, the supervised analyses by RDA and ANOSIM did not identify any significant differences between the microbiota community composition at BL and following any of the three dietary interventions (data not shown). Taken together, these results suggest that the dietary interventions do not result in dramatic changes in the overall biodiversity of the oral microbiome, but rather more subtle changes in community composition. As mentioned previously in the Methods section, eight athletes were recruited in both the camps, and the two baseline profiles (B1 vs. B2) of these eight athletes were compared. No substantive differences between the two microbiota profiles were apparent, as assessed by Shannon alpha-diversity and ANOSIM beta-diversity analyses. These tests indicate that the time between the study camps was sufficiently long to ensure a "washout" between the two camps, and, thereby, no potential carryover effects from the previous diet on the subsequent results/profiles.

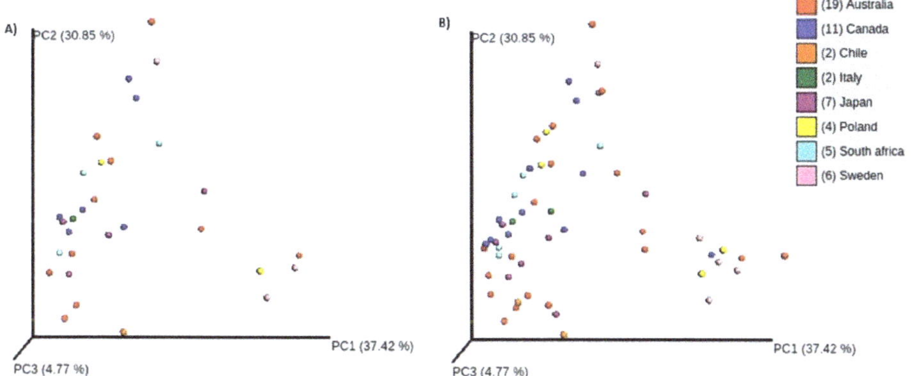

Figure 1. Principle component analysis of weighted UniFrac distances for the oral microbiomes of athletes at Baseline only (BL, **A**); and when combined with their profiles obtained after the diet-training intervention period (**B**). Samples are colored based on the athlete's country of origin and show no significant clustering indicative of a dietary and/or ethnic effect on the oral microbiomes.

3.1. Comparisons of Community Profiles of Saliva Samples between Baseline and Post Interventions

LefSe analyses was used to identify discriminating taxa between baseline (BL) and post diet-training interventions. OTU's affiliated to *Streptococcus*, *Peptostreptococcus*, *Actinomyces*, *Granulicatella*, *Atopobium*, *Veillonella* and *Prevotella* were found to be enriched following the consumption of HCHO diet, whereas *Parvimonas* was discriminatory and enriched for the BL samples from these same athletes (Supplementary Figure S1). Analysis by the sPLS-DA of the same athlete samples identified *Prevotella*, *Actinobacillus*, *Fusobacterium*, *Haemophilus* and *Gemella* to be associated and increased in BL samples (Figure 2). Pairwise comparisons (Supplementary Figure S2) of the oral microbiota profiles at BL and following HCHO diet training intervention was also examined using mixed effect linear regression, and the relative abundance of *Atopobium* was found to increase ($p = 0.015$), whereas *Capnocytophaga* ($p = 0.027$) and *Porphyromonas* ($p = 0.03$) were decreased after consumption of HCHO diet, when compared to the BL. However, no significant differences were observed once correction for multiple testing using false discovery rate was applied (FDR = 0.49).

LefSe analysis was then used to compare the microbiota profiles between BL and PCHO diet and revealed that the OTU's affiliated with *Leptotrichia*, *Neisseria*, *Moryella* and *Actinomyces* to be discriminatory and enriched for BL, whereas OTU's affiliated with *Streptococcus*, *Kingella*, unclassified members of *Neisseriaceae* and *Prevotella* were increased and discriminatory in the same athletes after consumption of the PCHO diet (Supplementary Figure S3). Analysis by sPLS-DA further showed that the relative abundances of *Haemophilus*, *Neisseria*, *Porphyromonas*, *Leptotrichia*, *Kingella*, *Prevotella*, *Unclassified Neisseriaceae*, *Rothia*, *Selenomonas* and *Tannerella* were increased following the consumption of the PCHO diet *whereas Unc. Aerococcaceae, Unc. CW040, Lautropia* and *Parvimonas* were distinct and increased in the BL samples of the same athletes who later received the PCHO dietary intervention (Figure 3). Repeated measures mixed effect linear regression analysis showed that the genus *Actinomyces* ($p = 0.04$), *Moryella* ($p = 0.05$), *Oribacterium* ($p = 0.04$), *Peptostreptococcus* ($p = 0.009$) and some unclassified *Erysipelotrichaceae* ($p = 0.04$) were reduced in response to the PCHO diet as compared to BL (Supplementary Figure S4). However, the statistical significance of all these differences was lost once correction for multiple testing using the false discovery rate was applied (FDR = 0.4) (Supplementary Figure S4).

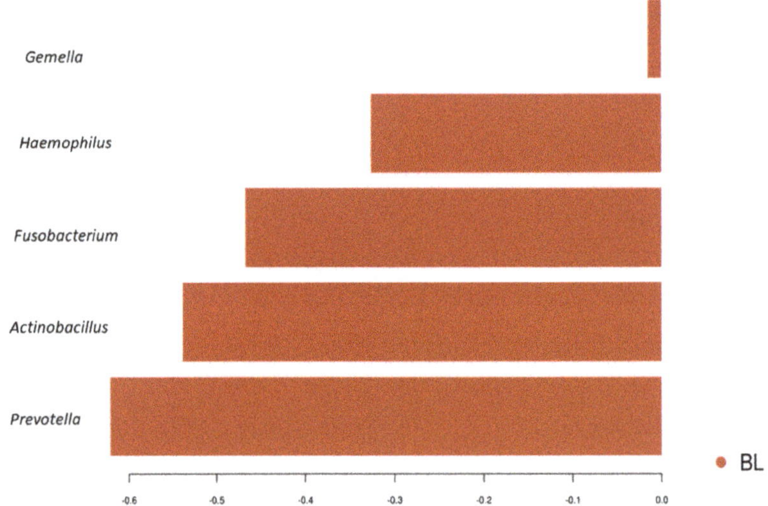

Figure 2. Genera differentiating between the oral microbiota profiles of athletes at baseline (BL, red) and after their consumption of the High Carbohydrate diet (HCHO) identified by sparse Partial Least Squares Discriminant Analysis (sPLS–DA).

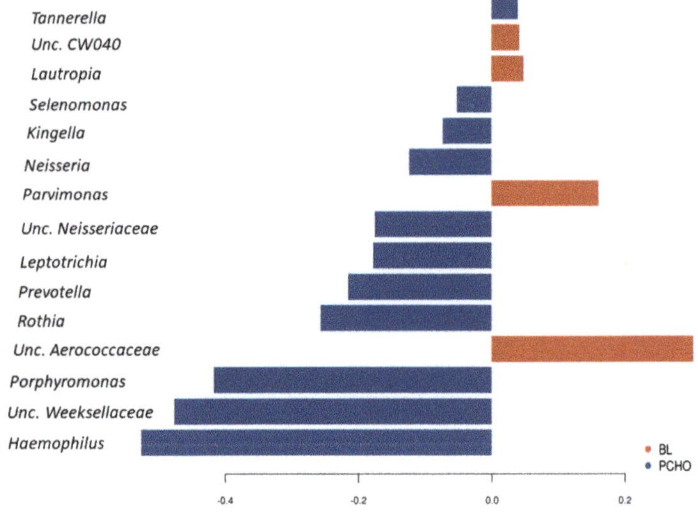

Figure 3. Genera differentiating between the oral microbiota profiles of athletes at baseline (BL, red) and after their consumption of the Periodised Carbohydrate diet (PCHO, blue) identified by sparse Partial Least Squares Discriminant Analysis (sPLS–DA).

Discriminating taxa for BL and for the same samples following the LCHF diet. Training intervention was also identified using LefSE and sPLS-DA. LefSe analysis revealed *Leptotrichia*, *Lachnospiraceae* and *TM-7* affiliated OTU's to be increased in the BL samples, whereas *Lactobacillales*, *Streptococcus*, *Neisseria* affiliated OTU's were discriminatory and increased in the LCHF group (Supplementary Figure S5). The sPLS-DA identified *Selenomonas*, *Unc. Planococcaceae*, *Unc. Enterobacteriaceae*, *Peptostreptococcus*, *Gemella*, *Granulicatella*, *Parvimonas Unc. Clostridiaceae* to increase following LCHF diet training intervention, whereas *Unc. F16*, *Unc. Neisseriaceae*, *Leptotrichia*, *Lactobacillus*, *Lautropia* and *Kingella* to be distinct and enriched in the BL samples of same athletes (Figure 4). According to repeated measures analysis using mixed effect linear regression, the genus *Fusobacterium* ($p = 0.02$), *Lautropia* ($p = 0.05$), *Aggregatibacter* ($p = 0.04$), *Leptotrichia* ($p = 0.040$) and some unclassified *F16* ($p = 0.04$) were reduced, whereas *Granulicatella* ($p = 0.03$), some unclassified Planococcaceae ($p = 0.03$) and *Streptococcus* ($p = 0.048$) were increased in response to LCHF diet when compared to microbiota profiles at their BL (Supplementary Figure S6). However, no significant differences were observed once correction for multiple testing using false discovery rate was applied (FDR = 0.4).

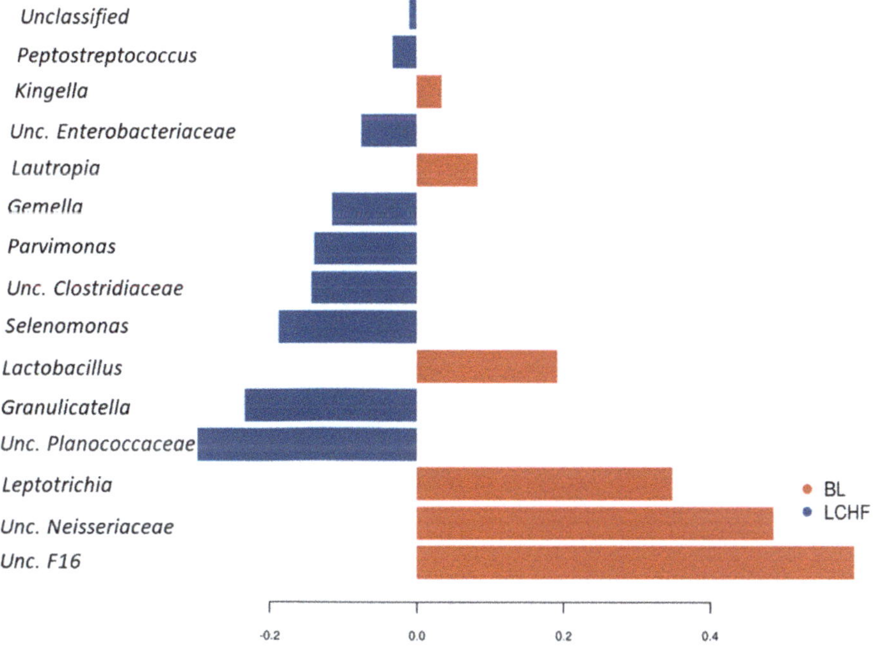

Figure 4. Genera differentiating between the oral microbiota profiles of athletes at baseline (BL, red) and after their consumption of the Low Carbohydrate High Fat diet (LCHF, blue) identified by sparse Partial Least Squares Discriminant Analysis (sPLS–DA).

3.2. Comparisons of Community Profiles of Saliva Samples at the Conclusion of Dietary Interventions

Figure 5 summarises the results of these analyses, showing the community profiles present in saliva samples at the conclusion of the dietary intervention periods, with annotations around some key genera and their inferred nitrate reductase capacity (Figure 5A). These profiles were compared with each other using sPLS-DA, which can be used to extract those taxa that most strongly discriminate the community structure between treatment groups (Figure 5B). Longitudinal comparison of the taxonomic profiles in the samples using sPLS-DA (Figure 5B) showed the strongest effect of the LCHF dietary intervention and in particular increase in the abundance of Gram-positive (Firmicutes) bacteria such as *Streptococcus*, *Peptostreptococcus*, and *Rothia*.

LefSe analyses was also used to examine the differences in oral microbiomes post intervention and these analyses showed that the discriminating and enriched taxa (at the OTU level) were *Streptococcus* affiliated OTUs for the LCHF diet intervention, whereas Gram-negative bacteria (e.g., *Haemophilus* and *Leptotrichia* spp.) were among the enriched and discriminating taxa for the HCHO/PCHO diets (Figure 6A,B).

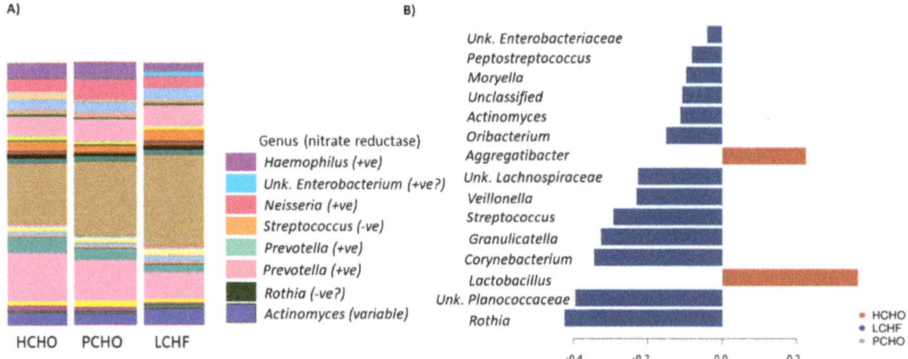

Figure 5. Oral microbiome profiles (genus-level) of athletes consuming either a high carbohydrate (HCHO), periodised carbohydrate (PCHO) or a low-carbohydrate high-fat diet (LCHF) after the diet-training intervention where bar plots represent: (**A**) relative abundance of genera in saliva samples after dietary interventions and their inferred nitrate reductase activity; (**B**) microbial families associated with different diets as identified by sparse Partial Least Squares Discriminant Analysis (sPLS-DA).

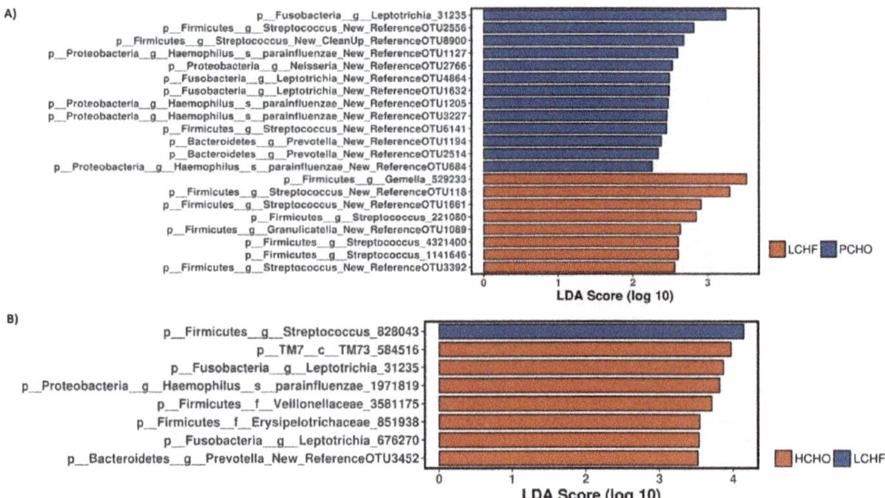

Figure 6. Linear Discriminant Analysis (LDA) Effect Size (LefSe) analysis at Operational Taxonomic Unit (OTU) level to compare the oral microbiome profiles of athletes post training diet interventions between Periodised Carbohydrate/Low Carbohydrate High Fat diet (PCHO/LCHF) (**A**) and High carbohydrate-Low Carbohydrate high-fat diet (HCHO/LCHF) (**B**), respectively.

4. Discussion

Despite the relatively small number of participants in this study, the dietary pattern consumed by these elite athletes during intensified training was shown to invoke remarkable effects on specific oral bacterial taxa—the bacterial communities in the mouth. The lack of a matching cohort of non-athletes (non-race walkers) and the lack of comprehensive data on the habitual dietary intake of athletes (i.e., BL samples) are acknowledged but were beyond the logistical and financial scope of the trial design. Furthermore, but understandably, the elite nature of the athletes ensured the group size is quite small, which also reduces the power needed for stringent statistical tests of significance, or the further subgrouping of the athletes according to ethnicity, etc. However, and despite these limitations, this is the first study of its type with elite endurance athletes, and one of the very few studies of the diet/nutrients × oral microbiota interactions affecting the physiology and/or metabolism of healthy human subjects. Furthermore, while fluctuations in the abundance and/or activities of nitrate-reducing bacteria in the oral microbiome are recognised to affect an individual's responsiveness to nitrate supplementation [16], this is the first study of the effects of dietary manipulation on this specific microbiome.

The bacterial taxa found in this study are similar to those represented in the Human Oral Microbiome Database (HOMD), as well as those typically reported in other studies of mainstream human subjects [17,18]. The beta-diversity UniFrac principal coordinates analysis showed no apparent clustering of the oral microbial communities based on the ethnicity of the athletes nor any distinct effects of the dietary interventions under investigation in this study. This is similar to the findings of other studies of healthy individuals in which no significant clustering and bacterial taxa changes in the oral cavity have been reported [17]. Nevertheless, more subtle changes within the bacterial communities in association with the diets were observed, with some of these representing potential alterations in community–host symbiosis. Here, the comparisons of the oral microbiome collected after three weeks of consuming one of three widely used diets by elite athletes during intensified training revealed that, unlike the CHO-rich diets, a ketogenic-LCHF diet appears to shift the balance of bacterial taxa that are widely considered to be key governors of the enterosalivary nitrate-nitrite-nitric oxide (NO) axis within the oral cavity. This is an important finding since previous studies have demonstrated functional effects on host health when alterations to the oral microbiome interfere with this pathway [19]. Facultative anaerobic bacteria in the mouth reduce salivary gland concentrated nitrate to nitrite, which is then swallowed and absorbed into the bloodstream, before further reduction to NO [3,4,20].

The critical role of the oral microbiome in this effect has been recently demonstrated in healthy non-athletes, where a seven-day period of antiseptic mouth wash treatment was shown to disrupt the oral microbiota and, in the absence of any dietary modifications, was associated with reductions in plasma and oral nitrite levels and an increase in blood pressure [20]. Taken together, the changes seen following consumption of the LCHF diet, with respect to reductions in the relative abundances of well-known Gram-negative nitrate/nitrite reducers such as *Haemophilus*, *Prevotella* and *Neisseria*; and an increase in *Streptococcus* spp., which are not recognised to be directly involved in nitrate/nitrite reduction; raises the spectre that consumption of the LCHF diet can impair the enterosalivary nitrate-nitrite-NO axis. Indeed, further indirect support for this hypothesis can be found in a recent brief report that a three-day LCHF diet was associated with an impaired plasma nitrate/nitrite conversion following supplementation with potassium nitrate, compared with the response observed when people consumed a HCHO diet [21]. This suggests that a LCHF diet might alter the baseline contribution of the nitrate-nitrite-NO pathway to NO-related health and performance benefits in athletes, as well as reduce their responsiveness to nitrate/beetroot juice supplementation as a performance aid [22]. Further supporting evidence comes from the major outcome of the Supernova 1 study, which is the primary study to the current project and from which these saliva samples were derived [6]. The study found a reduction in exercise economy (i.e., an increased oxygen cost of exercise) across a range of walking speeds in the LCHF group. It was originally hypothesised that this contributed to the failure of the LCHF group to improve their performance of a 10,000 m race walking event, despite the improvement in aerobic capacity that was seen across each of the study groups in response to the three week block of intensified training It is plausible to attribute this loss of economy to the substantial increase in the contribution of fat oxidation to exercise substrate needs in the LCHF group, noting the longstanding recognition that CHO oxidation is slightly more economical in generating ATP than fat oxidation per unit of oxygen utilization [23]. Additionally, the increase in exercise tolerance and performance following acute and/or chronic nitrate supplementation include improved oxygen delivery to the muscle via the vasodilatory effects of NO, as well as a direct effect on mitochondria to reduce proton leak [22]. However, these benefits are not universally observed across and within studies, and this variability is partially attributed to individual responsiveness, in addition to the more obvious contribution of unsuitable study protocols in relation to both the supplementation and exercise elements. However, based on the findings reported here, it is also plausible that part of the reduced exercise efficiency observed in the Supernova 1 study might be attributed to an altered oral microbiome, resulting in a reduction in nitrate/nitrite reducing activity and NO generation, with coordinate effects on circulation and mitochondrial function.

5. Conclusions

In conclusion, the results presented here are the first direct comparison of the oral microbiota profiles of elite athletes, and the effects of the dietary pattern consumed during intensified training for race-walking. The LCHF diet resulted in the most dramatic effects on the oral microbiota, with reductions in the relative abundance (*Haemophilus, Neisseria* and *Prevotella*), and with a coincident increase in the relative abundance of *Streptococcus* spp. The athletes participating in this study following consumption of the LCHF diet also showed a loss of exercise economy (i.e., an increased oxygen cost of exercise) across a range of walking speeds compared to athletes consuming the carbohydrate rich diets [6]. The findings reported here therefore justify the need to examine how diet x oral microbiome interactions affect elite athlete performance; and, particularly, NO homeostasis, and any coordinate impacts on cardiovascular and circulatory physiology.

Supplementary Materials: The following are available online http://www.mdpi.com/2072-6643/11/3/614/s1, Figure S1: Genera differentiating between the oral microbiota profiles of athletes at baseline (BL, red) and after their consumption of the High Carbohydrate diet (HCHO, blue) identified by LefSE; Figure S2: Mixed effect linear regression identified significant reductions in the relative abundances of *Capnocytophaga* ($p = 0.027$) and *Porphyromonas* ($p = 0.032$) whereas significant increase in the relative abundance of *Atopobium* ($p = 0.015$) after consumption of the High carbohydrate (HCHO) diet. Relative abundance was compared by mixed effect linear regression, including sampling time point as fixed effect and athlete as random effect. BL: baseline. Samples collected from the same individual are connected by lines; Figure S3: Genera differentiating between the oral microbiota profiles of athletes at baseline (BL, red) and after their consumption of the Periodised Carbohydrate diet (PCHO, blue) identified by LefSE; Figure S4: Mixed effect linear regression identified significant reductions in the relative abundances of *Actinomyces* ($p = 0.043$), *Moryella* ($p = 0.053$), *Oribacterium* ($p = 0.042$), *Peptostreptococcus* ($p = 0.009$) and *Unc. Erysipelotrichaceae* ($p = 0.042$) after consumption of the Periodised carbohydrate (PCHO) diet. Relative abundance was compared by mixed effect linear regression, including sampling time point as fixed effect and athlete as random effect. BL: baseline. Samples collected from the same individual are connected by lines; Figure S5: Genera differentiating between the oral microbiota profiles of athletes at baseline (BL, red) and after their consumption of the Low Carbohydrate High Fat diet (LCHF, blue) identified by LefSE; Figure S6: Mixed effect linear regression identified significant reductions in the relative abundances of *Fusobacterium* ($p = 0.020$), *Lautropia* ($p = 0.048$), *Leptotrichia* ($p = 0.040$), *Aggregatibacter* ($p = 0.044$) and *Unc. F16* ($p = 0.040$) whereas significant increase in the relative abundances of *Granulicatella* ($p = 0.038$), *Streptococcus* ($p = 0.048$) and *Planococcaceae* ($p = 0.038$) after consumption of the low carbohydrate high fat (LCHF) diet. Relative abundance was compared by mixed effect linear regression, including sampling time point as fixed effect and athlete as random effect. BL: baseline. Samples collected from the same individual are connected by lines.

Author Contributions: Conceptualization, L.M.B. and M.M.; methodology, L.M.B., M.M., and L.K.; formal analysis, N.M.; investigation, N.M.; resources, L.M.B. and M.M.; original draft preparation, N.M.; manuscript review and editing, L.M.B., M.M., L.K., N.V., B.C., H.M.O., M.L.R., and K.L.C.; visualization, N.M.; supervision, M.M.; project administration, L.M.B. and M.M.; funding acquisition, L.M.B. and M.M.

Funding: This research has been supported by funds from Collaborative Research Network for Advancing Exercise and Sports Science CRN AESS, a Program Grant from Australian Catholic University Research Fund

(No. 201300800) and a grant from the Australian Institute of Sport's High-Performance Sport Research Fund. N.M. is supported by a University of Queensland International Fellowship.

Acknowledgments: The authors thank all the subjects for their participation in this study and N.M. especially acknowledges the support and guidance from Alicia Kang, Erin Shanahan, Naoki Fukuma and Luisa Gomez Arango.

Conflicts of Interest: The authors declare no conflict of interest.

References

1. Dewhirst, F.E.; Chen, T.; Izard, J.; Paster, B.J.; Tanner, A.C.; Yu, W.H.; Lakshmanan, A.; Wade, W.G. The Human Oral Microbiome. *J. Bacteriol.* **2010**, *192*, 5002–5017. [CrossRef]

2. He, J.; Li, Y.; Cao, Y.; Xue, J.; Zhou, X. The oral microbiome diversity and its relation to human diseases. *Folia Microbiol.* **2015**, *60*, 69–80. [CrossRef]

3. Lundberg, J.O.; Weitzberg, E.; Cole, J.A.; Benjamin, N. Nitrate, bacteria and human health. *Nat. Rev. Microbiol.* **2004**, *2*, 593. [CrossRef] [PubMed]

4. Lundberg, J.O.; Weitzberg, E.; Gladwin, M.T. The nitrate-nitrite-nitric oxide pathway in physiology and therapeutics. *Nat. Rev. Drug Discov.* **2008**, *7*, 156–167. [CrossRef] [PubMed]

5. Petersson, J.; Carlström, M.; Schreiber, O.; Phillipson, M.; Christoffersson, G.; Jägare, A.; Roos, S.; Jansson, E.A.; Persson, A.E.; Lundberg, J.O.; et al. Gastroprotective and blood pressure lowering effects of dietary nitrate are abolished by an antiseptic mouthwash. *Free Radic. Biol. Med.* **2009**, *46*, 1068–1075. [CrossRef] [PubMed]

6. Burke, L.M.; Ross, M.L.; Garvican-Lewis, L.A.; Welvaert, M.; Heikura, I.A.; Forbes, S.G.; Mirtschin, J.G.; Cato, L.E.; Strobel, N.; Sharma, A.P.; et al. Low carbohydrate, high fat diet impairs exercise economy and negates the performance benefit from intensified training in elite race walkers. *J. Physiol.* **2017**, *595*, 2785–2807. [CrossRef]

7. Mirtschin, J.G.; Forbes, S.F.; Cato, L.E.; Heikura, I.A.; Strobel, N.; Hall, R.; Burke, L.M. Organization of Dietary Control for Nutrition-Training Intervention Involving Periodized Carbohydrate Availability and Ketogenic Low-Carbohydrate High-Fat Diet. *Int. J. Sport Nutr. Exerc. Metab.* **2018**, 1–10. [CrossRef] [PubMed]

8. Burke, L.M.; Hawley, J.A.; Jeukendrup, A.; Morton, J.P.; Stellingwerff, T.; Maughan, R.J. Toward a Common Understanding of Diet-Exercise Strategies to Manipulate Fuel Availability for Training and Competition Preparation in Endurance Sport. *Int. J. Sport Nutr. Exerc. Metab.* **2018**, *28*, 451–463. [CrossRef] [PubMed]

9. Ó Cuív, P.; Aguirre de Cárcer, D.; Jones, M.; Klaassens, E.S.; Worthley, D.L.; Whitehall, V.L.; Kang, S.; McSweeney, C.S.; Leggett, B.A.; Morrison, M. The effects from DNA extraction methods on the evaluation of microbial diversity associated with human colonic tissue. *Microb. Ecol.* **2011**, *61*, 353–362. [CrossRef]

10. Shanahan, E.R.; Zhong, L.; Talley, N.J.; Morrison, M.; Holtmann, G. Characterisation of the gastrointestinal mucosa-associated microbiota: A novel technique to prevent cross-contamination during endoscopic procedures. *Aliment. Pharmacol. Ther.* **2016**, *43*, 1186–1196. [CrossRef]

11. Caporaso, J.G.; Kuczynski, J.; Stombaugh, J.; Bittinger, K.; Bushman, F.D.; Costello, E.K.; Fierer, N.; Peña, A.G.; Goodrich, J.K.; Gordon, J.I.; et al. QIIME allows analysis of high-throughput community sequencing data. *Nat. Methods* **2010**, *7*, 335–336. [CrossRef] [PubMed]

12. Haas, B.J.; Gevers, D.; Earl, A.M.; Feldgarden, M.; Ward, D.V.; Giannoukos, G.; Ciulla, D.; Tabbaa, D.; Highlander, S.K.; Sodergren, E.; et al. Chimeric 16S rRNA sequence formation and detection in Sanger and 454 pyrosequenced PCR amplicons. *Genome Res.* **2011**, *21*, 494–504. [CrossRef]

13. DeSantis, T.Z.; Hugenholtz, P.; Larsen, N.; Rojas, M.; Brodie, E.L.; Keller, K.; Huber, T.; Dalevi, D.; Hu, P.; Andersen, G.L. Greengenes, a chimera-checked 16S rRNA gene database and workbench compatible with ARB. *Appl. Environ. Microbiol.* **2006**, *72*, 5069–5072. [CrossRef] [PubMed]

14. Cao, K.-A.L.; Costello, M.-E.; Lakis, V.A.; Bartolo, F.; Chua, X.-Y.; Brazeilles, R.; Rondeau, P. MixMC: A Multivariate Statistical Framework to Gain Insight into Microbial Communities. *PLoS ONE* **2016**, *11*, e0160169. [CrossRef]

15. Zakrzewski, M.; Proietti, C.; Ellis, J.J.; Hasan, S.; Brion, M.-J.; Berger, B.; Krause, L. Calypso: A user-friendly web-server for mining and visualizing microbiome–environment interactions. *Bioinformatics* **2017**, *33*, 782–783. [CrossRef] [PubMed]

16. Burleigh, M.C.; Liddle, L.; Monaghan, C.; Muggeridge, D.J.; Sculthorpe, N.; Butcher, J.P.; Henriquez, F.L.; Allen, J.D.; Easton, C. Salivary nitrite production is elevated in individuals with a higher abundance of oral nitrate-reducing bacteria. *Free Radic. Biol. Med.* **2018**, *120*, 80–88. [CrossRef] [PubMed]

17. Bik, E.M.; Long, C.D.; Armitage, G.C.; Loomer, P.; Emerson, J.; Mongodin, E.F.; Nelson, K.E.; Gill, S.R.; Fraser-Liggett, C.M.; Relman, D.A. Bacterial diversity in the oral cavity of 10 healthy individuals. *ISME J.* **2010**, *4*, 962. [CrossRef]

18. Aas, J.A.; Paster, B.J.; Stokes, L.N.; Olsen, I.; Dewhirst, F.E. Defining the normal bacterial flora of the oral cavity. *J. Clin. Microbiol.* **2005**, *43*, 5721–5732. [CrossRef] [PubMed]

19. Vanhatalo, A.; Blackwell, J.R.; L'Heureux, J.E.; Williams, D.W.; Smith, A.; van der Giezen, M.; Winyard, P.G.; Kelly, J.; Jones, A.M. Nitrate-responsive oral microbiome modulates nitric oxide homeostasis and blood pressure in humans. *Free Radic. Biol. Med.* **2018**, *124*, 21–30. [CrossRef]

20. Qu, X.M.; Wu, Z.F.; Pang, B.X.; Jin, L.Y.; Qin, L.Z.; Wang, S.L. From Nitrate to Nitric Oxide: The Role of Salivary Glands and Oral Bacteria. *J. Dent. Res.* **2016**, *95*, 1452–1456. [CrossRef]

21. Piatrikova, E. The Impact of High-Carbohydrate and High-Fat Diets in Combination with Nitrate on O_2 Uptake Kinetics and Performance during High-Intensity Exercise. Published Online First: 16 March 2017. Available online: https://ore.exeter.ac.uk/repository/handle/10871/28215 (accessed on 16 December 2017).

22. Jones, A.M. Influence of dietary nitrate on the physiological determinants of exercise performance: A critical review. *Appl. Physiol. Nutr. Metab. Physiol. Appl. Nutr. Metab.* **2014**, *39*, 1019–1028. [CrossRef] [PubMed]

23. Krogh, A.; Lindhard, J. The Relative Value of Fat and Carbohydrate as Sources of Muscular Energy. *Biochem. J.* **1920**, *14*, 290–363. [CrossRef] [PubMed]

© 2019 by the authors. Licensee MDPI, Basel, Switzerland. This article is an open access article distributed under the terms and conditions of the Creative Commons Attribution (CC BY) license (http://creativecommons.org/licenses/by/4.0/).

Article

Energy Status and Body Composition Across a Collegiate Women's Lacrosse Season

Hannah A. Zabriskie [1], Bradley S. Currier [1], Patrick S. Harty [1], Richard A. Stecker [1], Andrew R. Jagim [2] and Chad M. Kerksick [1,*]

[1] Exercise and Performance Nutrition Laboratory, Department of Exercise Science, Lindenwood University, St. Charles, MO 63301, USA; hzabriskie@lindenwood.edu (H.A.Z.); bc129@lindenwood.edu (B.S.C.); pharty@lindenwood.edu (P.S.H.); rstecker@lindenwood.edu (R.A.S.)

[2] Sport Medicine Research, Mayo Clinic Health Systems, Onalaska, WI 54650, USA; jagim.andrew@mayo.edu

*   Correspondence: ckerksick@lindenwood.edu; Tel.: +1-636-627-4629

Received: 15 January 2019; Accepted: 18 February 2019; Published: 23 February 2019

Abstract: Little data is available regarding the energy and nutritional status of female collegiate team sport athletes. Twenty female NCAA Division II lacrosse athletes (mean $\pm$ SD: 20.4 $\pm$ 1.8 years; 68.8 $\pm$ 8.9 kg; 168.4 $\pm$ 6.6 cm; 27.9 $\pm$ 3% body fat) recorded dietary intake and wore a physical activity monitor over four consecutive days at five different time points (20 days total) during one academic year. Body composition, bone health, and resting metabolic rate were assessed in conjunction with wearing the monitor during off-season, pre-season, and season-play. Body fat percentage decreased slightly during the course of this study ($p = 0.037$). Total daily energy expenditure (TDEE) ($p < 0.001$) and activity energy expenditure (AEE) ($p = 0.001$) energy were found to change significantly over the course of the year, with pre-season training resulting in the highest energy expenditures (TDEE: 2789 $\pm$ 391 kcal/day; AEE: 1001 $\pm$ 267 kcal/day). Caloric (2124 $\pm$ 448 kcal/day), carbohydrate (3.6 $\pm$ 1.1 g/kg), and protein (1.2 $\pm$ 0.3 g/kg) intake did not change over the course of the year ($p > 0.05$). Athletes self-reported a moderate negative energy balance (366–719 kcal/day) and low energy availability (22.9–30.4 kcal/kg FFM) at each measurement period throughout the study. Reported caloric and macronutrient intake was low given the recorded energy expenditure and macronutrient intake recommendations for athletes. Athletic support staff should provide athletes with appropriate fueling strategies, particularly during pre-season training, to adequately meet energy demands.

Keywords: female athletes; energy balance; nutrition; RED-S; calories; recommendations; gender; health; energy availability

---

1. Introduction

Energy is required for all bodily functions. The total amount of energy expended in one day, total daily energy expenditure (TDEE), is the sum of resting metabolic rate (RMR), activity energy expenditure (AEE), non-exercise activity thermogenesis (NEAT), and the thermic effect of food (TEF). While RMR is a consistent contributor to TDEE, making up about 60–65% of daily energy expenditure, AEE can vary widely from day to day within a single person and between different individuals [1,2]. In order to maintain energy balance, individuals must attempt to match energy intake with the amount of energy expended each day. Thus, regulation of energy balance is a primary focus of athletes and athletic professionals [3] to ensure optimal energy is available to support training, recovery, and lean body mass.

In instances when excess calories are consumed relative to TDEE, weight gain oftentimes occurs. Conversely, if insufficient calories are consumed, a state of negative energy balance exists during which athletes can experience undesirable loss of fat free mass (FFM) [4] and be at increased risk of injuries [5] and illness [6]. Energy availability, or the number of calories available to each kilogram of FFM after accounting for AEE is an emerging measure of energy status that is easier to assess, as opposed to TDEE or energy balance, from a time and logistics perspective, as it focuses solely on activity-related energy expenditure [7]. Low energy availability is correlated with a negative energy balance and Relative Energy Deficiency in Sport (RED-S) among female athletes [8]. RED-S is a comprehensive syndrome which includes the three components originally described as the female athlete triad [9] and extends to include a multifactorial state of physiological dysfunction that can have a profound impact on athlete health. It is important for athletes, coaches, athletic trainers, and all other athletic personnel to understand how to avoid these low energy states and to be aware of factors that may increase these risks.

Much of the research regarding athletic energy status has focused on male athletes. While efforts have established the energy status of female athletes performing endurance [8,10] and team sports [11–16], the generalizability of these results is hampered by the unique demands of each sport. Furthermore, the energy needs of an athlete can be dependent on sport type (e.g., aesthetic, continuous, and intermittent), position, level of competition, and seasonal training demands. For instance, in collegiate lacrosse, the non-stop clock, large field of play, and frequent changes in acceleration and direction create different energy requirements from those for athletes competing in other sports. These nuances are further complicated for athletes who are required to travel or perform frequently. Because of these distinctions, more research is needed to provide relevant information on energy requirements as it pertains to many types of female athletes.

Few researchers have undertaken the task of tracking energy status in a single cohort of female athletes over an entire annual training calendar. Reed et al. [13] has measured energy availability during pre-, mid-, and post-season periods in NCAA Division I female soccer players and Woodruff et al. [14] has compared energy availability from pre-season to post-season in elite female volleyball players. However, while these protocols assessed energy demands during or surrounding in-season play, these reports do not provide a long-term view of changes in energy status. Zanders et al. [17] has monitored collegiate (NCAA Division II) women's basketball players longitudinally across a complete academic calendar (September–April); however, the demands facing basketball players are likely different from females participating in field-based team sports. Therefore, more research is needed to assess off-season energy status in addition to pre-season and in-season status in field-based team sports, as energy demands may differ due to variation in training practices and body composition goals. To meet these needs, the following investigation sought to document the fluctuations in energy expenditure, energy balance, and body composition over the course of an academic year in Division II collegiate female lacrosse players.

2. Materials and Methods

2.1. Research Design

Data collection commenced at the beginning of the fall academic semester and finished at the end of the spring academic semester. Participants were observed and assessed for changes in RMR, body composition, bone density, energy intake, TDEE, and perceived recovery. Participants were assessed during five phases, with each testing period separated by four to six weeks. During each phase, participants were monitored for a period of four days (two weekdays, two weekend days), during which they wore a physical activity monitor (Actiheart, CamNTech, Cambridge, UK) and recorded all food consumed using a commercially available food and nutrition tracking application (MyFitnessPal, Under Armour, Baltimore, MD, USA). Monitoring of daily energy expenditure and energy intake occurred during all five phases. During the first, third, and fifth phases, body composition and bone density were assessed using dual energy X-ray absorptiometry (DEXA) scan and RMR was measured within two weeks of the monitoring period. Figure 1 provides an overview of each testing phase. Phase I consisted of practices and conditioning, with one group of six athletes participating in a game. Phase

II consisted of off-season conditioning and captain-led practices. Phase III consisted of pre-season training and practices. Phase IV consisted of the first half of season-play and associated conditioning and practices. Phase V covered the second half of the season and finished just before the start of post-season tournament play.

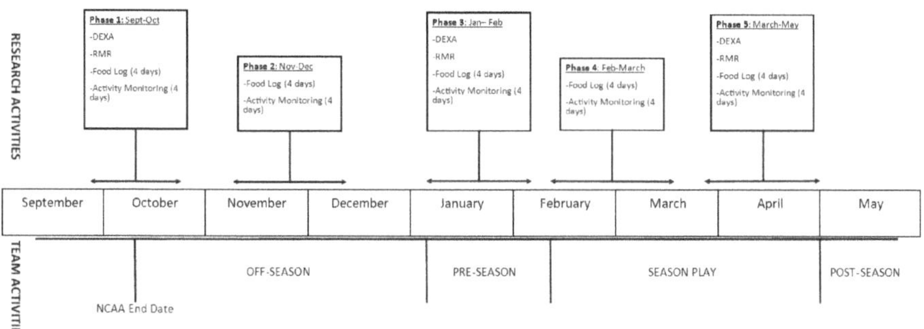

Figure 1. Overview of research and team activities. DEXA, dual energy X-ray absorptiometry; RMR, resting metabolic rate.

2.2. Subjects

A total of 20 NCAA Division II female lacrosse athletes completed this investigation (mean $\pm$ SD: 20.4 $\pm$ 1.8 years; 68.8 $\pm$ 8.9 kg; 168.4 $\pm$ 6.6 cm; 27.9 $\pm$ 3% body fat). Athletes were recruited during a team meeting, during which researchers presented the study design and answered questions regarding participation. Initially, 22 NCAA athletes were recruited for this study, though any athlete who became injured and was unable to fully participate in team activities was excluded from participation. One athlete withdrew during Phase IV due to noncompliance (not wearing physical activity monitor) and another was excluded for medical reasons before the start of Phase III. The final sample included 20 athletes without significant injury who were fully participating in team activities during each assessment phase. All subjects were informed of the risks associated with participation in this study and signed an informed consent document approved by the Lindenwood University IRB (Protocol # IRB-19-L0012, approved 25 August 2017).

2.3. Methodology

2.3.1. Anthropometrics, Body Composition and Bone Health

Body weight and height were measured during each phase of the study. Shoes and excess clothing were removed and then body weight was obtained to the nearest 0.1 kg (Digital Scale BWB-627A Class III, Tanita, Tokyo, Japan). Height was measured to the nearest 0.25 in using a stadiometer (HR-200, Tanita, Tokyo, Japan) and subsequently converted to cm.

Body composition was measured during phases I, III, and V. Participants arrived at the laboratory after following an overnight (8–10 h) fast and having abstained from exercise and caffeine for 24 h. Participants provided a urine sample to determine hydration status. In cases where the urine specific gravity was greater than 1.02, 24 fluid ounces of water were provided to achieve a more standardized hydration status for each testing period. A whole-body DEXA scan (Hologic Discovery A, Hologic, Bedford, MA, USA) was then performed to obtain body composition and bone density parameters using manufacturer provided software (Hologic APEX Software, Version 4.5.3, Hologic, Bedford, MA, USA) with the NHANES correction factor applied. However, it should be noted that only a full body scan was obtained and was used to calculate bone mineral content (BMC), bone mineral density (BMD), and Z-score.

2.3.2. Resting Metabolic Rate

RMR assessment was performed during phases I, III, and V on the same day as the DEXA scan. A metabolic cart (TrueMax 2400 Metabolic Measurement System, ParvoMedics, Sandy, UT, USA) was calibrated to within less than 2% of the previous day's calibration factor. All RMR assessments during Phase I were performed using the same cart. However, the laboratory purchased a second metabolic cart which was also used in phases III and V, and each athlete was assigned to one of the two carts for phases III and V. Therefore, half of study participants used the same cart for the entire study, while the other half used a different cart for two of the three phases.

Following completion of the DEXA and 15 min of quiet and seated rest, participants were directed to lay supine on a padded exam table. Participants then had a plastic hood and attached drape placed over their head and shoulders. A blanket was positioned over the participant to ensure a comfortable resting temperature and to eliminate any expired air from leaking out of the closed system. Gases were collected and analyzed for 20 min. RMR was calculated by identifying five consecutive min in the last 10 min of assessment with a less than 5% change in oxygen use. The average of these five min was considered to be the RMR (kcal/day).

2.3.3. Diet Record

Diet records were kept during each phase for four consecutive days, which included two weekdays and two weekend days. Participants were asked to download the MyFitnessPal application to their mobile device and share their diary with a designated member of the study team. In one instance, the participant was unable to download the application and therefore kept a paper record during the day and uploaded her intake to the MyFitnessPal website each evening. Participants were educated on dietary reporting strategies by a trained research assistant. Study participants were provided educational materials about serving sizes (written and pictorial comparisons of common household items and respective serving sizes) and were instructed on how to alter serving sizes in the MyFitnessPal application. Further education about the functionality of the application was provided and participants were instructed to record all food and drink they consumed, including alcohol. Energy (kilocalorie) and macronutrients (carbohydrate, fat, and protein) data was retrieved from each participant's diary for each phase and expressed as a daily average for total and relative intakes.

2.3.4. Physical Activity Monitoring

Physical activity monitoring occurred during each phase of this study on the same days for which diet information was collected. Physical activity monitoring occurred on two weekdays and two weekend days which were consecutive. Each participant was outfitted with a physical activity monitor (Actiheart, CamNtech, Cambridge, UK) which was to be worn at all times, except when showering. Accelerometry data provided by the Actiheart device was used for all energy calculations within this study. The monitor was attached using two electrodes (Kendall 230 Foam Electrodes, Covidien, Mansfield, MA, USA), with one positioned over the xiphoid process and the other on the left side of the chest between and the fifth and sixth rib. When the monitor was initially applied by a study team member, each electrode location was outlined with permanent marker and participants were allowed to replace the electrodes as they lost adhesion. Additionally, participants were provided with an Actiheart-compatible chest strap (Polar, Kempele, Finland) to be used during heavy exercise instead of the electrodes.

The Actiheart devices have been previously shown to accurately calculate TDEE and AEE [18]. RMR was calculated via the manufacturer-provided software using the Schofield equation for use in the TDEE estimation. The software also automatically estimated TEF to be 10% of TDEE. Each of these values was recorded for each of the four monitoring days during each phase expressed as a daily average by collapsing the data collected during the four-day monitoring period.

Only six Actiheart devices were available for use during this study. Therefore, each phase took four weeks to complete, as the sample was divided into four groups of five to six athletes.

2.3.5. Markers of Energy Deficiency

Energy balance (kcal/day) was calculated as the difference between energy intake and TDEE during each phase. Energy availability (kcal/kg FFM) was calculated during each phase using the following equation [19]:

$$\text{Energy Availability} = (\text{Energy intake} - \text{AEE})/\text{kg FFM} \tag{1}$$

For phases when FFM was not directly measured (i.e., for Phase II and Phase IV), the FFM measured in the previous phase was used.

RMR ratio, or the ratio of measured RMR to predicted RMR, was calculated per the findings of Staal et al. [20]. RMR ratio was calculated twice; it was calculated once using the Cunningham equation and once with the Schofield equation. The Cunningham equation has been utilized in prior research of RMR ratio [20] and has been validated in athletic populations [21]. The Schofield equation is [22]

$$(\text{RMR} = [14.808 \times \text{weight (kg)}] + 486.3) \tag{2}$$

It was also utilized for RMR ratio calculations due to it being the manufacturer default for RMR prediction in the Actiheart software. Fat-free mass index (FFMI) was calculated with no height correction as kilograms of FFM per meter of height squared (kg/m^2). Both RMR ratio and FFMI were only calculated for phases when they were directly measured (Phases I, III, and V).

2.3.6. Assessment of Recovery

During each phase, athletes were provided with a paper questionnaire featuring visual analog scales (VAS). Perceived rest, soreness, and training satisfaction were all assessed using VAS. Each VAS was 100 mm in length and responses are reported as a number between 0 and 100.

2.4. Statistical Analysis

All analysis was completed in SPSS v25 (Chicago, IL, USA). All data are presented as means $\pm$ SD. The Shapiro-Wilk test was used to identify non-normal data. Several of the body composition variables were moderately skewed, though transformations did not improve normality. Therefore, because the sample sizes were consistent throughout each phase, analysis was performed using a repeated measures ANOVA, relying upon the robustness of the procedure to overcome the moderate skew. Bonferroni *post hoc* corrections were utilized to identify significantly different phases, when indicated. Each ANOVA model was fit with $N = 20$ as that was the number of participants who completed every phase of the study. That was therefore the final sample size utilized for all results. Pearson correlations were used to quantify relationships between variables. A significance level of $\alpha = 0.05$ was used.

3. Results

3.1. Body Composition and Bone Health

An overview of body composition and bone health for each phase is found in Table 1. This population of female lacrosse players had an average body mass index (BMI) of 24.23 ± 2.3 kg/m^2 and an average weight of 69.3 ± 9.5 kg over the entire course of the study. The average percent body fat was $27.4 \pm 3.0\%$ throughout the year.

Weight (kg) did not significantly change over any of the five phases ($p = 0.201$). FM ($p = 0.118$) did not change between phases I, III, and V, while FFM trended toward a change ($p = 0.054$). However, body fat percentage decreased slightly over the year ($p = 0.037$), though Bonferroni *post hoc* testing only identified a trend ($p = 0.076$) towards athletes having significantly higher body fat percentage in Phase

I compared to Phase V. Bone mineral content ($p < 0.001$) significantly increased over the season with Phase V being significantly greater than Phase I ($p = 0.007; p = 0.028$) and Phase 2 ($p < 0.001; p = 0.014$). However, bone mineral density did not significantly change ($p = 0.17$).

Table 1. Body Composition and Bone Health over the Academic Year.

	Phase 1 (Off-Season)	Phase 2 (Off-Season)	Phase 3 (Pre-Season)	Phase 4 (In-Season)	Phase 5 (In-Season)	Overall *p*-Value
Body Weight (kg)	68.8 ± 8.9	69.6 ± 9.5	69.6 ± 10.0	69.3 ± 10.0	68.9 ± 10.1	0.20
FFM (kg)	47.0 ± 5.3	--	47.9 ± 5.4	--	47.4 ± 5.6	0.054
FM (kg)	18.4 ± 4.4	--	18.3 ± 4.7	--	17.8 ± 4.8	0.118
Percent Fat (%) *	27.9 ± 3.0	--	27.3 ± 2.7	--	27.0 ± 3.2	**0.037**
BMC (g) *	2575 ± 230 [5]	--	2572 ± 230 [5]	--	2610 ± 247 [1,3]	**<0.001**
BMD (g/cm^3)	1.20 ± 0.07	--	1.20 ± 0.07	--	1.24 ± 0.14	0.167
z-Score *	1.30 ± 0.76 [5]	--	1.28 ± 0.80 [5]	--	1.46 ± 0.75 [1,3]	**0.004**

Data is presented as mean ± SD. Legend: BMC, bone mineral content; BMD, bone mineral density; FFM, fat-free mass; FM, fat mass. * Significant effect of Phase via Repeated Measures ANOVA. [1] Significantly different from Phase I. [3] Significantly different from Phase III. [5] Significantly different from Phase V. The bold indicates statistical significance.

3.2. Metabolic Rate and Energy Expenditure

RMR was elevated ($p < 0.001$) during Phases III ($p = 0.002$) and V ($p = 0.001$) compared to Phase I. RMR was found to be moderately correlated with weight ($r = 0.50$–0.68, $p < 0.05$), FM ($r = 0.48$–0.67, $p < 0.05$), and FFM ($r = 0.51$–0.61, $p < 0.05$), during Phases I and III. Stronger correlations were noted between RMR and weight ($r = 0.78$, $p < 0.001$), FM ($r = 0.78$, $p < 0.001$), and FFM ($r = 0.71$, $p < 0.001$) in Phase V. When combining all phases, total body weight was most strongly correlated with RMR ($r = 0.61$, $p < 0.001$). A moderate correlation was also identified between RMR and FM ($r = 0.58$, $p < 0.001$) and FFM ($r = 0.58$, $p < 0.001$).

Energy expenditures from each phase can be found in Table 2. TDEE was significantly different during the phases of this study ($p < 0.001$), with Phase III resulting in greater EE than Phases I, IV, and V ($p < 0.05$). AEE changed significantly over the course of the season ($p = 0.001$), with AEE recorded in Phase III being significantly higher than that recorded during Phase I ($p = 0.004$), Phase IV ($p = 0.002$), and Phase V ($p = 0.04$), while trending towards being significantly greater than that in Phase II ($p = 0.065$). Physical activity level (PAL) was also significantly higher in Phase III than in any other phase ($p < 0.05$), as depicted in Table 2.

Table 2. Resting Metabolism and Daily Energy Expenditure over the Academic Year.

	Phase 1 (Off-Season)	Phase 2 (Off-Season)	Phase 3 (Pre-Season)	Phase 4 (In-Season)	Phase 5 (In-Season)	Overall *p*-Value
RMR (kcal/day) *	1536 ± 152 [3,5]	--	1683 ± 162 [1]	--	1732 ± 244 [1]	**<0.001**
TDEE (kcal/day) *	2608 ± 378 [3]	2579 ± 376	2798 ± 391 [1]	2513 ± 248 [3]	2582 ± 303 [3]	**<0.001**
AEE (kcal/day) *	842 ± 267 [3]	804 ± 244	1001 ± 267 [1]	749 ± 161 [3]	817 ± 235 [3]	**0.001**
PAL *	1.75 ± 0.19 [3]	1.72 ± 0.14 [3]	1.87 ± 0.15 [1]	1.69 ± 0.15 [3]	1.73 ± 0.18 [3]	**0.001**

Data is presented as mean ± SD. Legend: AEE, activity energy expenditure; PAL, physical activity level; RMR, resting metabolic rate; TDEE, total daily energy expenditure. * Significant effect of Phase via Repeated Measures ANOVA. [1] Significantly different from Phase I. [3] Significantly different from Phase III. [5] Significantly different from Phase V. The bold indicates statistical significance.

3.3. Energy Intake

Self-reported caloric intake, including food, beverages, and alcohol, did not significantly change over the season ($p = 0.247$), and the athletes recorded ingesting 2124 ± 448 kcals on average throughout all phases (see Table 3). The ingestion of both absolute ($p = 0.262$) and relative ($p = 0.146$) carbohydrate intake was constant across the year. The same was also true for absolute ($p = 0.168$) and relative ($p = 0.163$) protein intake. Absolute fat ingestion, however, did change during the five phases ($p = 0.03$); specifically, Phase II resulted in significantly lower absolute fat ingestion than Phase V.

Table 3. Self-Reported Caloric and Macronutrient Intake across the Academic Year.

	Phase 1 (Off-Season)	Phase 2 (Off-Season)	Phase 3 (Pre-Season)	Phase 4 (In-Season)	Phase 5 (In-Season)	Overall *p*-Value
Calories (kcal/day)	2242 ± 462	2015 ± 451	2079 ± 435	2124 ± 505	2161 ± 392	0.247
Carbohydrate (g/day)	262 ± 61	231 ± 59	247 ± 74	248 ± 66	236 ± 74	0.262
Protein (g/day)	80 ± 19	72 ± 20	82 ± 22	84 ± 16	79 ± 20	0.168
Fat (g/day) *	78 ± 20	70 ± 25 [5]	74 ± 23	81 ± 26	88 ± 23 [2]	**0.03**
Relative Carbohydrate (g/kg/day)	3.9 ± 1.1	3.4 ± 0.9	3.6 ± 1.2	3.6 ± 0.9	3.5 ± 1.2	0.146
Relative Protein (g/kg/day)	1.2 ± 0.3	1.1 ± 0.3	1.2 ± 0.4	1.2 ± 0.3	1.2 ± 0.4	0.163

Data is presented as mean ± SD. kcal = Kilocalorie; g = gram; kg = kilogram; * Significant effect of Phase via Repeated Measures ANOVA. [2] Significantly different from Phase II. [5] Significantly different from Phase V.

3.4. Energy Balance, Energy Availability and Surrogate Markers of Energy Deficiency

As presented in Table 4, energy availability changed significantly during the course of the academic year ($p = 0.017$). *Post hoc* comparisons showed that Phase III trended toward a lower energy availability than in Phase I ($p = 0.058$) and Phase IV ($p = 0.057$). Energy balance also changed across the course of this investigation ($p = 0.01$), with Phase III trending towards a more negative energy balance than Phase I ($p = 0.053$) and significantly a poorer balance than Phase IV ($p = 0.029$).

Table 4. Energy Balance, Energy Availability and Surrogate Markers of Energy Deficiency.

	Phase 1 (Off-Season)	Phase 2 (Off-Season)	Phase 3 (Pre-Season)	Phase 4 (In-Season)	Phase 5 (In-Season)	Overall *p*-Value
Energy Balance (kcal/day) *	−366 + 527 [‡]	−564 + 484	−719 + 440	−389 + 432 [3]	−421 ± 418	**0.01**
Energy Availability (kcal/kg FFM) *	30.4 ± 11.0 [‡]	26.2 ± 10.5	22.9 ± 8.5	28.7 ± 9.5 [‡]	28.9 ± 9.2	**0.017**
RMR Ratio, Schofield (Measured/Predicted) *	1.02 ± 0.7	--	1.11 ± 0.1 [1]	--	1.15 ± 0.1 [1]	**<0.001**
RMR Ratio, Cunningham (Measured/Predicted) *	1.0 ± 0.1	--	1.08 ± 0.1 [1]	--	1.1 ± 0.1 [1]	**0.001**
Free Mass Index (kg/m^2)	16.6 ± 1.2	--	16.7 ± 1.2	--	16.5 ± 1.3	0.175

Data is presented as mean ± SD. kcal = Kilocalorie; g = gram; kg = kilogram; * Significant effect of Phase via Repeated Measures ANOVA. [1] Significantly different from Phase I. [3] Significantly different from Phase III. [‡] Trend toward difference from Phase III. The bold indicates statistical significance.

RMR ratio was significantly different during the three phases in which it was assessed ($p \leq 0.001$) for both the Schofield and Cunningham equations. RMR ratio was found to be lowest during Phase I, which was lower than both Phase III ($p < 0.02$) and Phase V ($p = 0.001$). FFMI did not significantly change over the season ($p = 0.175$), with the average FFMI being 16.6 ± 1.2 kg/m^2. Only during Phase III was a significant correlation noted between RMR ratio (Cunningham equation) and energy availability ($r = -0.445$, $p = 0.049$). All other time points revealed no significant correlations between surrogate markers (both Schofield and Cunningham RMR ratios or FFMI) and energy balance or energy availability ($p > 0.05$).

3.5. Recovery Measures

Few relationships between energy balance, energy availability, and perceived measures of recovery were identified. Energy balance ($r = 0.581$, $p = 0.007$) and energy availability ($r = 0.6$, $p = 0.005$) were positively correlated with sleep quality during Phase II. During Phase V, energy balance ($r = 0.514$, $p = 0.02$) and energy availability ($r = 0.496$, $p = 0.026$) were found to be significantly positively correlated with perceived rest. In addition, energy availability was positively correlated with training satisfaction during Phase V ($r = 0.45$, $p = 0.046$).

4. Discussion

The purpose of this investigation was to assess the body composition, energy expenditure, and dietary habits of NCAA Division II female lacrosse players during an academic year. While body composition variables and diet were relatively stable, energy expenditure changed significantly during the course of the study. However, the athletes appeared to be in a consistent state of negative energy balance due to their self-reported energy intake. Energy availability was also shown to be low, hovering near the clinical energy deficiency threshold of 30 kcal/kg FFM [19]. It should be noted that objective measures of energy status such as body weight and body composition did not significantly change, which may offer evidence of some level of underreporting of dietary intake as has been previously reported.

While the athletes in the present study demonstrated a negative energy balance at every time point, this is not unprecedented. Hill et al. [10] has described low self-reported caloric intake resulting in significant energy deficiency in lightweight rowers, though it was noted that underreporting of food was problematic in this population. Also in agreement with our findings, negative energy balance and similar levels of TDEE have been reported in elite synchronized swimmers [11], junior elite female soccer players [12], and Division II female basketball players [17]. Additionally, a study recording only energy expenditure and not energy intake has identified similar levels of TDEE in elite female soccer players [15]. Despite the presented cohort having a significant energy deficiency and low energy availability with no significant changes in body composition, it is reasonable to suspect, based upon prior reports detailed above, that this is due to underreporting of dietary intakes. Food logs can present a substantial participant burden and, unfortunately, are frequently fraught with underreported and misreported information [23]. Thus, until more accessible and simpler methods of accurately recording dietary intake are available, collecting a valid representation of energy intake will continue to be a barrier in energy balance and energy availability investigations. The only diet-related variable that changed over the course of the season was absolute fat intake, and this change, specifically low intake during Phase II compared to the end of the season, is likely a reflection of schedule demands and food availability while travelling for competition during Phase V.

The greatest energy deficiency was reported during Phase III, which occurred during pre-season training immediately following the academic winter break. In accordance with our findings, previous studies have also reported the highest energy expenditure during pre-season preparation [13,17,24]. However, there was no concomitant increase in energy or macronutrient intake in Phase III. Fat intake during Phase V was increased in comparison to other phases. We are not able to specifically identify why fat intake was increased during this phase, but it is tempting to speculate that due to the rigors of in-season travel and classes, etc. that athletes were left with less time to consider their food choices and consequently selected foods that were higher in fat content. Altogether, these changes and suggestions create the need for athletic support staff to emphasize proper nutrition, and specifically greater caloric intake, while also focusing on how to become competition ready during pre-season training when risk of energy deficiency may be higher. Athletes may also need to be better educated regarding how to anticipate travel and training schedules that should require a greater energy intake.

The assessment of energy availability or energy balance can be challenging, as it relies upon accurate dietary records and participant compliance with energy expenditure estimation methods. Therefore, some alternative markers of energy deficiency or low energy availability have been suggested to offer insight into energy status while minimizing participant burden. One of these markers is the ratio of measured RMR to predicted RMR, with values less than 0.90 representing suppressed RMR and energy deficiency [20]. Because body mass and body composition are also influenced by energy status, some metrics of body composition have been advocated to offer insight into energy status. While low BMI is generally associated with RED-S [9,25], it may not be a sufficient surrogate marker as it fails to account for body composition. Unpublished data (currently in review) from our research group in a large cohort of nearly 400 female athletes has identified and proposed a lower limit of fat-free mass index as a simple way of estimating energy status with minimal testing required (DEXA only). In this paper, we proposed a lower limit FFMI value of 16.92 kg/m^2 as evaluated by a whole-body DEXA scan utilizing the TBAR 1209 correction factor. Whether or not this value holds true against future investigation and whether or not this FFMI value is actually correlated to low energy availability remains to be seen. As reported, no correlation was noted between FFMI and energy balance or availability. However, a different correction factor (NHANES) was utilized. Results from the present study provide interesting insight into the use of surrogate markers of energy deficiency longitudinally. For example, there were no changes noted in RMR ratio or FFMI over the course of this study despite significant changes in energy deficiency between phases. In addition, no meaningful correlations were identified between RMR ratio or FFMI and energy balance or energy availability. However, both of these measures simply may not respond to changes in energy deficiency very rapidly, suggesting that these measures may be more appropriate for acute assessments of high-risk athletes, particularly if screening a large group of athletes at one time. The utility of RMR ratio is also dependent upon the use of accurate and appropriate prediction equations. Previous research has documented that many accepted equations do not accurately predict RMR in Division II female athletes [26]. The observed increase in RMR during Phase III and V without any change in calorie intake, body mass, or fat-free mass is challenging to interpret. While not fully assessed within our current paper, it is possible that some level of adaptive thermogenesis was occurring that resulted in RMR being elevated due to the increased energy expended secondary to the recovery demands of in-season activity. While possible, all RMR measures followed standardized protocols whereby each person had refrained from exercise for at least 24 h and observed an overnight fast. Therefore, it is not likely the change was due to a deviation in our measurement approach. In addition, it is also possible the activity levels outside of documented training activities may have occurred which may account for our reported changes in RMR. Nonetheless, future research should address these considerations.

Despite the robust scope of this investigation, it was not without limitations. One of the primary limitations was the lack of confirmatory data via questionnaires or logs, primarily with regards to menstrual function and physical activity. It was known, anecdotally, that it was not uncommon for this athletic team to participate in self-directed exercise in addition to scheduled team activities. Including a daily physical activity log would have aided in the validation of the AEE and TDEE measurements and would likely help to better explain participants' energy balances. Administration of an assessment for disordered eating would have provided greater insight into RED-S in this population. Additionally, it would have been beneficial to monitor menstrual function. Tomten and colleagues [27] have published a study investigating weight stable runners with and without menstrual disorders. Despite similar energy expenditures and stable body weights, runners who exhibited menstrual dysfunction also tended to have lower energy intake and therefore also a negative energy balance. While the present study similarly evidenced weight stability despite an estimated energy deficit, menstrual function was not assessed. Hence, identifying athletes with menstrual irregularities could potentially explain how the body altered energy expenditure to accommodate the deficient energy intake, particularly considering the consistency of body weight measurements that were observed across our study design. Future studies should include both physical activity and menstrual function questionnaires.

5. Conclusions

Despite the limitations previously described, the scope of this study is unprecedented in female lacrosse athletes. This is the first investigation to have assessed energy expenditure, energy status, and body composition with this volume over an entire academic year in female athletes playing a field-based team sport. Overall, the female athletes in this study expended significant amounts of energy each day and consistently failed to match their levels of energy expenditure with adequate caloric intake. Because of this, athletes exhibited a negative energy balance and a low energy availability.

Consequently, female team sport athletes should not be overlooked as a population at risk of negative energy balance and low energy availability. The results of this study also help to provide insight into PAL values of female team sport athletes which can be used by sport nutrition practitioners to identify energy requirements of comparable athletes undergoing similar training and potential changes throughout a season.

Author Contributions: Conceptualization, C.M.K. and A.R.J.; methodology, C.M.K.; software, H.A.Z. and C.M.K.; formal analysis, H.A.Z.; investigation, H.A.Z., B.S.C., P.S.H., and R.A.S.; resources, C.M.K. and H.A.Z.; data curation, H.A.Z.; writing—original draft preparation, H.A.Z.; writing—review and editing, H.A.Z, B.S.C., P.S.H., R.A.S., A.R.J., and C.M.K.; visualization, H.A.Z.; supervision, C.M.K., R.A.S., and A.R.J.; project administration, H.A.Z. and R.A.S.

Funding: This research received no external funding.

Conflicts of Interest: The authors declare no conflict of interest.

References

1. Scheers, T.; Philippaerts, R.; Lefevre, J. Variability in physical activity patterns as measured by the SenseWear Armband: How many days are needed? *Eur. J. Appl. Physiol.* **2012**, *112*, 1653–1662. [CrossRef] [PubMed]

2. Drenowatz, C.; Hand, G.A.; Shook, R.P.; Jakicic, J.M.; Hebert, J.R.; Burgess, S.; Blair, S.N. The association between different types of exercise and energy expenditure in young nonoverweight and overweight adults. *Appl. Physiol. Nutr. Metab.* **2015**, *40*, 211–217. [CrossRef] [PubMed]

3. Wilson, P.B.; Madrigal, L.A.; Burnfield, J.M. Weight control practices of Division I National Collegiate Athletic Association athletes. *Phys. Sportsmed.* **2016**, *44*, 170–176. [CrossRef] [PubMed]

4. Pikosky, M.A.; Smith, T.J.; Grediagin, A.; Castaneda-Sceppa, C.; Byerley, L.; Glickman, E.L.; Young, A.J. Increased protein maintains nitrogen balance during exercise-induced energy deficit. *Med. Sci. Sports Exerc.* **2008**, *40*, 505–512. [CrossRef] [PubMed]

5. Joy, E.A.; Campbell, D. Stress fractures in the female athlete. *Curr. Sports Med. Rep.* **2005**, *4*, 323–328. [CrossRef] [PubMed]

6. Novas, A.; Rowbottom, D.; Jenkins, D. Total daily energy expenditure and incidence of upper respiratory tract infection symptoms in young females. *Int. J. Sports Med.* **2002**, *23*, 465–470. [CrossRef] [PubMed]

7. Logue, D.; Madigan, S.M.; Delahunt, E.; Heinen, M.; Mc Donnell, S.J.; Corish, C.A. Low Energy Availability in Athletes: A Review of Prevalence, Dietary Patterns, Physiological Health, and Sports Performance. *Sports Med.* **2018**, *48*, 73–96. [CrossRef] [PubMed]

8. Melin, A.; Tornberg, A.B.; Skouby, S.; Moller, S.S.; Sundgot-Borgen, J.; Faber, J.; Sidelmann, J.J.; Aziz, M.; Sjodin, A. Energy availability and the female athlete triad in elite endurance athletes. *Scand. J. Med. Sci. Sports* **2015**, *25*, 610–622. [CrossRef] [PubMed]

9. Mountjoy, M.; Sundgot-Borgen, J.; Burke, L.; Carter, S.; Constantini, N.; Lebrun, C.; Meyer, N.; Sherman, R.; Steffen, K.; Budgett, R.; et al. The IOC consensus statement: Beyond the Female Athlete Triad–Relative Energy Deficiency in Sport (RED-S). *Br. J. Sports Med.* **2014**, *48*, 491–497. [CrossRef] [PubMed]

10. Hill, R.J.; Davies, P.S. Energy intake and energy expenditure in elite lightweight female rowers. *Med. Sci. Sports Exerc.* **2002**, *34*, 1823–1829. [CrossRef] [PubMed]

11. Ebine, N.; Feng, J.Y.; Homma, M.; Saitoh, S.; Jones, P.J. Total energy expenditure of elite synchronized swimmers measured by the doubly labeled water method. *Eur. J. Appl. Physiol.* **2000**, *83*, 1–6. [CrossRef] [PubMed]

12. Gibson, J.C.; Stuart-Hill, L.; Martin, S.; Gaul, C. Nutrition status of junior elite Canadian female soccer athletes. *Int. J. Sport Nutr. Exerc. Metab.* **2011**, *21*, 507–514. [CrossRef] [PubMed]

13. Reed, J.L.; De Souza, M.J.; Williams, N.I. Changes in energy availability across the season in Division I female soccer players. *J. Sports Sci.* **2013**, *31*, 314–324. [CrossRef] [PubMed]

14. Woodruff, S.J.; Meloche, R.D. Energy availability of female varsity volleyball players. *Int. J. Sport Nutr. Exerc. Metab.* **2013**, *23*, 24–30. [CrossRef] [PubMed]

15. Mara, J.K.; Thompson, K.G.; Pumpa, K.L. Assessing the Energy Expenditure of Elite Female Soccer Players: A Preliminary Study. *J. Strength Cond. Res.* **2015**, *29*, 2780–2786. [CrossRef] [PubMed]

16. Sell, K.M.; Ledesma, A.B. Heart Rate and Energy Expenditure in Division I Field Hockey Players During Competitive Play. *J. Strength Cond. Res.* **2016**, *30*, 2122–2128. [CrossRef] [PubMed]

17. Zanders, B.R.; Currier, B.S.; Harty, P.S.; Zabriskie, H.A.; Smith, C.R.; Stecker, R.A.; Richmond, S.R.; Jagim, A.R.; Kerksick, C.M. Changes in Energy Expenditure, Dietary Intake, and Energy Availability Across an Entire Collegiate Women's Basketball Season. *J. Strength Cond. Res.* **2018**. [CrossRef] [PubMed]

18. Crouter, S.E.; Churilla, J.R.; Bassett, D.R., Jr. Accuracy of the Actiheart for the assessment of energy expenditure in adults. *Eur. J. Clin. Nutr.* **2007**, *62*, 704. [CrossRef] [PubMed]

19. Loucks, A.B.; Kiens, B.; Wright, H.H. Energy availability in athletes. *J. Sports Sci.* **2011**, *29* (Suppl. 1), S7–S15. [CrossRef]

20. Staal, S.; Sjodin, A.; Fahrenholtz, I.; Bonnesen, K.; Melin, A. Low RMRratio as a Surrogate Marker for Energy Deficiency, the Choice of Predictive Equation Vital for Correctly Identifying Male and Female Ballet Dancers at Risk. *Int. J. Sport Nutr. Exerc. Metab.* **2018**, 1–24. [CrossRef] [PubMed]

21. Ten Haaf, T.; Weijs, P.J. Resting energy expenditure prediction in recreational athletes of 18–35 years: Confirmation of Cunningham equation and an improved weight-based alternative. *PLoS ONE* **2014**, *9*, e108460. [CrossRef] [PubMed]

22. Schofield, W.N. Predicting basal metabolic rate, new standards and review of previous work. *Hum. Nutr. Clin. Nutr.* **1985**, *39* (Suppl. 1), 5–41. [PubMed]

23. Capling, L.; Beck, K.L.; Gifford, J.A.; Slater, G.; Flood, V.M.; O'Connor, H. Validity of Dietary Assessment in Athletes: A Systematic Review. *Nutrients* **2017**, *9*, 1313. [CrossRef] [PubMed]

24. Reed, J.L.; De Souza, M.J.; Kindler, J.M.; Williams, N.I. Nutritional practices associated with low energy availability in Division I female soccer players. *J. Sports Sci.* **2014**, *32*, 1499–1509. [CrossRef] [PubMed]

25. Barrack, M.T.; Gibbs, J.C.; De Souza, M.J.; Williams, N.I.; Nichols, J.F.; Rauh, M.J.; Nattiv, A. Higher Incidence of Bone Stress Injuries with Increasing Female Athlete Triad–Related Risk Factors: A Prospective Multisite Study of Exercising Girls and Women. *Am. J. Sports Med.* **2014**, *42*, 949–958. [CrossRef] [PubMed]

26. Watson, A.D.; Zabriskie, H.A.; Witherbee, K.E.; Sulavik, A.; Gieske, B.T.; Kerksick, C.M. Determining a Resting Metabolic Rate Prediction Equation for Collegiate Female Athletes. *J. Strength Cond. Res.* **2019**, in press. [CrossRef] [PubMed]

27. Tomten, S.E.; Hostmark, A.T. Energy balance in weight stable athletes with and without menstrual disorders. *Scand. J. Med. Sci. Sports* **2006**, *16*, 127–133. [CrossRef] [PubMed]

© 2019 by the authors. Licensee MDPI, Basel, Switzerland. This article is an open access article distributed under the terms and conditions of the Creative Commons Attribution (CC BY) license (http://creativecommons.org/licenses/by/4.0/).

Article

Satiating Effect of High Protein Diets on Resistance-Trained Individuals in Energy Deficit

Justin Roberts [1,*], Anastasia Zinchenko [2,3], Krishnaa T. Mahbubani [4], James Johnstone [1], Lee Smith [1], Viviane Merzbach [1], Miguel Blacutt [3], Oscar Banderas [3], Luis Villasenor [3], Fredrik T. Vårvik [3] and Menno Henselmans [3]

[1] School of Psychology and Sport Science, Cambridge Centre for Sport and Exercise Sciences, Anglia Ruskin University, East Road, Cambridge CB1 1PT, UK; james.johnstone@anglia.ac.uk (J.J.); lee.smith@anglia.ac.uk (L.S.); viviane.merzbach@anglia.ac.uk (V.M.)

[2] Department of Biochemistry, Kings College, University of Cambridge, Kings Parade, Cambridge CB2 1ST, UK; a.zinchenko@live.de

[3] International Scientific Research Foundation for Fitness and Nutrition, 1073 LC Amsterdam, The Netherlands; miguelblacutt@hotmail.com (M.B.); oscarbanderastk@gmail.com (O.B.); luis@villasenor.net (L.V.); ftvaarvik@gmail.com (F.T.V.); menno.henselmans@gmail.com (M.H.)

[4] Department of Surgery, Addenbrookes Hospital, Cambridge CB2 0QQ, UK; ktam2@cam.ac.uk

* Correspondence: Justin.roberts@anglia.ac.uk; Tel.: +44-1223-695-154

Received: 30 November 2018; Accepted: 26 December 2018; Published: 28 December 2018

Abstract: Short-term energy deficit strategies are practiced by weight class and physique athletes, often involving high protein intakes to maximize satiety and maintain lean mass despite a paucity of research. This study compared the satiating effect of two protein diets on resistance-trained individuals during short-term energy deficit. Following ethical approval, 16 participants (age: 28 ± 2 years; height: 1.72 ± 0.03 m; body-mass: 88.83 ± 5.54 kg; body-fat: $21.85 \pm 1.82\%$) were randomly assigned to 7-days moderate (PRO_{MOD}: 1.8 $g \cdot kg^{-1} \cdot d^{-1}$) or high protein ($PRO_{HIGH}$: 2.9 $g \cdot kg^{-1} \cdot d^{-1}$) matched calorie-deficit diets in a cross-over design. Daily satiety responses were recorded throughout interventions. Pre-post diet, plasma ghrelin and peptide tyrosine tyrosine (PYY), and satiety ratings were assessed in response to a protein-rich meal. Only perceived satisfaction was significantly greater following PRO_{HIGH} (67.29 ± 4.28 v 58.96 ± 4.51 mm, $p = 0.04$). Perceived cravings increased following PRO_{MOD} only (46.25 ± 4.96 to 57.60 ± 4.41 mm, $p = 0.01$). Absolute ghrelin concentration significantly reduced post-meal following PRO_{MOD} (972.8 ± 130.4 to 613.6 ± 114.3 $pg \cdot mL^{-1}$; $p = 0.003$), remaining lower than PRO_{HIGH} at 2 h (-0.40 ± 0.06 v -0.26 ± 0.06 $pg \cdot mL^{-1}$ normalized relative change; $p = 0.015$). Absolute PYY concentration increased to a similar extent post-meal (PRO_{MOD}: 84.9 ± 8.9 to 147.1 ± 11.9 $pg \cdot mL^{-1}$, PRO_{HIGH}: 100.6 ± 9.5 to 143.3 ± 12.0 $pg \cdot mL^{-1}$; $p < 0.001$), but expressed as relative change difference was significantly greater for PRO_{MOD} at 2 h ($+0.39 \pm 0.20$ $pg \cdot mL^{-1}$ v -0.28 ± 0.12 $pg \cdot mL^{-1}$; $p = 0.001$). Perceived hunger, fullness and satisfaction post-meal were comparable between diets ($p > 0.05$). However, desire to eat remained significantly blunted for PRO_{MOD} ($p = 0.048$). PRO_{HIGH} does not confer additional satiating benefits in resistance-trained individuals during short-term energy deficit. Ghrelin and PYY responses to a test-meal support the contention that satiety was maintained following PRO_{MOD}, although athletes experiencing negative symptoms (i.e., cravings) may benefit from protein-rich meals as opposed to over-consumption of protein.

Keywords: dietary protein; satiety; ghrelin; peptide YY; resistance training

---

1. Introduction

Weight loss is an essential component of success for many strength athletes, as well as individuals involved in general strength training. Weight class athletes restrict dietary energy intake typically for 7–21 days prior to a competition [1], or during phases of training, to enhance the strength to body-mass ratio, improve body composition, and increase the competitive advantage in weight class events [2]. For physique/ bodybuilding athletes, similar practices of caloric restriction often occurs for longer time periods i.e., >12 weeks to reach a very low body fat percentage, often going below 5% to improve aesthetical appearance and succeed in bodybuilding competitions [2,3]. During dieting phases many athletes may substantially increase their protein intake, as this practice has been shown to be beneficial to maintain lean mass whilst reducing body-fat [4–13]. Although the recommended daily intake of protein for healthy adults has been indicated at 0.83 $g \cdot kg^{-1} \cdot d^{-1}$ ([14] with habitual protein intakes being reportedly greater at ~1.2 $g \cdot kg^{-1} \cdot d^{-1}$ [15]), protein requirements for resistance-trained individuals to achieve maximal fractional synthetic rate and strength performance may exceed these recommendations ranging from 1.3 to 1.8 $g \cdot kg^{-1} \cdot d^{-1}$ [16,17]. During acute energy deficit, protein requirements may be subtly higher at 1.8–2.0 $g \cdot kg^{-1} \cdot d^{-1}$ to offset lean muscle losses. There is, however, controversy as to whether substantially higher protein intakes ranging from 2.3 to 3.1 $g \cdot kg^{-1} \cdot d^{-1}$ may in fact yield optimal results [13] with evidence that resistance-trained athletes consume protein intakes as high as 4.3 $g \cdot kg^{-1} \cdot d^{-1}$ [18].

Higher protein intakes may also provide an additional benefit of increased satiety. A recent meta-analysis by Dhillon et al. (2016) concluded that higher protein meals (ranging from 97 to 188 g) increased fullness ratings more than lower protein meals [19]. However, this analysis focused on short-term studies (up to 10 h) in untrained individuals and can therefore not necessarily be extrapolated to longer-term satiety over several days/ weeks. There may be a possibility that a person becomes accustomed to a high protein intake and that a further increase in protein intake does not result in additional satiety. This point can be particularly important for strength athletes who generally consume a higher protein diet beyond the recommended levels for maximal performance [16,17].

Very few long-term, randomized, controlled trials have examined satiety and *ad libitum* energy intake in normal weighted individuals following a low-protein vs a high-protein diet [20–22]. Only three studies have been conducted on resistance-trained individuals showing that a higher protein intake may be beneficial for muscle retention and well-being following a weight loss diet [4,6,23]. Walberg et al. have shown that even though a higher protein diet (1.6 $g \cdot kg^{-1} \cdot d^{-1}$) during an extreme weight loss intervention (18 $kcal \cdot kg^{-1}$) led to positive nitrogen balance compared to a lower protein group (0.8 $g \cdot kg^{-1} \cdot d^{-1}$); it also led to reduced maximal isometric muscular endurance in weight lifters [6]. In a different study, 2.3 $g \cdot kg^{-1} \cdot d^{-1}$ protein intake was shown to be beneficial for muscle retention in 40% energy deficit compared to the 1.0 $g \cdot kg^{-1} \cdot d^{-1}$ control group [4]. However, higher protein intake was also associated with increased fatigue for participants in this study, although this has been contested elsewhere [23].

The finding of increased fatigue is particularly important considering that many athletes increase their protein intake during caloric restriction in order to limit muscle loss, reduce hunger and increase the feeling of satiety [13]. An increase in fatigue can result in decreased motivation to train [4], reduced training efficiency and thus have a negative effect on athletic performance [6] particularly during the competition preparation phase, in which many weight-class strength athletes undergo weight loss interventions. It is important to note that the control condition examined in this study [4] represented an unusually low protein intake (1.0 $g \cdot kg^{-1} \cdot d^{-1}$) for strength athletes that is below the established recommendations [16]. Equally so, very high protein intakes, exceeding these recommendations, may not only negatively impact on an athletes' well-being and reduce the effort given to training, but also carries the risk of displacement of other nutrients.

Therefore, the aim of this study was to compare the satiating effect of two diets with a different protein content in resistance-trained subjects in energy deficit, participant well-being and training motivation. A moderately high protein diet (PRO$_{MOD}$) with an intake corresponding to the recommendation for maximal performance (1.8 $g \cdot kg^{-1} \cdot d^{-1}$) [16,17] and a very high protein diet (PRO$_{HIGH}$) with a protein intake of 2.9 $g \cdot kg^{-1} \cdot d^{-1}$ were used as main interventions. It was hypothesised that a protein intake of 2.9 $g \cdot kg^{-1} \cdot d^{-1}$ would not impact on satiety levels more so than an intake of 1.8 $g \cdot kg^{-1} \cdot d^{-1}$ but may reduce the motivation to train.

2. Materials and Methods

2.1. Participants

This study was conducted in accordance with the Declaration of Helsinki, and the protocol was approved by the Faculty of Science and Technology Ethics Committee, Anglia Ruskin University (approval number: FST/FREP/15/595). A priori sample size based on G*power software ($\alpha = 0.05$ and $1\text{-}\beta = 0.80$) using satiety data from Mettler et al. (2010) [4] based on resistance training individuals consuming 2.3 $g \cdot kg^{-1} \cdot d^{-1}$ vs. 1 $g \cdot kg^{-1} \cdot d^{-1}$, estimated 14 in each group. Participants were required to have a resistance training background of >6 months, while actively training >3 h per week, in a similar manner to previous research [4]. This was to ensure participants were undergoing habitual training at the point of study inclusion, and beyond the time-frame typically employed during acute training studies [24–29].

Informed consent was obtained from all individual participants prior to study inclusion. All participants satisfactorily completed a health screen questionnaire, and had no known history of cardiovascular or metabolic abnormalities (e.g., diabetes); or recent viral infections or injuries which would prevent them from maintaining habitual training sessions. As part of the inclusion criteria, participants were required to not have any blood related disorders, no known eating disorders and no known adverse reactions to whey protein supplementation, gluten or coconut oil. Additionally, all participants were required to not be taking any medication/supplementation which could influence satiety levels. Lacto-vegetarians were eligible for the study.

Twenty-two individuals (13 men, 9 women) volunteered for study inclusion. However, five participants withdrew prior to first dietary intervention (due to personal factors conflicting with the study requirements) and one participant did not complete the final assessment, resulting in 16 resistance trained participants in this randomised, controlled trial (9 men, 7 women). None of the participants reported using any anabolic drugs, and were required to refrain from taking additional supplementation (e.g., creatine, beta-alanine) for 4 weeks prior to and during the study. Participant characteristics are shown in Table 1.

Table 1. Participant characteristics and baseline measurements.

Variable	All Participants (n = 16)	Male (n = 9)	Female (n = 7)
Age (years)	28 ± 2	26 ± 2	30 ± 3
Height (m)	1.72 ± 0.03	1.79 ± 0.02	1.63 ± 0.04 *
Body-mass (kg)	88.83 ± 5.54	95.68 ± 5.73	80.01 ± 6.21
Body-fat (%)	21.85 ± 1.82	18.33 ± 1.91	26.37 ± 2.57 *
BIA FM (kg)	25.52 ± 3.37	23.12 ± 3.02	28.60 ± 6.82
BIA FM (%)	27.53 ± 2.46	23.40 ± 1.99	32.84 ± 4.40 *
BIA FFM (kg)	63.31 ± 3.50	72.56 ± 3.05	51.41 ± 3.48 *
BIA FFM (%)	72.47 ± 2.46	76.60 ± 1.99	67.16 ± 4.40 *
PA (°)	8.27 ± 0.16	8.71 ± 0.10	7.70 ± 0.19 *
TBW (%)	52.26 ± 1.77	55.23 ± 1.43	48.43 ± 3.17 *
ICW (%)	58.49 ± 0.81	59.48 ± 0.48	57.23 ± 1.43
ECW (%)	41.48 ± 0.81	40.52 ± 0.84	42.71 ± 1.45

Data presented as mean ± SE. Body-fat (%) refers to estimation via skinfold measures. BIA denotes bioelectrical impedance analysis. FM = fat-mass; FFM = fat-free mass; PA = phase angle; TBW = total body water; ICW = intracellular water; ECW = extracellular water. Hydration markers from BIA. * denotes significantly different to male cohort, $p \leq 0.05$.

2.2. Pre-Intervention Measures

All testing took place within the Cambridge Centre for Sport and Exercise Science, Human Physiology Laboratory, Anglia Ruskin University, Cambridge. Prior to main trial procedures, participants were required to visit the laboratory ~72–96 h prior to start of the first and second trials to ascertain body composition measures required for energy intake calculations. As such, participants were requested to arrive acutely fasted (i.e., no food within 3 h of assessment, and maintain habitual hydration patterns) with last consumption of fluid (~0.5 L water) 1 h prior to assessment to standardise procedures. Body-mass (Seca 799, Hamburg, Germany), height (Seca CE123 stadiometer, Hamburg, Germany) and body composition were assessed under temperature controlled conditions.

Following a standardized 5-min resting period in a supine position, single frequency bioelectrical impedance (Impedimed DF50, Carlsbad, CA, USA) was undertaken for initial assessment of body composition measures, with particular emphasis on hydration indices (to confirm adherence to the pre-testing requirements) and phase angle (a proxy marker of muscle quality to confirm trained status (with >7.0 reported for female athletes, and >8.0 for male athletes from previous research [30]). In addition, estimated body fat was evaluated using an 8 site skinfold calliper assessment (using guidelines outlined by the International Society for the Advancement of Kinanthropometry (ISAK)). All body composition measures were undertaken by the same researcher. Within the same timeframe, participants completed habitual food intake and training diaries.

2.3. Dietary Assessment

As part of pre-intervention measures, and throughout the intervention, participants were requested to maintain habitual food/activity diaries (following individual guidance in diary collation, with emphasis on meal content, portion size and weight, and fluid intake) using the smart phone app/browser program (www.MyFitnessPal.com). This method has been shown to be a reliable food tracking method used in previous studies [31–33]. Diaries were assessed using Nutritics Professional Dietary Analysis software (Nutritics Ltd., Co. Dublin, Ireland). Initial assessment was used to monitor typical food choices and habitual caloric balance (including criterion assessment of training levels). From this, assessment of individual maintenance caloric intake was undertaken using the formula of Katch-McArdle (1996) based on an estimated resting daily energy expenditure (RDEE) of 370 + (21.6 * lean body weight in kg) and adjusted against training requirements and non-exercise adaptive thermogenesis [34], based on previous research [35].

2.4. Experimental Design and Intervention Measures

This study employed an experimental, randomised controlled, counter-balanced, crossover design. Participants were randomly assigned to a 7-day matched calorie-restricted diet (20% from estimated maintenance calories) of either moderate (1.8 $g \cdot kg^{-1} \cdot d^{-1}$) or high (2.9 $g \cdot kg^{-1} \cdot d^{-1}$) total protein intake ($PRO_{MOD}$ and PRO_{HIGH} respectively) with a fixed frequency of 4 meals per day. Guidance on meal intake, food options in line with habitual patterns and portion size was provided to each participant, along with provision of additional whey protein to supplement daily intake where required. Participants visited the laboratory prior to starting the dietary intervention (day 0), and again at on the completion of the 7 days (day 8) for assessment of satiety measures to a test meal (see below). After each diet, participants were required to undertake a 3-day period in which the same meal frequency (4 meals per day) was maintained, and protein intake was fixed (1.8 $g \cdot kg^{-1} \cdot d^{-1}$). However, energy, carbohydrate and fat intake was permitted *ad libitum* at each meal until participants felt full. Perception of satiety was noted by participants across each day during this period. This was included to determine the dieting condition less likely to result in acute post-diet overeating.

Throughout this 10-day period, participants recorded all food and fluid intakes using an individually allocated MyFitnessPal account, and maintained habitual training patterns. Compliance of dietary intake was checked daily by the research team, and further individual support was provided as required to meet expected intakes. At the end of the first dietary period, participants undertook a washout period and returned to habitual intake patterns for 4 weeks before returning to carry out the opposing dietary condition. Mean dietary intakes at baseline and for each intervention are shown in Table 2.

Table 2. Mean dietary intake at baseline and across intervention periods.

Variable	Category	PRO$_{MOD}$		PRO$_{HIGH}$	
		PRE	**POST**	**PRE**	**POST**
Energy Intake	(kcal·d^{-1})	2359.44 ± 220.84 *	2057.44 ± 104.47	2270.71 ± 150.41 *	2119.44 ± 122.79
	(kcal·kg^{-1}·d^{-1})	26.89 ± 1.94	23.83 ± 0.93	25.53 ± 1.17	24.35 ± 0.84
Protein Intake	(g·d^{-1})	158.19 ± 13.55	162.94 ± 10.14	166.71 ± 18.33	256.13 ± 16.64 [a,b]
	(g·kg^{-1}·d^{-1})	1.80 ± 0.12	1.84 ± 0.02	1.84 ± 0.15	2.89 ± 0.01 [a,b]
	(%EI)	27.79 ± 1.98	31.68 ± 1.33	28.79 ± 2.13	48.44 ± 1.93 [a,b]
Carbohydrate Intake	(g·d^{-1})	220.75 ± 25.34	238.75 ± 13.32	221.79 ± 12.07	160.88 ± 13.88 [a,b]
	(g·kg^{-1}·d^{-1})	2.57 ± 0.25	2.81 ± 0.18	2.55 ± 0.16	1.90 ± 0.16 [a,b]
	(%EI)	38.08 ± 2.52	46.55 ± 1.32 [a]	40.28 ± 2.45	30.32 ± 1.82 [a,b]
Fat Intake	(g·d^{-1})	90.94 ± 11.40	46.88 ± 2.62 [a]	73.79 ± 7.68	47.31 ± 2.80 [a]
	(g·kg^{-1}·d^{-1})	1.02 ± 0.11	0.55 ± 0.03 [a]	0.83 ± 0.08 [b]	0.55 ± 0.03 [a]
	(%EI)	14.83 ± 0.77	9.12 ± 0.25 [a]	12.86 ± 0.90 [b]	8.94 ± 0.21 [a]

Data presented as mean ± SE, and for macronutrient categories expressed in grams per day, grams per kg per day, and percentage of energy intake (EI). PRO$_{MOD}$ denotes moderate protein condition (target: 1.80 g·kg^{-1}·d^{-1}); PRO$_{HIGH}$ denotes high protein condition (target: 2.90 g·kg^{-1}·d^{-1}). * denotes main effect for time (p = 0.04), but no significant post-hoc findings. [a] denotes significantly different to PRE within condition ($p \leq 0.005$). [b] denotes significantly different to PRO$_{MOD}$ at same time-point ($p \leq 0.024$).

2.5. Laboratory Measures

Laboratory assessment took place on day 0 and 8 of each main intervention period, in addition to pre-trial body composition measures. Participants were instructed to refrain from strenuous physical activity 24–48 h prior to all laboratory visits, arrive strictly overnight fasted (>12 h) and avoid any undue exertion travelling to the laboratory. Upon arrival, participants were required to rest in a seated position for ~30 min without any undue distractions and complete an online satiety questionnaire (see below). Following this, a venous whole blood sample (T0) was collected from participants by a qualified phlebotomist into duplicate 4 mL K3EDTA vacutainers (Greiner Bio-One GmbH, Kremsmunster, Austria). Once collected, stabilisers were added to each of the samples prior to being centrifuged for 15 min at 3000 g. The plasma layer was pipetted and aliquoted into sterile, non-pyrogenic, polypropylene cryovials (Fisherbrand, Fisher Scientific, Loughborough, UK) and immediately frozen at −20 °C for later assessment of satiety hormones: ghrelin and peptide YY (PYY). Body composition measures (height, weight, bioelectrical impedance) were assessed at the end of each 7-day intervention period as previously described.

2.6. Test Meal and Satiety Measures

At each visit for both intervention conditions, following resting measures, participants consumed a standardised high-protein breakfast meal (Asda Scottish porridge oats (Asda Stores Ltd., Leeds, UK) comprising (per 100g): 367 kcal; 12.0 g protein; 61.0 g carbohydrate; 6.2 g fat), added whey protein (Pure Protein GF-1, USN UK Ltd., Longbridge, Birmingham, UK), coconut oil (Asda 100% raw extra virgin coconut oil, Asda Stores Ltd., Leeds, UK) with 200 mL water under supervision based on individual estimated caloric requirements for the PRO$_{HIGH}$ condition and a protein content of 0.725 g kg^{-1}). On immediate completion of the test meal, a stopwatch was started and participants captured their subjective post-prandial satiety responses using a previously validated satiety/hunger questionnaire based on a standard 100 mm visual analogue scale [36,37] at 15, 30, 60, 90 and 120 min. The questionnaire was based on immediate response to set questions ("How 'hungry' do you feel?"; "How 'full' do you feel?"; "How 'satisfied' do you feel?"; "How 'much' can you eat right now?"). Participants were required to remain in a quiet, seated position during this time with no activity except light reading, and moving to a designated couch (supine position) for collection of blood samples at T60 and T120 (minutes) for assessment of selected satiety hormones. Data was normalised to individual relative change and relative change differences in a similar manner to blood analyses (see Section 2.8 below).

2.7. Post-Laboratory Measures and Support

An online satiety questionnaire was additionally completed at the end of each day across the 7-day intervention, and during the post-diet 3-day *ad libitum* period. The questionnaire captured individual perceived responses to the following questions: (i) how hungry were you today?, (ii) how full were you today?, (iii) how satisfied were you today with your diet?, (iv) how much do you think you can eat right now?, (v) how high have your food cravings been today?, (vi) how often did you feel energetic and active today?, (vii) how often were you in a good mood today?, (viii) how much did you enjoy your training today?, (ix) how much do you feel you can give your best effort at training today?. Mean data was collated for days 1–3 and 5–7 across the intervention period, and for days 1–3 of the *ad libitum* period. To increase dietary compliance during each intervention period, participants were provided with additional whey protein (Pure Protein GF-1, whey protein concentrate/ soya protein isolate, USN UK Ltd., Longbridge, Birmingham, UK) comprising (per 100g): 370 kcal; 71 g protein; 9.8 g carbohydrate; 4.4 g fat.

2.8. Biochemical Analyses

All samples were analysed at the Department of Surgery, Addenbrookes Hospital, Cambridge. Plasma samples with 4-(2-Aminoethyl) benzenesulfonyl fluoride hydrochloride (AEBSF; Sigma, Dorset, UK) to a final concentration of 1 mg·mL^{-1} were frozen to $-20\,^{\circ}$C within 30 min of centrifuging (15 min at 3000 g). Samples were subsequently thawed aseptically in a $37\,^{\circ}$C water bath, centrifuged at 3000 g for 5 min to sediment any cryoprecipitates and then analysed in duplicate using a human total ghrelin ELISA kit or human PYY ELISA kit (EMD Millipore Corporation, Billerica, MA, USA) for ghrelin and PYY measurements according to manufacturers' instructions. ELISA plates were read using a FLUOstar OPTIMA plate reader (BMG Labtech, Aylesbury, UK). In addition to raw concentrations (pg·mL^{-1}), data were individually normalised as relative change (to reflect normalised post-intervention effects) and relative change differences (to reflect normalised intervention effects taking into consideration pre-intervention results) according to the following equations:

Relative change (relative Δ pg·mL^{-1}) = $\frac{y-x}{x}$ where x = the pre-meal resting sample, y = the post-meal at the respective sample time-points (i.e., 60, 120 min);

Relative change difference (relative Δ difference pg·mL^{-1}) = $\left(\frac{y_{post}-x_{post}}{x_{post}} \right) - \left(\frac{y_{pre}-x_{pre}}{x_{pre}} \right)$ where pre is the pre-intervention results, and post is the post-intervention results.

2.9. Statistical Analyses

Statistical analyses were performed using SPSS (v24, IBM, Armonk, NY, USA). Dependent variable distributions were assessed for normality using a Shapiro-Wilk test as well as manual inspections of M-estimators, histograms, stem-and-leaf plots and boxplots. Potential order and treatment effects were assessed prior to main analyses using a paired samples t-test. Baseline participant characteristics (gender) were assessed using an independent samples t-test. A mixed design repeated measures ANOVA (diet, time) was performed for main analyses, with Bonferonni post-hoc comparisons where applicable. Where pertinent, relative change questionnaire scores were assessed using a paired samples t-test. An alpha level of ≤ 0.05 was employed for statistical significance. Data are reported as mean $\pm$ S.E.

3. Results

3.1. Nutrition Intake and Body Composition Data

Mean dietary intake across both interventions is shown in Table 2. Prior to starting the first intervention average caloric intake for all participants was significantly greater

(2379.81 ± 201.48 kcal·d^{-1}, $F = 5.07$, $p = 0.013$, $\eta p^2 = 0.25$) in comparison to main intervention intakes, although similar to predicted habitual intakes when taking into consideration rest days (2381.25 ± 131.31 kcal·d^{-1}). Across intervention periods, targeted caloric deficit was achieved in relation to estimated maintenance calories (based on individual requirements for both training and rest days) for both PRO$_{MOD}$ ($-22.9 \pm 1.0\%$) and PRO$_{HIGH}$ ($-21.0 \pm 1.1\%$) conditions.

Whilst protein was generally maintained during PRO$_{MOD}$ compared with pre-intervention intakes, PRO$_{HIGH}$ resulted in an expected increase in relative protein from 1.84 ± 0.15 to 2.89 ± 0.01g·kg^{-1}·d^{-1} (F = $78.29_{(diet \times time)}$, $p < 0.0001$, $\eta p^2 = 0.84$). Post-intervention protein intakes were significantly different between conditions ($p < 0.0001$), with mean intakes demonstrating excellent compliance with target amounts. Relative carbohydrate intake was reduced during PRO$_{HIGH}$ by $25.37 \pm 6.16\%$ from 2.55 ± 0.16 g·kg^{-1}·d^{-1} to 1.90 ± 0.16g·kg^{-1}·d^{-1} (F = $11.54_{(diet \times time)}$, $p = 0.004$, $\eta p^2 = 0.44$), and was significantly lower than PRO$_{MOD-post}$ ($p < 0.0001$). To meet caloric deficit requirements, during both interventions total fat intake was significantly reduced by $39.42 \pm 5.76\%$ ($p < 0.0001$) and $27.63 \pm 6.35\%$ ($p < 0.0001$) for PRO$_{MOD}$ and PRO$_{HIGH}$ respectively (F = $4.64_{(diet \times time)}$, $p = 0.048$, $\eta p^2 = 0.24$). Following interventions, no differences for fat intake were reported between conditions ($p > 0.05$). However, relative fat intakes pre-intervention were significantly higher for PRO$_{MOD}$ compared to PRO$_{HIGH}$ ($p = 0.008$).

Mean body-mass significantly decreased within both interventions (F = $24.00_{(time)}$, $p < 0.0001$, $\eta p^2 = 0.62$) by $1.28 \pm 0.30\%$ with PRO$_{MOD}$ (from 88.37 ± 5.42 to 87.18 ± 5.31 kg, $p = 0.001$) and by $1.49 \pm 0.36\%$ with PRO$_{HIGH}$ (from 88.48 ± 5.54 to 87.14 ± 5.42 kg, $p = 0.002$) demonstrating compliance with acute energy deficit based on estimated maintenance calories. This corresponded with a significant within-condition reduction in phase angle (F = $26.02_{(time)}$, $p < 0.0001$, $\eta p^2 = 0.63$) for both PRO$_{MOD}$ ($-0.41 \pm 0.09°$, $p < 0.0001$) and PRO$_{HIGH}$ ($-0.29 \pm 0.12°$, $p = 0.04$). Short-term energy deficit resulted in a reduction in hydration indices, with both total body water (PRO$_{MOD}$: 53.14 ± 1.81 to $51.16 \pm 1.59\%$, $p = 0.002$; PRO$_{HIGH}$: 53.04 ± 1.70 to $51.24 \pm 1.80\%$, $p = 0.018$) and intracellular water (PRO$_{MOD}$: 59.07 ± 0.75 to $57.88 \pm 0.67\%$, $p = 0.001$; PRO$_{HIGH}$: 58.89 ± 0.71 to $57.88 \pm 0.81\%$, $p = 0.035$) decreasing within-condition. It was noted that mean training frequency was comparable between dietary conditions (PRO$_{MOD}$: 6.38 ± 0.63 sessions; PRO$_{HIGH}$: 7.00 ± 0.50 sessions, $p = 0.44$) across the 10-day intervention period.

3.2. Dietary Intervention Perceived Satiety Responses

Across both dietary interventions, participants maintained a daily satiety questionnaire as previously described. Data, based on absolute values (using a 100 mm visual analogue scale), were averaged for the beginning (days 1–3) and end of each intervention (days 5–7). Mean responses are shown in Table 3. Perception of satisfaction was significantly greater at the end of PRO$_{HIGH}$ (F = $4.52_{(diet)}$, $p = 0.05$, $\eta p^2 = 0.23$) compared with PRO$_{MOD}$ over days 5–7 (67.29 ± 4.28 mm v 58.96 ± 4.51 mm respectively, $p = 0.04$). Participants also reported a significant mean increase in perception of cravings (F = $5.93_{(time)}$, $p = 0.028$, $\eta p^2 = 0.28$) within PRO$_{MOD}$ only from days 1–3 (46.25 ± 4.96 mm) to days 5–7 (57.60 ± 4.41 mm, $p = 0.01$). No other differences were reported between conditions, including perceived training enjoyment and motivation to train.

Table 3. Satiety and well-being questionnaire responses during dietary intervention phases.

Question	PRO$_{MOD1-3}$	PRO$_{MOD5-7}$	PRO$_{HIGH1-3}$	PRO$_{HIGH5-7}$
Hunger	48.54 ± 3.48	51.46 ± 4.89	43.33 ± 4.22	41.67 ± 3.68
Fullness	62.29 ± 4.11	58.65 ± 4.64	64.38 ± 3.81	65.52 ± 3.99
Satisfaction	65.42 ± 4.38	58.96 ± 4.51	68.65 ± 2.34	67.29 ± 4.28 [a]
Desire to eat	50.94 ± 5.53	57.92 ± 5.45	45.00 ± 5.13	46.15 ± 5.58
Cravings	46.25 ± 4.96	57.60 ± 4.41 [b]	40.83 ± 4.59	47.19 ± 4.60
Energy	65.00 ± 2.81	67.81 ± 4.54	64.48 ± 3.18	69.06 ± 4.64
Mood	72.92 ± 3.35	67.92 ± 5.33	67.71 ± 3.26	71.88 ± 3.89
Training enjoyment	72.29 ± 4.35	68.23 ± 4.73	71.67 ± 4.48	73.02 ± 4.19
Training motivation	71.67 ± 4.97	66.56 ± 4.68	70.21 ± 3.93	70.83 ± 5.92

Data represent mean scores over days 1–3 and days 5–7 during moderate (PRO$_{MOD}$) or high (PRO$_{HIGH}$) diet periods. Data based on arbitrary units (a.u.) from participant responses using a visual analogue scale (VAS, 0–100 mm) and presented as mean ± SE. Question categories paraphrased. [a] denotes significant difference to PRO$_{MOD5-7}$ ($p = 0.04$). [b] denotes significant increase within group compared to days 1–3 ($p = 0.01$).

3.3. Test-Meal Satiety Hormone Responses

Absolute values: Mean plasma ghrelin and PYY concentrations in response to the test-meal are shown in Table 4. No differences were reported for pre- or post-intervention concentrations between conditions ($p > 0.05$) for either analyte prior to consuming the test-meal. However, both ghrelin and PYY were notably lower (albeit non-significant, $p = 0.06$ and $p = 0.11$ respectively) at T0 following PRO$_{MOD}$ compared with pre-intervention data. Plasma ghrelin reduced as expected by T60 in all assessments, but typically increased by T120 (F = 28.77$_{(time)}$, $p < 0.001$, $\eta p^2 = 0.63$), remaining significantly lower than pre-meal concentrations ($p \leq 0.036$). However, following PRO$_{MOD}$, plasma ghrelin continued to decrease to 613.57 ± 114.26 pg·mL^{-1} by T120 ($p = 0.003$ within condition compared to T0) and was significantly different overall compared to pre intervention responses ($p = 0.015$). Across all assessments, plasma PYY significantly increased from T0 to T120 (F = 33.05$_{(time)}$, $p < 0.001$, $\eta p^2 = 0.70$), with no differences reported between pre- or post-intervention concentrations.

Table 4. Mean plasma ghrelin and peptide YY concentrations in response to test meal.

Hormone	Time (mins)	PRO$_{MOD}$		PRO$_{HIGH}$	
		PRE	POST	PRE	POST
Ghrelin (pg·mL^{-1})	0	1125.97 ± 125.85	972.81 ± 130.42	920.12 ± 143.43	1088.17 ± 158.77
	60	696.42 ± 96.61 *	659.73 ± 86.39 *	672.27 ± 119.29 *	786.61 ± 117.33 *
	120	758.91 ± 129.38 *	613.57 ± 114.26 *, a	714.75 ± 136.56 *	850.60 ± 147.68 *
PYY (pg·mL^{-1})	0	103.62 ± 10.15	84.87 ± 8.94	87.94 ± 11.45	100.65 ± 9.54
	60	130.83 ± 9.09 *	129.38 ± 10.49 *	133.84 ± 15.14 *	141.02 ± 12.23 *
	120	137.60 ± 8.71 *	147.14 ± 11.94 *	142.11 ± 13.29 *	143.34 ± 11.98 *

Data presented as mean ± SE. PRO$_{MOD}$ denotes moderate protein condition; PRO$_{HIGH}$ denotes high protein condition. Time-point 0 refers to resting concentrations pre-feeding; other time-points refer to corresponding time post-breakfast meal. * denotes significant differences to 0 time-point within condition only ($p \leq 0.036$). [a] denotes overall difference to pre intervention responses ($p = 0.015$).

Relative values: Plasma ghrelin and PYY expressed as normalized relative change (individual results and mean data, reflective of intervention diet effect) and normalized relative change difference (mean data, reflective of overall change in comparison to habitual intake) are shown in Figures 1 and 2 respectively. Normalized relative change in ghrelin concentration was significantly lower (F = 5.90$_{(diet \times time)}$, $p = 0.029$, $\eta p^2 = 0.30$) at T120 following PRO$_{MOD}$ ($-0.40 ± 0.06$ pg·mL^{-1}) compared to PRO$_{HIGH}$ ($-0.26 ± 0.06$ pg·mL^{-1}, $p = 0.015$), with no prior differences reported at T60 between interventions ($p > 0.05$). Relative change difference in ghrelin concentration was comparable between conditions ($p > 0.05$).

Normalized relative change for PYY increased from 0.59 ± 0.14 pg·mL^{-1} at T60 to 0.79 ± 0.16 pg·mL^{-1} at T120 with PRO$_{MOD}$, but was not deemed significant ($p=0.07$, main effect for time) in comparison to PRO$_{HIGH}$ (0.46 ± 0.10 pg·mL^{-1} at T60 to 0.51 ± 0.12 pg·mL^{-1} at T120). However, relative change difference for PYY was significantly greater for PRO$_{MOD}$ at both T60 (0.26 ± 0.18 pg·mL^{-1}) and T120 (0.39 ± 0.20 pg·mL^{-1}) compared with negative findings for PRO$_{HIGH}$ at the same time-points (T60: -0.15 ± 0.11 pg·mL^{-1}, T120: -0.28 ± 0.12 pg·mL^{-1}; $p \leq 0.018$).

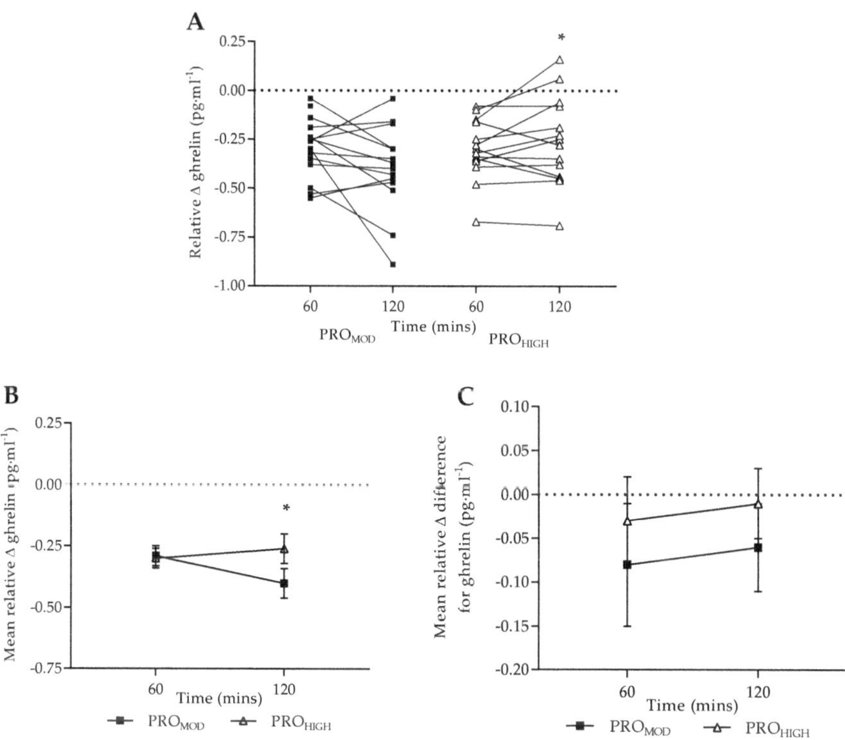

Figure 1. Plasma ghrelin concentrations following test-meal (pg·mL^{-1}; mean $\pm$ SE): (**A**) individual values expressed as normalised relative change to baseline levels; (**B**) mean values expressed as normalised relative change to baseline levels; (**C**) mean relative difference change (taking into consideration pre-intervention results). PRO$_{MOD}$ PRO$_{HIGH}$ denote moderate and high protein conditions. Dashed line provides reference point to pre-meal normalisation. * denotes significant difference overall to PRO$_{MOD}$ at corresponding time-point ($p = 0.015$).

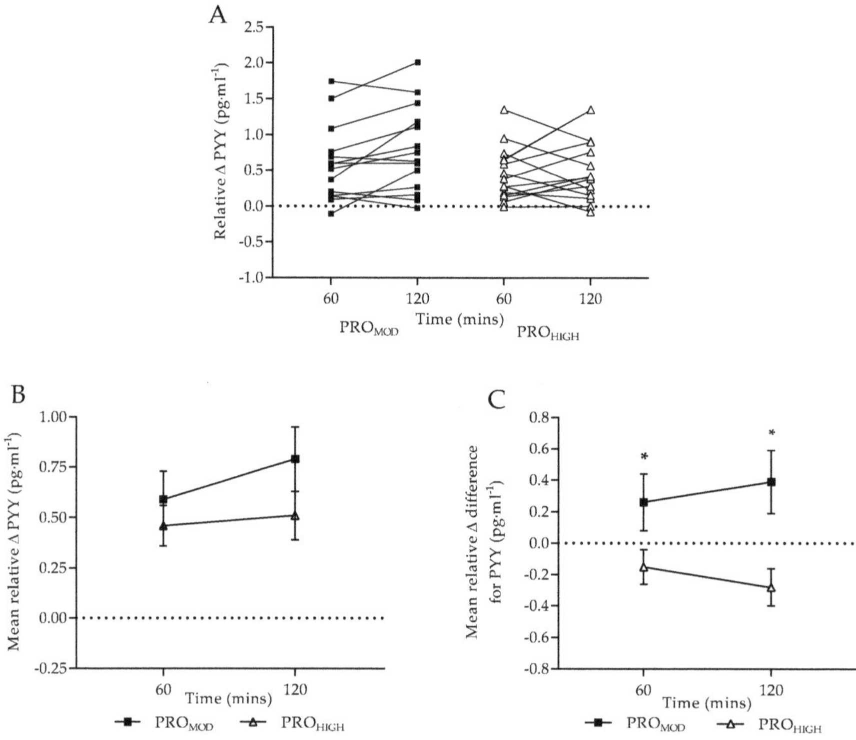

Figure 2. Plasma peptide YY concentrations following test-meal (pg·mL^{-1}; mean ± SE): (**A**) individual values expressed as normalised relative change to baseline levels; (**B**) mean values expressed as normalised relative change to baseline levels; (**C**) mean relative difference change (taking into consideration pre-intervention results). PRO$_{MOD}$ and PRO$_{HIGH}$ denote moderate and high protein conditions. Dashed line provides reference point to pre-meal normalisation. * denotes significant difference between dietary conditions at each time-point ($p \leq 0.018$).

3.4. Test-Meal Satiety Questionnaire Responses

Satiety questionnaire responses to the test-meal following each intervention (normalized relative change) are shown in Figure 3. Perception of hunger was significantly reduced (F = 15.34$_{(time)}$, $p < 0.0001$, $\eta p^2 = 0.51$) for 60 min post test-meal following both interventions ($p \leq 0.024$) with no differences reported between diets ($p > 0.05$). In a similar pattern, perception of fullness significantly increased post-meal (F = 16.52$_{(time)}$, $p < 0.0001$, $\eta p^2 = 0.52$) and remained above pre-meal values at T120 for both interventions ($p \leq 0.05$), with no differences reported between diets ($p > 0.05$). Interestingly, following immediate completion of the test-meal satisfaction was only significantly increased with PRO$_{MOD}$ ($p = 0.006$), beyond which perception of satisfaction remained elevated across all time-points within each dietary condition (F = 13.65$_{(time)}$, $p < 0.0001$, $\eta p^2 = 0.48$). Perception of desire to eat reduced following consumption of the test meal as expected (F = 16.90$_{(time)}$, $p < 0.0001$, $\eta p^2 = 0.53$) and remained lower than pre-meal values for 60 min in both interventions ($p \leq 0.02$). However, perception of desire to eat only remained blunted at T90 and T120 within-PRO$_{MOD}$ only ($p \leq 0.048$).

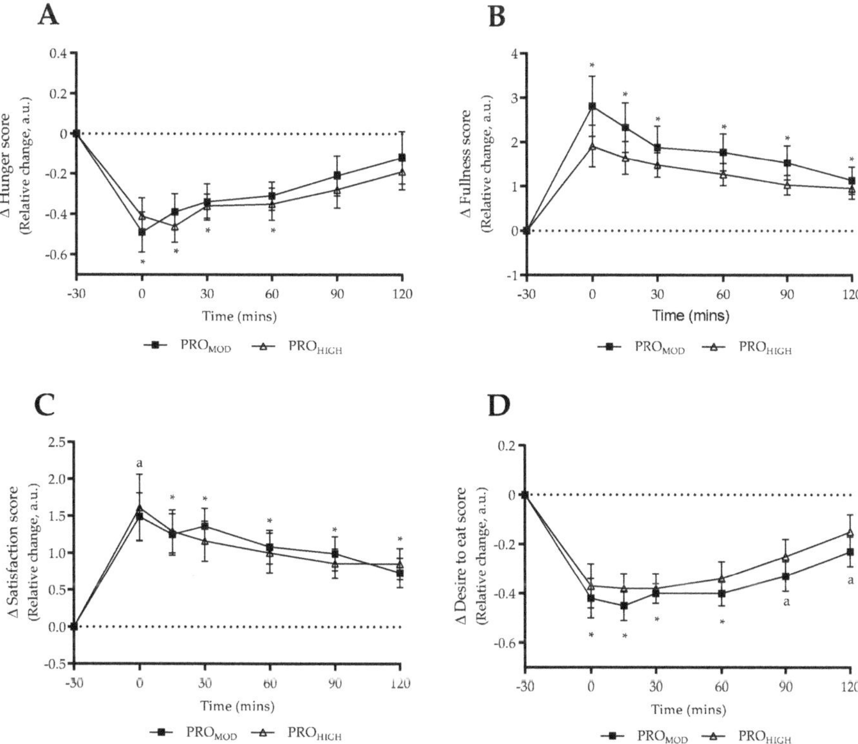

Figure 3. Questionnaire responses in relation to test meal under laboratory conditions. Data expressed as relative change (post diet, normalised) for both moderate (PRO$_{MOD}$) and high (PRO$_{HIGH}$) protein conditions (mean $\pm$ SE). (**A**) Perception of hunger; (**B**) Perception of fullness; (**C**) Perception of satisfaction; (**D**) Perception of desire to eat. Responses in relation to pre-meal (-30 min) time-point. 0 min denotes immediate completion of test meal. Dashed line provides reference point to pre-meal perceived state. [*] denotes significant difference compared to pre-meal time-point ($p \leq 0.05$) within both dietary conditions. [a] denotes significant difference compared to pre-meal time-point for PRO$_{MOD}$ only ($p \leq 0.048$). No differences reported between dietary conditions.

Satiety questionnaire responses to the test-meal expressed as normalized relative change difference (taking into consideration pre-intervention responses) are shown in Figure 4. No significant differences were report within or between dietary conditions for perception of hunger, fullness or satisfaction ($p > 0.05$). For perception of desire to eat, a significant main effect for time was reported (F = 5.43$_{(time)}$, $p < 0.0001$, $\eta p^2 = 0.27$), with post-hoc analysis demonstrating a significant increase on immediate completion of the test-meal for PRO$_{HIGH}$ only ($p = 0.028$).

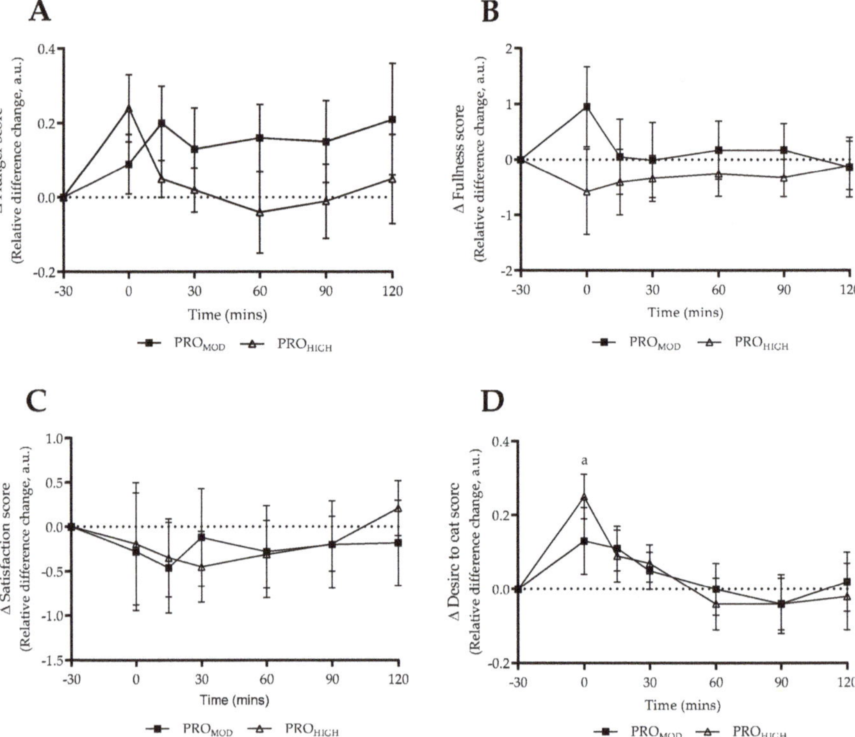

Figure 4. Questionnaire responses in relation to test meal under laboratory conditions. Data expressed as relative difference change (baseline to post diet) for both moderate (PRO$_{MOD}$) and high (PRO$_{HIGH}$) protein conditions (mean $\pm$ SE). (**A**) Perception of hunger; (**B**) Perception of fullness; (**C**) Perception of satisfaction; (**D**) Perception of desire to eat. Responses in relation to pre-meal (-30 min) time-point. 0 min denotes immediate completion of test meal. Dashed line provides reference point to pre-meal perceived state. [a] denotes difference to pre-meal time-point within PRO$_{HIGH}$ only ($p = 0.028$). No differences between dietary conditions observed.

3.5. Ad libitum Intake and Satiety Responses

Mean dietary intake during the *ad libitum* phase following each dietary intervention is shown in Table 5. Despite higher absolute energy and fat intakes with PRO$_{HIGH}$, no significant differences were reported between conditions for any variable ($p > 0.05$). Compliance with target protein intake was met for both conditions. Satiety responses across days 1-3 of the *ad libitum* phase are shown in Table 6, along with delta scores in comparison to days 5-7 of the previous intervention. Perception of hunger was significantly reduced following transition to the *ad libitum* period (F = 5.18$_{(time)}$, $p = 0.038$, $\eta p^2 = 0.26$) within PRO$_{MOD}$ only (-9.67 ± 4.68 mm, $p = 0.046$). Likewise, perception of fullness (F = 10.33$_{(time)}$, $p = 0.006$, $\eta p^2 = 0.41$) and satisfaction (F = 5.72$_{(time)}$, $p = 0.03$, $\eta p^2 = 0.28$) were significantly improved within-condition relative to the end of the PRO$_{MOD}$ diet (fullness: 9.78 $\pm$ 3.74 mm, $p = 0.015$; satisfaction: 13.11 $\pm$ 4.33 mm, $p = 0.005$). Perceived desire to eat also reduced on transition to an *ad libitum* phase (F = 14.86$_{(time)}$, $p = 0.002$, $\eta p^2 = 0.50$) for PRO$_{MOD}$ only (-16.78 ± 4.54 mm, $p = 0.001$).

A significant interaction effect was reported for perceived cravings (F = 5.00$_{(dietxtime)}$, $p = 0.041$, $\eta p^2 = 0.25$), with post-hoc analysis indicating a significant reduction for PRO$_{MOD}$ (-13.00 ± 4.27 mm, $p = 0.004$). The relative change in perceived cravings for PRO$_{MOD}$ was significantly reduced in comparison to PRO$_{HIGH}$ ($p = 0.04$) following transition to an *ad libitum* phase. Participants reported improvements in general mood within-condition ($p = 0.05$) only. Perceived motivation to train was reportedly improved with PRO_{MOD} ($F = 4.52_{(time)}$, $p = 0.05$, $\eta p^2 = 0.23$) by 8.97 ± 5.08 mm ($p = 0.05$). No other differences were reported between dietary conditions during the *ad libitum* phase ($p > 0.05$).

Table 5. Mean dietary intake during *ad libitum* period.

Variable	Category	PRO_{MOD}	PRO_{HIGH}
Energy Intake	$(kcal \cdot d^{-1})$	2197.81 ± 177.58	2348.14 ± 232.46
	$(kcal \cdot kg^{-1} \cdot d^{-1})$	25.64 ± 1.62	26.52 ± 1.81
Protein Intake	$(g \cdot d^{-1})$	162.38 ± 11.18	165.07 ± 12.89
	$(g \cdot kg^{-1} \cdot d^{-1})$	1.85 ± 0.03	1.86 ± 0.07
	(%EI)	30.87 ± 2.13	29.40 ± 1.89
Carbohydrate Intake	$(g \cdot d^{-1})$	214.94 ± 19.61	230.61 ± 24.33
	$(g \cdot kg^{-1} \cdot d^{-1})$	2.57 ± 0.24	2.65 ± 0.23
	(%EI)	39.14 ± 1.98	39.66 ± 1.85
Fat Intake	$(g \cdot d^{-1})$	72.69 ± 9.71	81.29 ± 11.01
	$(g \cdot kg^{-1} \cdot d^{-1})$	0.84 ± 0.09	0.90 ± 0.09
	(%EI)	12.64 ± 0.76	13.29 ± 0.70

Data presented as mean $\pm$ SE, and for macronutrient categories expressed in grams per day, grams per kg per day, and percentage of energy intake (EI). PRO_{MOD} denotes post moderate protein condition; PRO_{HIGH} denotes post high protein condition. During *ad libitum* phase, meal frequency set to 4 meals, and target protein intake of $1.80 \, g \cdot kg^{-1} \cdot d^{-1}$, otherwise unrestricted. No significant differences reported between conditions.

Table 6. Questionnaire responses during *ad libitum* dietary phase.

Question Category	PRO_{MOD}	PRO_{HIGH}	ΔPRO_{MOD}	ΔPRO_{HIGH}
Hunger	41.67 ± 3.61	40.78 ± 3.48	-9.67 ± 4.68 [a]	-0.89 ± 2.74
Fullness	69.00 ± 4.39	73.39 ± 3.10	9.78 ± 3.74 [a]	7.87 ± 4.61
Satisfaction	73.11 ± 3.80	71.56 ± 3.00	13.11 ± 4.33 [a]	4.26 ± 4.89
Desire to eat	39.67 ± 2.96	38.03 ± 3.62	-16.78 ± 4.54 [a]	-8.12 ± 5.22
Cravings	42.89 ± 4.12	46.15 ± 4.23	-13.00 ± 4.27 [a]	-1.04 ± 5.08 *
Energy	71.33 ± 3.31	71.93 ± 2.91	2.99 ± 4.30	2.87 ± 3.34
Mood	76.11 ± 4.03	77.55 ± 2.09	8.33 ± 3.70 [a]	5.68 ± 2.72 [a]
Training enjoyment	79.35 ± 4.44	74.67 ± 4.51	9.23 ± 6.57	1.67 ± 4.43
Training motivation	78.85 ± 5.39	72.89 ± 4.99	8.97 ± 5.08 [a]	1.33 ± 3.93

Data represent mean scores over days 1–3 during *ad libitum* post diet period. Data based on arbitrary units (a.u.) from participant responses using a visual analogue scale (VAS) and presented as mean $\pm$ SE. Δ scores are relative to end of intervention diet. Question categories paraphrased. [a] denotes significant difference within group compared to end of intervention diet ($p \leq 0.05$).* denotes significant difference to ΔPRO_{MOD} ($p = 0.04$).

4. Discussion

The main finding from this study was that the perceived satiating effect of two diets with different protein content was generally comparable in resistance-trained participants undergoing a period of acute energy deficit. Therefore, consuming more protein during acute energy deficit does not appear to improve perceived satiety, as reported elsewhere [38–42]. However, when considering hormonal adaptations to an acute dietary intervention in response to a protein-rich test-meal, a moderate protein, energy deficit diet ($1.8 \, g \cdot kg^{-1} \cdot d^{-1}$) was deemed more satiating when compared to a high protein ($2.9 \, g \cdot kg^{-1} \cdot d^{-1}$) diet. These results suggest that whilst in acute energy deficit, a PRO_{MOD} diet may be sufficient to maintain training requirements in accordance with previous recommendations [5,43,44]. It is, however, important to note that a PRO_{HIGH} diet did not appear to disadvantage participants and may offer individual benefits discussed below.

Across the 7-day intervention period, participant responses to the daily satiety questionnaire were largely similar for all categories, with the exception of both cravings and satisfaction (Table 3). By the end of the dietary intervention, whilst consuming a PRO_{MOD} diet, participants reported increased cravings, which was not experienced to the same extent following PRO_{HIGH}. Additionally, perceived satisfaction was not only maintained during PRO$_{HIGH}$, but was significantly greater than PRO$_{MOD}$ at the end of the intervention period. Similar findings in response to a high protein meal have been observed elsewhere [45]. This implies that during short-term caloric deficit a substantial increase in protein ratio beyond recommended levels for maximal lean muscle gain may be better sustained by resistance trained athletes who are more prone to cravings. Participant well-being, mood, training enjoyment and motivation to train were comparable between dietary conditions. As training frequency was equally maintained across dietary conditions, this inferred that an increase in protein intake beyond 1.8 g·kg^{-1}·d^{-1} did not compromise training responses, and did not result in reduced motivation or perceived energy during training as hypothesized.

In the current study subjects maintained their habitual training routine without any external motivation that was intended to increase their performance. As exercise fatigue has been reported to increase during acute caloric deficit [4], it is possible that participants in the current study subconsciously decreased training effort, despite maintaining session frequency. Interestingly, phase angle (PA) assessed via bioelectrical impedance indicated a significant reduction following both dietary conditions, more notably with PRO$_{MOD}$. PA has been reported to be a proxy measure of muscle "quality" and is associated with sex, age, body mass index (BMI) and fat mass percentage (FM%) [30,46–48]. A decrease in PA during both dietary conditions may indicate unfavorable changes to muscle integrity as a result of energy restriction. However, the comparable reduction in hydration indices likely infers decreased PA was transient based on acute energy deficit. Our previous findings demonstrated an increase in PA when resistance trainees underwent a period of supervised intensive training to volitional exhaustion during short-term energy deficit when consuming a PRO$_{HIGH}$ diet [30]. It is feasible, therefore, that in order to maintain muscle integrity during periods of acute caloric deficit more intensive training intensity and substantially increased protein intake is required, which is supported elsewhere [13].

Plasma ghrelin responses were comparable with previous research [45,49–52]. However, in response to a laboratory test meal, notable differences were reported between dietary conditions. Following PRO$_{MOD}$, plasma ghrelin remained significantly reduced at 2 h post-meal indicating a sustained satiating effect. When data was considered as normalized relative change (Figure 1B), this pattern was highlighted further with differences reported between dietary conditions in response to a protein-rich test meal. Furthermore, as no differences were observed between dietary conditions when taking into consideration pre-intervention ghrelin responses (Figure 1C), this would suggest that the satiating effect following PRO$_{MOD}$ was associated with the higher protein content of the test meal.

It should be noted that baseline protein intake for all participants was nearly identical to the protein requirements for the PRO$_{MOD}$ condition. Therefore, the sustained reduction in 2-h ghrelin response observed following PRO$_{MOD}$ may also be related to the energy balance of the individual (both a reduction in caloric and fat intake) across the intervention period [53,54]. It is also noteworthy, that following PRO$_{MOD}$, plasma ghrelin was reduced at T0 compared with pre-intervention levels which presented an interesting trend ($p = 0.06$, $\eta p^2 = 0.22$) considering the reported increase in cravings within-condition by day 7. As plasma ghrelin followed expected responses pre and post PRO$_{HIGH}$ [55], the findings indicate that satiety may be acutely enhanced by the consumption of a high-protein meal during periods of energy deficit when following a PRO$_{MOD}$ approach. Individuals experiencing periodic cravings when following a PRO$_{MOD}$ diet may therefore benefit by increasing the protein content of individual meals when required [56] without necessarily over-consuming protein.

This may be important considering the potential risks of longer-term consumption of very high protein intakes in some individuals (i.e., those with existing, or predisposition for, kidney disease [57]), and current recognition that high protein diets modulate intestinal microbiota production of tryptophan, leading to elevated metabolites (e.g., indoxyl sulphate) acting as uremic toxins [58,59]. There is also current suggestion that high protein diets may negatively impact intestinal function, potentially via bacterial metabolites produced as a result of undigested proteins [60]. High protein diets, in combination with reduced carbohydrate/fiber intake may reduce short-chain fatty acid concentrations, impacting on colonic environment/mucosa [15] which may influence individual health in the longer term. This supports the contention that a PRO_{MOD} approach may be sufficient for resistance-trained individuals during energy deficit.

On completion of the dietary interventions, absolute plasma PYY did not significantly differ between conditions in response to a standardized test meal, as observed elsewhere [49]. The expected increase in plasma PYY [52] in response to a protein-rich test meal was consistent, with a similar rate of change from T0-T60, following both dietary conditions. It is, however, noteworthy that following PRO_{MOD}, a greater rate of change in plasma PYY from T60-T120 was evident ($p = 0.07$), which is comparable with the sustained satiating effect observed with plasma ghrelin. When data were compared as normalized relative change (Figure 2B), no differences were reported between dietary conditions ($p = 0.06$) potentially explained by the mixed individual responses observed. However, when taking into consideration pre-intervention PYY responses (Figure 2C), an overall positive response was observed for PRO_{MOD} at both T60 and T120 in contrast to PRO_{HIGH}. This infers that a PRO_{MOD} intervention improved acute satiety to a greater extent than PRO_{HIGH}. Collectively, these results suggest that individuals may become accustomed to a higher protein intake, and therefore chronic consumption of a PRO_{HIGH} diet may lose its satiating effect over time [61].

Various factors modulate PYY concentrations, including acute/chronic energy balance, as well as nutrient composition of a meal. High protein meal content has been shown to elicit greater PYY secretion leading to short-term sensations of fullness [62,63]. PYY is an agonist at neuropeptide Y2 receptors and increasing levels leads to both reduced hunger perception and food intake [62,64]. In the current study, nutrient composition of the test-meals remained the same across all testing sessions. Therefore, although absolute changes in PYY were not reportedly different between dietary interventions, the relative change difference indicated that both the decrease in caloric intake, along with the higher protein content of the test meal resulted in acute increases in PYY concentrations over the 2-h monitoring period for PRO_{MOD}. Although not significant, the observed increase in absolute PYY concentration for PRO_{HIGH} at T0 may be explained by the increase in dietary protein over the 7-days. However, the hormonal response to a protein-rich meal for PRO_{HIGH} was not considered different to pre-intervention patterns. This further supports the contention that a moderate protein diet favorably improved acute physiological satiety in contrast to very high protein intakes.

Perceived satiety responses to the protein-rich test meal were largely comparable between dietary conditions in contrast to physiological responses. Only "desire to eat" remained significantly reduced until 2 h post-meal with PRO_{MOD} (Figure 3D), which in part supported the satiating effect observed for plasma ghrelin. However, this was not the case for perceived hunger, fullness or satisfaction. Whilst perceived satisfaction was comparable between conditions beyond T15, it was noted that participants reported being more satisfied on completion of the test-meal with PRO_{MOD} (T0, Figure 3C). However, as the mean values at T0 were similar between conditions, it was noted that a trend towards significance was observed for PRO_{HIGH} ($p = 0.06$) indicating that perceived responses to the test-meal were likely comparable for satisfaction. Overall, our results support previous findings that circulating levels of ghrelin and PYY do not correlate with the subjective perception of appetite [52,65].

Based on the acute energy deficit period undertaken it is feasible that perceived responses may not have changed dramatically compared to habitual intake. Interestingly when test-meal questionnaire responses were normalized in relation to pre-intervention results (Figure 4), no differences were reported between dietary conditions. The only within condition finding at T0 for "desire to eat" for PRO_{HIGH} (Figure 4D), may be explained by the fact that the protein content of the test-meal was based on the higher protein intervention. Therefore, following a 7-day high protein ($2.9 \text{ g·kg}^{-1}\text{·d}^{-1}$) energy deficit period, it is likely that participants had attained a degree of habituation to the protein intake and hence experienced a greater "desire to eat" on completion of the test-meal. This is, in part, supported by the hunger and fullness scores at T0, albeit non-significant.

On completion of the 7-day dietary intervention, participants then completed a 3-day period in which protein intake was fixed at $1.8 \text{ g·kg}^{-1}\text{·d}^{-1}$, with *ad libitum* consumption of carbohydrate, fat and energy intake until perceived fullness. This was included to provide insight as whether the preceding intervention increased the likelihood of over-eating on cessation of short-term energy deficit. Although energy and fat intakes were reportedly greater following PRO_{HIGH}, no differences were reported between groups during the *ad libitum* phase, as similarly observed elsewhere [64]. Despite having followed a monitored period of reduced (20%) energy intake, participants tended to only partially increase caloric intake mainly through increased total fat closer to pre-intervention levels. This suggests, at least in the short-term, that resistance-trained athletes are unlikely to over-consume calories on completion of an energy deficit period. This may reflect the tendency to follow habitual dietary intake patterns during periods of training.

In a similar manner, the satiety questionnaire responses during the *ad libitum* period were comparable between dietary conditions. However, when relative responses were compared to the end of the previous intervention period, the change from a PRO_{MOD} approach resulted in significant within-condition improvements in hunger, fullness, satisfaction, desire to eat, as well as general participants' mood and motivation to train. Notably, the reduction in cravings observed following PRO_{MOD} was significantly different to PRO_{HIGH}. Collectively this suggests that whilst a short-term PRO_{MOD} energy deficit period may result in improved hormonal satiety responses, periodic inclusion of higher protein intakes may support sustained periods in which athletes undergo caloric deficit. Alternatively, athletes undergoing periods of energy deficit may physiologically benefit from a PRO_{MOD} approach, but should be mindful of increasing protein content (either based on single meals, or as a dietary approach) when perceived cravings or a reduction in mood and/or training motivation occurs.

On completion of the study, several limitations were noted. Firstly, that in monitoring habitual dietary intake a longer lead in period would have provided clearer insight. All participants were resistance-trained individuals experienced in maintaining individual macronutrient ratios (i.e., $1.8\ g{\cdot}kg^{-1}{\cdot}d^{-1}$) and followed regular training routines. Dietary records were kept in the days leading into each testing period prior to each intervention. As such, the observed habitual intake (Table 2) included the requested rest days and therefore did not reflect maintenance intakes expected when taking into consideration training days over a typical week (i.e., $2628 \pm 144\ kcal{\cdot}d^{-1}$). However, based on the period collected, energy intakes were comparable to expected intakes for less active periods (e.g., $2381 \pm 131\ kcal{\cdot}d^{-1}$) for this cohort (comprising both male and female athletes). Therefore, although the habitual intakes were comparable prior to each intervention phase, they likely do not reflect accurate mean energy intakes. However, the decrease in body-mass observed across each intervention, along with monitored reduced energy intakes, indicated that participants achieved the 20% energy deficit as expected.

Although training frequency was maintained across both intervention periods, training intensity was not monitored. Therefore, although perceived energy, training motivation and training enjoyment was not different between dietary conditions, it is difficult to know whether participants maintained overall training intensity. It is feasible, as previously mentioned, that fatigue was offset through a subconscious reduction in training intensity during the energy deficit period. Anecdotally, most participants reported that the PRO_{HIGH} diet was challenging (i.e., reduced mental clarity) towards the end of the intervention phase, possibly indicating that such diets may be poorly tolerated in the longer term. Careful attention to monitoring training intensity, along with perceived well-being responses during periods of acute energy deficit may be warranted to ensure performance maintenance.

Another limitation observed was that other gut hormones (i.e., glucagon-like peptide 1 (GLP-1) and cholecystokinin (CCK)) were not assessed in this study. This would have supported our findings in determining that a PRO_{MOD} diet had a greater acute satiating effect after consumption of a high protein meal. GLP-1 for example is secreted in conjunction with PYY, and has been associated with appetite response and decreased food intake [62,66–68]. An increased GLP-1 response to PRO_{MOD} would provide clearer insight that the satiating effect of the test-meal was related to localized stomach/intestinal responses to a protein-rich breakfast meal, and confirmed our reasoning that athletes may become accustomed to chronic consumption of a high protein diet. Likewise, adjunct measurements of CCK released from the duodenum in response to meal consumption influences satiety acutely [62]. Along with reduced ghrelin responses, measures of CCK would have supported the contention that PRO_{MOD} resulted in greater satiety in this cohort.

Whilst the results from this study indicate that a PRO_{MOD} preferentially influences hormonal responses to a test-meal, it is evident that perceived responses were comparable between intervention diets. Further research is therefore warranted to assess whether perceived satiety responses are different when undertaking prolonged energy deficit periods. This is pertinent considering previous research has indicated that fasting ghrelin levels may be increased in response to longer term energy deficit [69,70]. Furthermore, with current recognition that high protein diets (including type of dietary protein) likely modulate gut microbiota, this may have inference to both neurotransmitter and satiety responses (e.g., GLP-1), especially considering inter-individual variance [15].

Although the sample size in the current study was deemed sufficient based on a priori power assessment (comparable to previous research [4]), determination of satiety responses in larger resistance-trained cohorts is also warranted. In addition, with the observation that athletes may benefit from consuming a PRO_{MOD} diet during energy deficit, but periodically include protein-high meals to offset negative symptoms (i.e., cravings), assessment of nutritional periodization during periods of caloric deficit would be beneficial to ascertain sustainability to such diets. Finally, maintenance of training performance and lean gains during periods of acute or chronic energy deficit should be considered to determine whether a PRO_{MOD} diet sustains beneficial responses in resistance-trained athletes.

5. Conclusions

During acute energy deficit in resistance-trained individuals, consuming protein intakes in excess of recommended athlete guidelines did not improve overall perceived satiety in comparison to more habitual, moderate protein intakes. Furthermore, a PRO_{MOD} diet favorably improved hormonal responses to a test-meal compared with a PRO_{HIGH} diet. Therefore, as long as training motivation/ performance is not compromised during short-term energy deficit, moderate protein intakes (1.8 $g \cdot kg^{-1} \cdot d^{-1}$) are likely to be adequate for resistance-trained individuals. The findings also suggest that where individuals experience negative symptoms (i.e., cravings), implementation of periodic high protein meals may be sufficient to maintain satisfaction, and diet sustainability.

Author Contributions: Conceptualization and design of the study was undertaken by M.H., F.T.V., A.Z., J.R. and K.T.M. J.R., A.Z., V.M., K.T.M. were involved with data collection, along with individual participant nutrition tracking support from M.B., O.B., and L.V.; blood collection and assay analyses were undertaken by J.R., V.M. and K.T.M. Formal analysis of data was undertaken by J.R., A.Z. and K.T.M. with support from L.S. and M.H. Original draft preparation was undertaken by J.R. and A.Z., with critical appraisal and editing by all authors. All authors reviewed the paper and approved the final version prior to submission.

Funding: Analytical costs for this research were funded by the Institute for Nutrition and Fitness Sciences (INFS) Private Ltd. The APC was funded by Anglia Ruskin University Open Access Fund.

Acknowledgments: This study was supported by Ultimate Sports Nutrition (USN) UK Ltd. with provision of whey protein.

Conflicts of Interest: The authors declare no conflict of interest. The funders had no role in the design of the study; in the collection, analyses, or interpretation of data; in the writing of the manuscript, or in the decision to publish the results.

References

1. Khodaee, M.; Olewinski, L.; Shadgan, B.; Kiningham, R.R. Rapid weight loss in sports with weight classes. *Curr. Sports Med. Rep.* **2015**, *14*, 435–441. [CrossRef]
2. Rossow, L.M.; Fukuda, D.H.; Fahs, C.A.; Loenneke, J.P.; Stout, J.R. Natural bodybuilding competition preparation and recovery: A 12-month case study. *Int. J. Sports Physiol. Perform.* **2013**, *8*, 582–592. [CrossRef] [PubMed]

3. Robinson, S.L.; Lambeth-Mansell, A.; Gillibrand, G.; Smith-Ryan, A.; Bannock, L. A nutrition and conditioning intervention for natural bodybuilding contest preparation: Case study. *J. Int. Soc. Sports Nutr.* **2015**, *12*, 1–11. [CrossRef] [PubMed]
4. Mettler, S.; Mitchell, N.; Tipton, K.D. Increased protein intake reduces lean body mass loss during weight loss in athletes. *Med. Sci. Sports Exerc.* **2010**, *42*, 326–337. [CrossRef] [PubMed]
5. Tarnopolsky, M.A.; Atkinson, S.A.; MacDougall, J.D.; Chesley, A.; Phillips, S.; Schwarcz, H.P. Evaluation of protein requirements for trained strength athletes. *J. Appl. Physiol.* **1992**, *73*, 1986–1995. [CrossRef] [PubMed]
6. Walberg, J.L.; Leidy, M.K.; Sturgill, D.J.; Hinkle, D.E.; Ritchey, S.J.; Sebolt, D.R. Macronutrient content of a hypoenergy diet affects nitrogen retention and muscle function in weight lifters. *Int. J. Sports Med.* **1988**, *9*, 261–266. [CrossRef] [PubMed]
7. Helms, E.R.; Aragon, A.A.; Fitschen, P.J. Evidence-based recommendations for natural bodybuilding contest preparation: Nutrition and supplementation. *J. Int. Soc. Sports Nutr.* **2014**, *11*, 20. [CrossRef] [PubMed]
8. Lemon, P.W.R. Beyond the zone: Protein needs of active individuals. *J. Am. Coll. Nutr.* **2000**, *19*, 513S–521S. [CrossRef]
9. Phillips, S.M. Dietary protein for athletes: From requirements to metabolic advantage. *Appl. Physiol. Nutr. Metab.* **2006**, *31*, 647–654. [CrossRef]
10. Phillips, S.M.; Moore, D.R.; Tang, J.E. A critical examination of dietary protein requirements, benefits, and excesses in athletes. *Int. J. Sport Nutr. Exerc. Metab.* **2007**, *17*, 58–76. [CrossRef]
11. Layman, D.K.; Boileau, R.A.; Erickson, D.J.; Painter, J.E.; Shiue, H.; Sather, C.; Christou, D.D. A reduced ratio of dietary carbohydrate to protein improves body composition and blood lipid profiles during weight loss in adult women. *J. Nutr.* **2003**, *133*, 411–417. [CrossRef] [PubMed]
12. Stiegler, P.; Cunliffe, A. The role of diet and exercise for the maintenance of fat-free mass and resting metabolic rate during weight loss. *Sports Med.* **2006**, *36*, 239–262. [CrossRef] [PubMed]
13. Helms, E.R.; Zinn, C.; Rowlands, D.S.; Brown, S.R. A systematic review of dietary protein during caloric restriction in resistance trained lean athletes: A case for higher intakes. *Int. J. Sport Nutr. Exerc. Metab.* **2014**, *24*, 127–138. [CrossRef] [PubMed]
14. Rand, W.M.; Pellett, P.L.; Young, V.R. Meta-analysis of nitrogen balance studies for estimating protein requirements in healthy adults. *Am. J. Clin. Nutr.* **2003**, *77*, 109–127. [CrossRef] [PubMed]
15. Blachier, F.; Beaumont, M.; Portune, K.J.; Steuer, N.; Lan, A.; Audebert, M.; Khodorova, N.; Andriamihaja, M.; Airinei, G.; Benamouzig, R.; et al. High-protein diets for weight management: Interactions with the intestinal microbiota and consequences for gut health. A position paper by the my new gut study group. *Clin. Nutr.* **2018**. [CrossRef] [PubMed]
16. Phillips, S.M.; Van Loon, L.J.C. Dietary protein for athletes: From requirements to optimum adaptation. *J. Sports Sci.* **2011**, *29*, S29–S38. [CrossRef] [PubMed]
17. Phillips, S.M. A brief review of higher dietary protein diets in weight loss: A focus on athletes. *Sports Med.* **2014**, *44* (Suppl. 2), S149–S153. [CrossRef]
18. Spendlove, J.; Mitchell, L.; Gifford, J.; Hackett, D.; Slater, G.; Cobley, S.; O'Connor, H. Dietary intake of competitive bodybuilders. *Sport Med.* **2015**, *45*, 1041–1063. [CrossRef]
19. Dhillon, J.; Craig, B.A.; Leidy, H.J.; Amankwaah, A.F.; Osei-Boadi Anguah, K.; Jacobs, A.; Jones, B.L.; Jones, J.B.; Keeler, C.L.; Keller, C.E.; et al. The effects of increased protein intake on fullness: A meta-analysis and its limitations. *J. Acad. Nutr. Diet.* **2016**, *116*, 968–983. [CrossRef]
20. Weigle, D.S.; Breen, P.A.; Matthys, C.C.; Callahan, H.S.; Meeuws, K.E.; Burden, V.R.; Purnell, J.Q. A high-protein diet induces sustained reductions in appetite, ad libitum caloric intake, and body weight despite compensatory changes in diurnal plasma leptin and ghrelin concentrations. *Am. J. Clin. Nutr.* **2005**, *82*, 41–48. [CrossRef]
21. Martens, E.A.; Lemmens, S.G.; Westerterp-Plantenga, M.S. Protein leverage affects energy intake of high-protein diets in humans. *Am. J. Clin. Nutr.* **2013**, *97*, 86–93. [CrossRef] [PubMed]
22. Martens, E.A.; Tan, S.Y.; Dunlop, M.V.; Mattes, R.D.; Westerterp-Plantenga, M.S. Protein leverage effects of beef protein on energy intake in humans. *Am. J. Clin. Nutr.* **2014**, *99*, 1397–1406. [CrossRef] [PubMed]
23. Helms, E.R.; Zinn, C.; Rowlands, D.S.; Naidoo, R.; Cronin, J. High-protein, low-fat, short-term diet results in less stress and fatigue than moderate-protein moderate-fat diet during weight loss in male weightlifters: A pilot study. *Int. J. Sport Nutr. Exerc. Metab.* **2015**, *25*, 163–170. [CrossRef] [PubMed]

24. DeFreitas, J.M.; Beck, T.W.; Stock, M.S.; Dillon, M.A.; Kasishke, P.R. An examination of the time course of training-induced skeletal muscle hypertrophy. *Eur. J. Appl. Physiol.* **2011**, *111*, 2785–2790. [CrossRef] [PubMed]

25. Brown, N.; Bubeck, D.; Haeufle, D.F.B.; Weickenmeier, J.; Kuhl, E.; Alt, W.; Schmitt, S. Weekly time course of neuro-muscular adaptation to intensive strength training. *Front. Physiol.* **2017**, *8*, 329. [CrossRef] [PubMed]

26. Sale, D.G. Neural adaptation to resistance training. *Med. Sci. Sports Exerc.* **1988**, *20* (Suppl. 5), S135–S145. [CrossRef] [PubMed]

27. Trappe, S.; Williamson, D.; Godard, M. Maintenance of whole muscle strength and size following resistance training in older men. *J. Gerontol.* **2002**, *57*, B138–B143. [CrossRef]

28. American College of Sports Medicine. American College of Sports Medicine position stand. Progression models in resistance training for healthy adults. *Med. Sci. Sports Exerc.* **2009**, *41*, 687–708. [CrossRef]

29. Utter, A.C.; Kang, J.; Nieman, D.C.; Brown, V.A.; Dumke, C.L.; McAnulty, S.R.; McAnulty, L.S. Carbohydrate supplementation and perceived exertion during resistance exercise. *J. Strength Cond. Res.* **2005**, *19*, 939–943. [CrossRef]

30. Roberts, J.; Zinchenko, A.; Suckling, C.; Smith, L.; Johnstone, J.; Henselmans, M. The short-term effect of high versus moderate protein intake on recovery after strength training in resistance-trained individuals. *J. Int. Soc. Sports Nutr.* **2017**, *14*, 44. [CrossRef]

31. Antonio, J.; Peacock, C.; Ellerbroek, A.; Fromhoff, B.; Silver, T. The effects of consuming a high protein diet (4.4 g/kg/d) on body composition in resistance-trained individuals. *J. Int. Soc. Sports Nutr.* **2014**, *11*, 19. [CrossRef] [PubMed]

32. Antonio, J.; Ellerbroek, A.; Silver, T.; Orris, S.; Scheiner, M.; Gonzalez, A.; Peacock, C. A high protein diet (3.4 g/kg/d) combined with a heavy resistance training program improves body composition in healthy trained men and women—A follow-up investigation. *J. Int. Soc. Sports Nutr.* **2015**, *12*, 39. [CrossRef] [PubMed]

33. Antonio, J.; Ellerbroek, A.; Silver, T.; Vargas, L.; Peacock, C. The effects of a high protein diet on indices of health and body composition—A crossover trial in resistance-trained men. *J. Int. Soc. Sports Nutr.* **2016**, *13*, 1–7. [CrossRef] [PubMed]

34. McArdle, W.; Katch, F.; Katch, V. *Exercise Physiology: Energy, Nutrition and Human Performance*, 5th ed.; Lippincott Williams and Wilkins: Philadelphia, PA, USA, 2001.

35. Cunningham, J. Body composition and resting metabolic rate: The myth of feminine metabolism. *Am. J. Clin. Nutr.* **1982**, *36*, 721–726. [CrossRef] [PubMed]

36. Stubbs, R.J.; Hughes, D.A.; Johnstone, A.M.; Rowley, E.; Reid, C.; Elia, M.; Stratton, R.; Delargy, H.; King, N.; Blundell, J.E. The use of visual analogue scales to assess motivation to eat in human subjects: A review of their reliability and validity with an evaluation of new hand-held computerized systems for temporal tracking of appetite ratings. *Br. J. Nutr.* **2000**, *84*, 405–415. [CrossRef] [PubMed]

37. Flint, A.; Raben, A.; Blundell, J.E.; Astrup, A. Reproducibility, power and validity of visual analogue scales in assessment of appetite sensations in single test meal studies. *Int. J. Obes. Relat. Metab. Disord.* **2000**, *24*, 38–48. [CrossRef] [PubMed]

38. Raben, A.; Agerholm-Larsen, L.; Flint, A.; Holst, J.J.; Astrup, A. Meals with similar energy densities but rich in protein, fat, carbohydrate, or alcohol have different effects on energy expenditure and substrate metabolism but not on appetite and energy intake. *Am. J. Clin. Nutr.* **2003**, *77*, 91–100. [CrossRef]

39. Bligh, H.F.; Godsland, I.F.; Frost, G.; Hunter, K.J.; Murray, P.; MacAulay, K.; Hyliands, D.; Talbot, D.C.; Casey, J.; Mulder, T.P.; et al. Plant-rich mixed meals based on Palaeolithic diet principles have a dramatic impact on incretin, peptide YY and satiety response, but show little effect on glucose and insulin homeostasis: An acute-effects randomised study. *Br. J. Nutr.* **2015**, *113*, 574–584. [CrossRef]

40. Blatt, A.D.; Roe, L.S.; Rolls, B.J. Increasing the protein content of meals and its effect on daily energy intake. *J. Am. Diet. Assoc.* **2011**, *111*, 290–294. [CrossRef]

41. Wiessing, K.R.; Xin, L.; Budgett, S.C.; Poppitt, S.D. No evidence of enhanced satiety following whey protein- or sucrose-enriched water beverages: A dose response trial in overweight women. *Eur. J. Clin. Nutr.* **2015**, *69*, 1238–1243. [CrossRef]

42. Ravn, A.M.; Gregersen, N.T.; Christensen, R.; Rasmussen, L.G.; Hels, O.; Belza, A.; Raben, A.; Larsen, T.M.; Toubro, S.; Astrup, A. Thermic effect of a meal and appetite in adults: An individual participant data meta-analysis of meal-test trials. *Food Nutr. Res.* **2013**, *57*. [CrossRef] [PubMed]

43. Phillips, S.M.; Chevalier, S.; Leidy, H.J. Protein "requirements" beyond the RDA: Implications for optimizing health. *Appl. Physiol. Nutr. Metab.* **2016**, *41*, 565–572. [CrossRef] [PubMed]

44. Lemon, P.W.; Tarnopolsky, M.A.; MacDougall, J.D.; Atkinson, S.A. Protein requirements and muscle mass/strength changes during intensive training in novice bodybuilders. *J. Appl. Physiol.* **1992**, *73*, 767–775. [CrossRef] [PubMed]

45. Jakubowicz, D.; Froy, O.; Wainstein, J.; Boaz, M. Meal timing and composition influence ghrelin levels, appetite scores and weight loss maintenance in overweight and obese adults. *Steroids* **2012**, *77*, 323–331. [CrossRef] [PubMed]

46. Barbosa-Silva, M.C.G.; Barros, A.J.D.; Wang, J.; Heymsfield, S.B.; Pierson, R.N. Bioelectrical impedance analysis: Population reference values for phase angle by age and sex. *Am. J. Clin. Nutr.* **2005**, *82*, 49–52. [CrossRef] [PubMed]

47. Norman, K.; Stobäus, N.; Pirlich, M.; Bosy-Westphal, A. Bioelectrical phase angle and impedance vector analysis—Clinical relevance and applicability of impedance parameters. *Clin. Nutr.* **2012**, *31*, 854–861. [CrossRef] [PubMed]

48. Souza, M.F.; Tomeleri, C.M.; Ribeiro, A.S.; Schoenfeld, B.J.; Silva, A.M.; Sardinha, L.B.; Cyrino, E.S. Effect of resistance training on phase angle in older women: A randomized controlled trial. *Scand. J. Med. Sci. Sports* **2016**, *8*, 1–9. [CrossRef]

49. Westerterp-Plantenga, M.S.; Lemmens, S.G.; Westerterp, K.R. Dietary protein—Its role in satiety, energetics, weight loss and health. *Br. J. Nutr.* **2012**, *108*, S105–S112. [CrossRef]

50. Gibbons, C.; Caudwell, P.; Finlayson, G.; Webb, D.L.; Hellström, P.M.; Näslund, E.; Blundell, J.E. Comparison of postprandial profiles of ghrelin, active GLP-1, and total PYY to meals varying in fat and carbohydrate and their association with hunger and the phases of satiety. *J. Clin. Endocrinol. Metab.* **2013**, *98*, E847–E855. [CrossRef]

51. Leidy, H.J.; Ortinau, L.C.; Douglas, S.M.; Hoertel, H.A. Beneficial effects of a higher-protein breakfast on the appetitive, hormonal, and neural signals controlling energy intake regulation in overweight/obese, "breakfast-skipping," late-adolescent girls. *Am. J. Clin. Nutr.* **2013**, *97*, 677–688. [CrossRef]

52. Doucet, É.; Laviolette, M.; Imbeault, P.; Strychar, I.; Rabasa-Lhoret, R.; Prud'homme, D. Total peptide YY is a correlate of postprandial energy expenditure but not of appetite or energy intake in healthy women. *Metab. Clin. Exp.* **2008**, *57*, 1458–1464. [CrossRef] [PubMed]

53. OConnor, K.L.; Scisco, J.L.; Smith, T.J.; Young, A.J.; Montain, S.J.; Price, L.L.; Lieberman, H.R.; Karl, J.P. Altered appetite-mediating hormone concentrations precede compensatory overeating after severe, short-term energy deprivation in healthy adults. *J. Nutr.* **2016**, *146*, 209–217. [CrossRef] [PubMed]

54. Moran, L.J.; Luscombe-Marsh, N.D.; Noakes, M.; Wittert, G.A.; Keogh, J.B.; Clifton, P.M. The satiating effect of dietary protein is unrelated to postprandial ghrelin secretion. *J. Clin. Endocrinol. Metab.* **2005**, *90*, 5205–5211. [CrossRef] [PubMed]

55. Leidy, H.J.; Armstrong, C.L.; Tang, M.; Mattes, R.D.; Campbell, W.W. The influence of higher protein intake and greater eating frequency on appetite control in overweight and obese men. *Obesity* **2010**, *18*, 1725–1732. [CrossRef] [PubMed]

56. MacKenzie-Shalders, K.L.; Byrne, N.M.; Slater, G.J.; King, N.A. The effect of a whey protein supplement dose on satiety and food intake in resistance training athletes. *Appetite* **2015**, *92*, 178–184. [CrossRef] [PubMed]

57. Cuenca-Sánchez, M.; Navas-Carrillo, D.; Orenes-Piñero, E. Controversies surrounding high-protein diet intake: Satiating effect and kidney and bone health. *Adv. Nutr.* **2015**, *6*, 260–266. [CrossRef] [PubMed]

58. Tan, X.; Cao, X.; Zou, J.; Shen, B.; Zhang, X.; Liu, Z.; Lv, W.; Teng, J.; Dign, X. Indoxyl sulfate, a valuable biomarker in chronic kidney disease and dialysis. *Hemodial. Int.* **2017**, *21*, 161–167. [CrossRef]

59. Vanholder, R.; Schepers, E.; Pletinck, A.; Nagler, E.V.; Glorieux, G. The uremic toxicity of indoxyl sulfate and p-cresyl sulfate: A systematic review. *J. Am. Soc. Nephrol.* **2014**, *25*, 1897–1907. [CrossRef]

60. Russell, W.R.; Gratz, S.W.; Duncan, S.H.; Holtrop, G.; Ince, J.; Scobbie, L.; Duncan, G.; Johnstone, A.M.; Lobley, G.E.; Wallace, R.J.; et al. High-protein, reduced-carbohydrate weight-loss diets promote metabolite profiles likely to be detrimental to colonic health. *Am. J. Clin. Nutr.* **2011**, *93*, 1062–1072. [CrossRef]

61. Long, S.J.; Jeffcoat, A.R.; Millward, D.J. Effect of habitual dietary-protein intake on appetite and satiety. *Appetite* **2000**, *35*, 79–88. [CrossRef]

62. Batterham, R.L.; Heffron, H.; Kapoor, S.; Chivers, J.E.; Chandarana, K.; Herzog, H.; Le Roux, C.W.; Thomas, E.L.; Bell, J.D.; Withers, D.J. Critical role for peptide YY in protein-mediated satiation and body-weight regulation. *Cell Metab.* **2006**, *4*, 223–233. [CrossRef] [PubMed]
63. Scheid, J.L.; De Souza, M.J. The Role of PYY in Eating Behavior and Diet. In *Handbook of Behavior, Food and Nutrition*; Preedy, V., Watson, R., Martin, C., Eds.; Springer: New York, NY, USA, 2011.
64. Karra, E.; Chandarana, K.; Batterham, R.L. The role of peptide YY in appetite regulation and obesity. *J. Physiol.* **2009**, *587*, 19–25. [CrossRef] [PubMed]
65. Blom, W.A.; Lluch, A.; Stafleu, A.; Vinoy, S.; Holst, J.J.; Schaafsma, G.; Hendriks, H.F. Effect of a high-protein breakfast on the postprandial ghrelin response. *Am. J. Clin. Nutr.* **2006**, *83*, 211–220. [CrossRef] [PubMed]
66. Campbell, M.D.; Gonzalez, J.T.; Rumbold, P.L.; Walker, M.; Shaw, J.A.; Stevenson, E.J.; West, D.J. Comparison of appetite responses to high- and low-glycemic index postexercise meals under matched insulinemia and fiber in type 1 diabetes. *Am. J. Clin. Nutr.* **2015**, *101*, 478–486. [CrossRef] [PubMed]
67. Clayton, D.J.; James, L.J. The effect of breakfast on appetite regulation, energy balance and exercise performance. *Proc. Nutr. Soc.* **2016**, *75*, 319–327. [CrossRef] [PubMed]
68. Clayton, D.J.; Stensel, D.J.; James, L.J. Effect of breakfast omission on subjective appetite, metabolism, acylated ghrelin and GLP-17-36 during rest and exercise. *Nutrition* **2016**, *32*, 179–185. [CrossRef]
69. Iepsen, E.W.; Lundgren, J.; Holst, J.J.; Madsbad, S.; Torekov, S.S. Successful weight loss maintenance includes long-term increased meal responses of GLP-1 and PYY3-36. *Eur. J. Endocrinol.* **2016**, *174*, 775–784. [CrossRef]
70. Sumithran, P.; Prendergast, L.A.; Delbridge, E.; Purcell, K.; Shulkes, A.; Kriketos, A.; Proietto, J. Long-term persistence of hormonal adaptations to weight loss. *N. Engl. J. Med.* **2011**, *365*, 1597–1604. [CrossRef]

 © 2018 by the authors. Licensee MDPI, Basel, Switzerland. This article is an open access article distributed under the terms and conditions of the Creative Commons Attribution (CC BY) license (http://creativecommons.org/licenses/by/4.0/).

Correction

Correction: Roberts et al. "Satiating Effect of High Protein Diets on Resistance-Trained Individuals in Energy Deficit" *Nutrients* 2019, 11(1), 56

Justin Roberts [1,*], Anastasia Zinchenko [2,3], Krishnaa Mahbubani [4], James Johnstone [1], Lee Smith [1], Viviane Merzbach [1], Miguel Blacutt [3], Oscar Banderas [3], Luis Villasenor [3], Fredrik T. Vårvik [3] and Menno Henselmans [3]

[1] Cambridge Centre for Sport and Exercise Sciences, School of Psychology and Sport Science, Anglia Ruskin University, East Road, Cambridge CB1 1PT, UK
[2] Kings College, Kings Parade, University of Cambridge, Cambridge CB2 1ST, UK
[3] International Scientific Research Foundation for Fitness and Nutrition, 1073 LC Amsterdam, The Netherlands
[4] Department of Surgery, Addenbrookes Hospital, Cambridge CB2 0QQ, UK
* Correspondence: Justin.roberts@anglia.ac.uk; Tel.: +44-1223-695-154

Received: 23 May 2019; Accepted: 3 July 2019; Published: 8 July 2019

The authors wish to make a correction to the published version of their paper [1]. In preparing a later manuscript, it was noted that the equations under Section 2.8 Biochemical Analyses should have reflected the post-meal hormone concentrations in relation to the pre-meal values (as undertaken during the analysis stage). Corrected versions of the relative change and relative change difference equations are shown below:

Relative change (relative Δ pg·mL^{-1}) $= \frac{y-x}{x}$ where x = the pre-meal resting sample, y = the post-meal at the respective sample time-points (i.e., 60, 120 min);

Relative change difference (relative Δ difference pg·mL^{-1}) $= \left(\frac{y_{post}-x_{post}}{x_{post}} \right) - \left(\frac{y_{pre}-x_{pre}}{x_{pre}} \right)$ where pre is the pre-intervention results, and post is the post-intervention results.

The data analysis undertaken previously already conforms to these corrected equations. We apologize for this change, which has no impact on the scientific outcomes or conclusions of the study. The original manuscript will remain online on the article webpage, with a reference to this correction.

Reference

1.   Roberts, J.; Zinchenko, A.; Mahbubani, K.; Johnstone, J.; Smith, L.; Merzbach, V.; Blacutt, M.; Banderas, O.; Villasenor, L.; Vårvik, F.; et al. Satiating Effect of High Protein Diets on Resistance-Trained Individuals in Energy Deficit. *Nutrients* **2019**, *11*, 56. [CrossRef]

© 2019 by the authors. Licensee MDPI, Basel, Switzerland. This article is an open access article distributed under the terms and conditions of the Creative Commons Attribution (CC BY) license (http://creativecommons.org/licenses/by/4.0/).

 *nutrients*

Article

The Effect of Branched-Chain Amino Acids, Citrulline, and Arginine on High-Intensity Interval Performance in Young Swimmers

Chun-Fang Hsueh [1], Huey-June Wu [2], Tzu-Shiou Tsai [3], Ching-Lin Wu [4] and Chen-Kang Chang [5],*

[1] Graduate Institute of Sport Coaching Science, Chinese Culture University, Taipei 114, Taiwan; swim_speedo@hotmail.com
[2] Department of Combat Sports and Martial Arts, Chinese Culture University, Taipei 114, Taiwan; wuhc0123@gmail.com
[3] Taipei Municipal Nan Gang High School, Taipei 115, Taiwan; Tsai481026@gmail.com
[4] Graduate Institute of Sports and Health Management, National Chung Hsing University, Taichung 402, Taiwan; psclw@dragon.nchu.edu.tw
[5] Department of Sport Performance, National Taiwan University of Sport, 16, Section 1, Shaun-Shih Road, Taichung 404, Taiwan
* Correspondence: wspahn@seed.net.tw; Tel.: +886-4-22213108 (ext. 2235)

Received: 29 October 2018; Accepted: 11 December 2018; Published: 14 December 2018

Abstract: High-intensity interval training has drawn significant interest for its ability to elicit similar training responses with less training volume compared to traditional moderate-intensity protocols. The purpose of this study was to examine the effect of co-ingestion of branched-chain amino acids (BCAA), arginine, and citrulline on 8 × 50 m high-intensity interval swim performance in trained young swimmers. This study used a randomized cross-over design. Eight male (age 15.6 ± 1.3 years) and eight female (age 15.6 ± 0.9 years) swimmers completed both amino acids (AA) and placebo (PL) trials. The participants ingested 0.085 g/kg body weight BCAA, 0.05 g/kg body weight arginine and 0.05 g/kg body weight citrulline before the swim test in the AA trial. The average 50 m time was significantly shorter in the AA trial than that in the PL trial. The AA trial was faster than the PL trial in the first, second, and the seventh laps. The AA trial showed significantly higher plasma BCAA concentrations and lower tryptophan/BCAA ratio. The other biochemical parameters and ratings of perceived exertion were similar between the two trials. The results showed that BCAA, arginine, and citrulline, allowed the participants to swim faster in a high-intensity interval protocol in young swimmers.

Keywords: central fatigue; tryptophan; ammonia; nitric oxide; stroke rate; stroke count

---

1. Introduction

High-intensity interval training (HIIT) has drawn significant interest from athletes of various sports, as well as general or less-fit populations [1,2]. The high-intensity nature of this protocol recruits both type-I and -II muscle fibers [3], resulting in significant improvement in cardiopulmonary and anaerobic capabilities [1,2,4,5]. Several studies have shown that HIIT can elicit similar training responses in competitive swimmers with less training volume compared to traditional moderate-intensity protocols [6,7]. One of the important factors for the success in HIIT is the ability to maintain the training intensity, especially at the later stages. However, the accumulation of peripheral and/or central fatigue, resulting from repeated high-intensity bouts, leads to declines in exercise intensity at the later stages of HIIT, and potentially reduce the training effect [1,2].

The peripheral factors for fatigue in HIIT include limitations in anaerobic and aerobic energy supply, and intramuscular accumulation of metabolic by-products such as H^+ and inorganic phosphate [8,9]. Moreover, the central nervous system may also be involved, as the capacity of the motor cortex to drive the knee extensors after high-intensity intermittent cycling was significantly decreased [10]. One of the mechanisms that contributes to central fatigue is the increase in cerebral concentration of the neurotransmitter serotonin (5-hydroxytryptamine) during exercise. The increased cerebral serotonin could lead to the feeling of fatigue and the loss of central drive and motivation [11]. An increase in serotonin concentration in presynaptic neutrons would lead to increased serotonin release and serotonin binding to postsynaptic receptors during nerve stimulation [12]. In addition, cerebral serotonin concentration was inversely correlated to running time to fatigue in rodents [13,14]. The rate of cerebral serotonin synthesis is regulated by the transport of free tryptophan, the precursor to serotonin, across the blood brain barrier [15]. Branched-chain amino acids (BCAA) have been hypothesized to alleviate central fatigue by competing with tryptophan in crossing the blood brain barrier through the L-system transporter [16]. Indeed, the decreased plasma free tryptophan/BCAA ratio would reduce the uptake of tryptophan, and subsequently, serotonin synthesis in the brain [14].

Nitric oxide (NO), a signaling molecule with a wide range of physiological functions, has been suggested to improve exercise performance by enhancing exercise-induced vasodilation [17], increasing the oxygenation status in the working muscles, and improving VO_2 kinetics [18]. Supplementations of arginine or citrulline, both of which are precursors to NO, have been suggested to improve performance in high-intensity exercise [19,20].

Previously, we revealed that supplementation of BCAA and arginine could improve repeated sprint performance in handball players [21]. We later added citrulline to the supplementation regimen [22,23] because the combined ingestion of citrulline and arginine may be more effective in increasing plasma arginine concentration than consuming either amino acid individually [24]. Although the nutritional strategies to support HIIT have been proposed, most results were obtained from land-based exercise [25]. In addition, the combination of BCAA, arginine, and citrulline on high-intensity interval swimming performance has not been investigated. With the increasing application of HIIT in swimming training, the aim of this study was to examine the effect of co-ingestion of BCAA, arginine, and citrulline on 8×50 m swim performance in trained young swimmers. The biochemical and stroke parameters were analyzed to investigate the potential mechanism of the supplementation.

2. Materials and Methods

2.1. Participants

Eight male (age: 15.6 ± 1.3 years; height: 1.74 ± 0.05 m; weight: 64.4 ± 7.6 kg) and eight female (age: 15.6 ± 0.9 years; height: 1.58 ± 0.06 m; weight: 54.0 ± 8.4 kg) swimmers were recruited from a high school in northern Taiwan. All participants have been participating in swimming training for at least 7 years and have competed at the national or international level. The characteristics and personal best performance of the participants are presented in Table 1. The exclusion criteria included cardiovascular disease risks, musculoskeletal injuries, smoking, or consumption of protein supplements in the past three months. After the experimental procedure and potential risks were explained, all participants and their legal guardians gave their written informed consent. The study protocol was approved by the Research Ethics Committee of China Medical University Hospital (CRREC-105-003).

Table 1. Basic characteristics of participants.

Gender *	Age (Year)	Height (m)	Weight (kg)	Body Mass Index	Best Style	Personal Best [†] 50 m (s)	Personal Best [†] 100 m (s)
M	15	1.78	67.1	21.18	Backstroke	29.8	62.2
M	17	1.78	71.1	22.44	Breaststroke	31.2	69.1
M	17	1.77	70.2	22.41	Front crawl	25.7	54.9
M	17	1.78	70.5	22.25	Front crawl	25.6	52.7
M	15	1.69	53.6	18.77	Front crawl	27.3	56.2
M	14	1.68	52.6	18.64	Butterfly	28.9	57.3
M	16	1.69	68.8	24.09	Butterfly	26.8	57.3
M	14	1.71	61.1	20.90	Front crawl	26.2	54.8
Mean [‡]	15.6	1.74	64.4	21.33			
SD [‡]	1.3	0.05	7.6	1.89			
F	16	1.46	46.3	21.72	Front crawl	29.7	65.2
F	15	1.59	45.4	17.96	Front crawl	29.8	66.1
F	17	1.60	51.4	20.08	Front crawl	28.2	59.9
F	16	1.64	60.9	22.64	Front crawl	28.0	61.2
F	15	1.53	44.1	18.84	Front crawl	29.5	61.1
F	16	1.58	56.5	22.63	Butterfly	31.3	69.2
F	16	1.65	60.9	22.37	Butterfly	33.5	71.8
F	14	1.60	66.6	26.02	Butterfly	31.7	68.9
Mean [§]	15.6	1.58	54.0	21.53			
SD [§]	0.9	0.06	8.4	2.55			

* M: male; F: female; [†] personal best record in their respective best style; [‡] mean and standard deviation of male participants; [§] mean and standard deviation of female participants.

2.2. Study Design

This study used a double-blind, placebo-controlled, randomized cross-over design. The study protocol is outlined in Figure 1. Each participant completed two trials, amino acids (AA) and placebo (PL), in a random order, separated by a wash-out period of at least seven days. The regular training schedule and dietary habits were maintained during the study period. The participants refrained from all training activity on the day prior to the trial. The study was conducted in January, 2017.

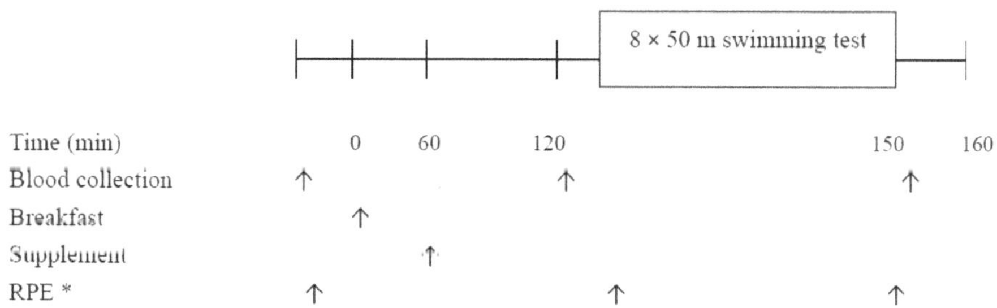

Figure 1. Study protocol. * RPE: ratings of perceived exertion. The symbols (↑) pointed out the actions at the specific time point.

2.3. Dietary Control

During the two days prior to each trial, the participants were provided with the same three meals per day, purchased from local convenience stores. The meals provided approximately

1624 kcal/day with 55% energy from carbohydrate, 25% from fat, and 18% from protein, according to the manufacturer's label. A standardized breakfast was given on the days of trials, including white bread 0.5 g/kg body weight, jam 0.1 g/kg body weight, butter 0.1 g/kg body weight, and soybean milk 5 mL/kg body weight (6.2 kcal/kg body weight, containing carbohydrate 1.0 g/kg body weight, protein 0.24 g/kg body weight, and fat 0.14 g/kg body weight).

2.4. Supplementation

On the days of the trials, the participants reported to a 25 m swimming pool in the morning after an overnight fast. After collecting blood samples from the antecubital vein, the participants consumed the standard breakfast, followed by one of the two interventions. In the AA trial, the participants ingested 0.085 g/kg body weight BCAA (leucine: isoleucine: valine = 10:7:3, containing vitamin E 6.67 IU/g BCAA, capsule, General Nutrition Corporation, Pittsburgh, PA, USA), 0.05 g/kg body weight arginine, and 0.05 g/kg body weight citrulline (arginine: citrulline = 1:1, tablet, General Nutrition Corporation). In the placebo (PL) trial, the participants consumed the identical number of empty capsules and tablets containing starch (Chung-Yu Biotech Co LTD, Taichung, Taiwan) and one capsule of vitamin E (100 IU, General Nutrition Corporation). All capsules and tablets were taken with water within 15 min.

2.5. High-Intensity Interval Swimming Test

Thirty min after consuming the amino acids or placebo, the participants began a controlled 1000 m warm-up. The warm-up included 8 × 50 m easy swim focusing on individual skills, 200 m mixed style, and 4 × 100 m free style. The warm-up lasted approximately 20 min. The 8 × 50 m high-intensity interval swimming test started 60 min after consuming the amino acids or placebo, following the international rules with push starts. There was a 3-min active recovery period between each sprint. The approximate work to rest ratio of 1:6 is chosen to allow the better recovery of acid/base balance and creatine phosphate resynthesis [2]. The participants were asked to swim with their best style with the best effort in each sprint. Four male and five female participants swam front crawl, two male and three female participants swam butterfly, one male participant swam breaststroke and another male participant swam backstroke. Participants from both trials were grouped into 3 or 4 according to their best record, and competed at the same time to encourage the best performance. No food or fluid was provided during the test. The ratings of perceived exertion (RPE) were recorded immediately before and after the test using Borg's 20-point scale [26].

2.6. Stroke Characteristics

The stroke characteristics were analyzed by an experienced swimmer/coach by reviewing the video files of the entire trials. The time to finish three consecutive strokes, measured by a stop watch, was recorded during approximately 35–40 m in each lap. The stroke rate was determined by dividing the time by three. For front crawl and backstroke, the timing started when the right hand entered the water, and ended immediately before the right hand entered the water for the second time. For breaststroke, the timing started when the arms started the outsweep, and ended when the arms completed the forward extension for the third time. For butterfly, the timing started when the arms entered the water, and ended immediately before the arms entered the water for the third time. The stroke count was measured during the entire lap, excluding the sculling or flipper movement after the start or turn.

2.7. Measurement of Blood Biochemical Parameters

The time point of blood sampling is shown in Figure 1. Venous blood samples were collected into tubes containing EDTA. After centrifugation at 1500× *g* for 15 min at 4 °C, the plasma samples were aliquoted and stored at −70 °C until further analysis. Plasma BCAA concentrations were measured enzymatically (Biovision, Milpitas, CA, USA) with a microplate spectrophotometer (Benchmark Plus, Bio-Rad, Hercules, CA, USA). Plasma tryptophan concentrations were analyzed with a fluorescence assay (Bridge-It, Mediomics, St. Louis, MO, USA). The fluorescence at excitation 485 nm and emission

665 nm was read by a microplate fluorescence reader (Plate Chameleon, Hidex, Turku, Finland). Plasma NO_x concentrations were determined using Griess reagent [27]. Plasma concentrations of urea, glucose, lactate, NH_3, glycerol, and non-esterified fatty acids (NEFA) were measured enzymatically with an automatic analyzer (Hitachi 7020, Tokyo, Japan) using commercial kits (Randox, Antrim, UK). Hemoglobin concentration and hematocrit in whole blood were measured by a blood cell analyzer (Sysmex Kx-21, Diamond Diagnostics, Holliston, MA, USA) to correct potential changes in plasma volume during the study periods [28].

2.8. Statistical Analysis

The results were initially analyzed by three-way (gender × trial × time) analysis of variance with repeated measurements. However, gender effect was insignificant in all performance, biochemical and stroke parameters. Therefore, the data from both genders were pulled together and analyzed by two-way (trial × time) analysis of variance with repeated measurements. If the interaction effect was to be found significant, the differences between the two trials were identified by Ryan-Holm-Bonferroni post hoc analysis [29]. If the time or lap effect was to be found significant, the differences within the same trial were identified with Bonferroni post hoc analysis. A $p < 0.05$ is considered statistically significant. With the power of 0.80 and sample size of 16, the minimal detectable difference is 1.06 standard deviation (SD). The data are presented as mean ± SD.

3. Results

The average 50 m time of the eight laps in the high-intensity interval swimming test was significantly shorter in the AA trial than that in the PL trial (AA: 30.50 ± 2.87 s vs. PL: 30.94 ± 3.02 s, $p < 0.001$, Figure 2a). Thirteen participants (6 male and 7 female) out of total 16 had faster average lap time in the AA trial (Figure 2a). When each lap was considered separately, there were significant trial, lap, and interaction effects (all $p < 0.001$, $\eta^2 = 0.795, 0.452, 0.236$, respectively). The post-hoc analysis showed that AA trial was faster than the PL trial in the first (AA: 30.47 ± 0.44 s vs. PL: 31.34 ± 0.57 s, $p < 0.001$), second (AA: 32.76 ± 2.54 s vs. PL: 33.58 ± 2.48 s, $p < 0.001$), and the seventh laps (AA: 31.64 ± 4.30 s vs. PL: 32.50 ± 4.60 s, $p < 0.001$) (Figure 2b).

The AA trial showed significantly higher plasma BCAA concentrations before and after the 8×50 m test (Figure 3a). Plasma tryptophan concentrations were decreased after the interval swim in both trials at the similar magnitude (Figure 3b). Nevertheless, the elevated plasma BCAA concentrations lead to significantly lower tryptophan/BCAA ratios before and after the interval swim in the AA trial (Figure 3c). Plasma NO_x levels showed a significant time effect ($p < 0.001$), but the post hoc analysis did not find significant difference among the three time points in either trials (Table 2). Plasma concentrations of NH_3, urea, lactate, glycerol, and NEFA were similar in both trials (Table 2). The participants also reported similar RPE before and after the 8×50 m test in both trials (Table 2).

(A)

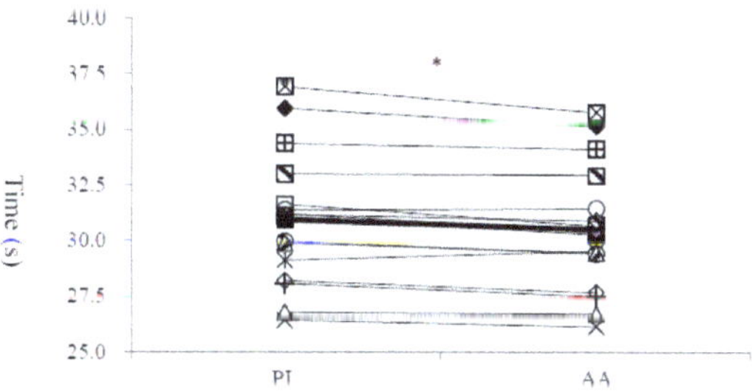

Figure 2. *Cont.*

(B)

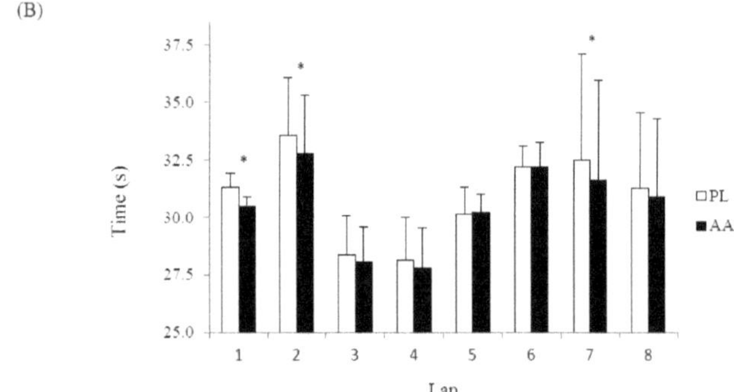

Figure 2. The performance of the 8 × 50 m high-intensity interval swimming test in the AA (amino acids supplemented) and PL (placebo) trials. (**A**) the average lap time of each participant.* Significant difference between the AA and PL trials, $p < 0.05$; (**B**) the average time in each lap. Trial effect: $p < 0.001$; lap effect: $p < 0.001$; interaction effect: $p < 0.001$; * Significant difference between the AA and PL trials, $p < 0.05$.

(A)

(B)

(C)

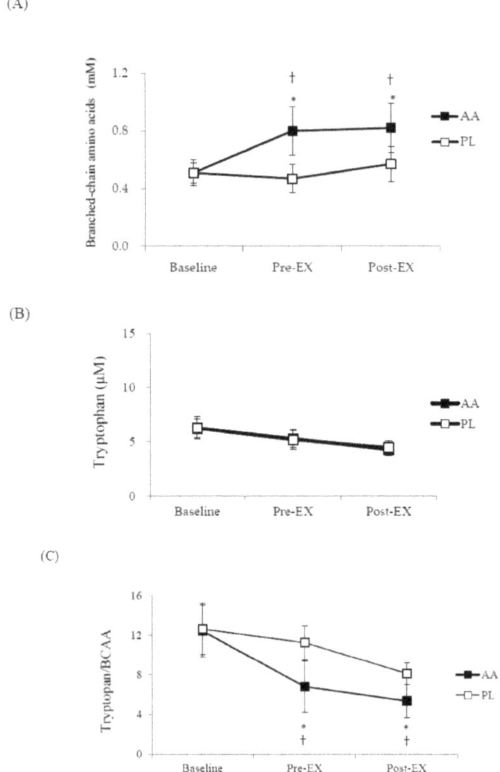

Figure 3. Plasma amino acid concentrations at baseline, before, and after the 8 × 50 m high-intensity interval swimming test in the AA (amino acids supplemented) and PL (placebo) trials. (**A**) Branched-chain amino acids, trial effect: $p < 0.001$; time effect: $p < 0.001$; interaction effect: $p < 0.001$. (**B**) Tryptophan, trial effect: $p = 0.821$; time effect: $p < 0.001$; interaction effect: $p = 0.383$. (**C**) Tryptophan/branched-chain amino acids ratio, trial effect: $p < 0.001$; time effect: $p < 0.0001$; interaction effect: $p < 0.001$; * Significantly different from the baseline in the same trial, $p < 0.05$; † AA trial vs. PL trial at the same time point, $p < 0.05$. Pre-Ex: before 8 × 50 m swim test; Post-Ex: immediately after 8 × 50 m swim test.

Table 2. Biochemical parameters and ratings of perceived exertion at the baseline, before, and after 8 × 50 m test in the in the AA (amino acids supplemented) and PL (placebo) trials.

	Trial	Baseline	Pre-Ex	Post-Ex
NO_x (µM)	AA	11.61 ± 4.94	31.65 ± 29.56	23.06 ± 15.28
	PL	12.68 ± 7.42	25.84 ± 29.39	18.36 ± 11.71
NH_3 (µM)	AA	115.20 ± 103.49	123.72 ± 66.29	123.86 ± 65.35
	PL	116.59 ± 40.69	109.02 ± 50.89	114.81 ± 50.31
Urea (mM)	AA	4.64 ± 0.77	5.38 ± 0.90	5.11 ± 0.91
	PL	4.77 ± 0.65	5.01 ± 0.87	4.63 ± 0.73
Lactate (mM)	AA	1.36 ± 0.56	1.91 ± 0.59	14.75 ± 4.43 *,[†]
	PL	1.43 ± 0.34	1.71 ± 0.38	14.12 ± 2.79 *,[†]
Glycerol (µM)	AA	46.00 ± 21.00	81.89 ± 28.83 *	240.89 ± 96.61 *,[†]
	PL	44.63 ± 25.27	71.35 ± 26.45	227.44 ± 58.36 *,[†]
NEFA (mM) [1]	AA	0.49 ± 0.27	0.42 ± 0.16	0.22 ± 0.09 [†]
	PL	0.44 ± 0.25	0.38 ± 0.10	0.20 ± 0.08 [†]
RPE [2]	AA	9.1 ± 2.4	15.0 ± 1.8 *	17.2 ± 2.3 *
	PL	9.7 ± 2.4	14.8 ± 2.2 *	17.3 ± 1.8 *

[1] NEFA. non-esterified fatty acids; [2] RPE: ratings of perceived exertion; * Significantly different from the baseline in the same trial, $p < 0.001$; [†] Significantly different from Pre-Ex in the same trial, $p < 0.05$. Pre-Ex: before 8 × 50 m swim test; Post-Ex: immediately after 8 × 50 m swim test.

The data of stroke rate and stroke count are shown in Table 3. None of the main effects was significant in stroke rate. However, the participants in the AA trial showed a trend of faster stroke rate in the first ($p = 0.011$) and second lap ($p = 0.011$ and 0.002, respectively, according to paired t test), compared to those in the PL trial. The last lap showed the highest stroke count in both trials. In the AA trial, the participants swam the last lap with significantly higher stroke count than in the third lap ($p = 0.015$), while in the PL trial, the stroke count in the last lap was significantly higher than in the third ($p = 0.038$) and fourth lap ($p = 0.041$).

Table 3. Stroke rate and stroke count in the 8×50 m high-intensity interval swimming test in the AA and PL trials.

Trial		Lap								
		1	2	3	4	5	6	7	8	Mean
Stroke rate [1]	PL	70.6 ± 19.5	71.2 ± 19.3	72.3 ± 19.0	71.4 ± 20.2	70.2 ± 18.7	69.2 ± 18.5	69.4 ± 18.2	69.4 ± 18.3	70.5 ± 19.0
(count/min)	AA	74.5 ± 18.5	76.0 ± 18.3	72.7 ± 19.5	72.4 ± 18.4	70.7 ± 16.5	71.4 ± 18.0	70.1 ± 17.4	71.9 ± 18.2	72.7 ± 18.1
Stroke count [2]	PL	30.7 ± 9.0 [a,b]	30.6 ± 9.5 [a,b]	30.8 ± 9.1 [a]	30.8 ± 9.2 [a,b]	31.1 ± 9.4 [a,b]	31.7 ± 9.5 [a,b]	31.6 ± 9.4 [a,b]	31.9 ± 9.0 [b]	31.1 ± 9.3
(count/lap)	AA	31.1 ± 8.5 [a,b]	30.6 ± 8.2 [a,b]	30.6 ± 8.2 [a]	31.1 ± 8.6 [a]	31.2 ± 8.3 [a,b]	31.3 ± 8.1 [a,b]	31.5 ± 8.4 [a,b]	32.1 ± 8.4 [b]	31.2 ± 8.3

[1] Trial effect $p = 0.102$, lap effect $p = 0.489$, interaction effect $p = 0.513$; [2] Trial effect $p = 0.910$, lap effect $p < 0.001$, interaction effect $p = 0.643$; The values with different superscripts (a, b) were significantly different within the same trial.

4. Discussion

The results of this study suggested that the combined supplementation of BCAA, arginine, and citrulline significantly improved performance in 8 × 50 m high-intensity interval swims in well-trained young swimmers. The participants in the AA trial showed significantly lower plasma tryptophan/BCAA ratio before and after the interval swims. In addition, the participants were able to swim faster under a similar degree of perceived effort, indicating the possibility of alleviated central fatigue.

The participants in the AA trial showed significantly higher plasma BCAA concentrations and lower tryptophan/BCAA ratio after the interval swims, compared to the PL trial. This is in agreement with our previous studies using the same supplements [22,23]. It has been revealed that central nervous system plays a role in the development of fatigue in repeated high-intensity exercise [9]. Several studies using functional magnetic resonance imaging have found increased activations in sensory processing and motor-related brain regions such as primary motor cortex, supplementary motor area and pre-motor cortex while performing fatiguing exercise tasks [30]. These activations suggest a greater perception of effort under fatigue and the need for higher motor output to sustain the same physical workload [31]. The decreased plasma tryptophan/BCAA ratio in the AA trial would reduce the cerebral uptake of tryptophan, leading to lower cerebral serotonin synthesis [14].

Although previous studies have reported that oral supplementation of BCAA could reduce RPE and mental fatigue in maximal exercise in untrained participants [32], our results suggest that the participants in the AA trial had better performance under the same level of RPE. This contradiction may result from the different protocols to measure physical performance. Blomstrand et al. used a fixed-rate protocol, followed by a 20-min maximal exercise on a bicycle treadmill [32]. On the other hand, the time-trial style of the present protocol required the participants to swim at their best effort in each lap. Therefore, the participants in both trials reached similar RPE, which is in agreement with our previous study [33].

A recent study revealed that a short-term citrulline supplementation could increase peak power output by 9% and total power output by 7% in 60-s all-out sprint that followed the 6-min bout of severe-intensity exercise [20]. In addition, a single dose of citrulline malate could improve performance in repeated high-intensity anaerobic resistance exercises [34]. Acute beetroot juice supplementation, which is rich in citrulline, also improved peak and mean power output, while reducing the time required to reach peak power output in the Wingate test [35]. A recent review summarized that acute beetroot juice could improve the performance in repeated high-intensity exercise by increasing phosphocreatine resynthesis and muscle shortening velocity [36]. These ergogenic effects could be mediated by the increased tissue oxygenation and improved O_2 kinetics. In the first and second lap during which central fatigue may not have been accumulated, it is possible that ergogenic effect was partially the result of the role of citrulline in improving short-term anaerobic performance. At the later stage of repeated high-intensity exercise, the aerobic system becomes an important energy source despite the all-out effort [9]. For example, the contribution of aerobic energy increased from 25% in the first sprint to approximately 50% in the second to forth sprints in 30 s × 4 swimming with 30 s rest in between [37]. Our participants also showed significantly elevated plasma glycerol concentrations after 8 × 50 m swim (Table 2), indicating greater lipolysis. The increased phosphocreatine resynthesis and muscle work efficiency, combining with delayed central fatigue, may lead to the better performance in the later stage of the high-intensity interval swim.

Our results showed that plasma ammonia and urea were similar between the two trials. In addition, ammonia concentrations remained unchanged after exercise in both trials, indicating that BCAA oxidation was not significantly increased in the AA trial. It is possible that the breakfast and the appropriate warm up before 8 × 50 m swim had already increased plasma ammonia levels. This is in agreement with Peyrebrune et al., who reported that plasma ammonia concentration was decreased after 8 × 50 yards sprint swimming [38]. These results agree with the notion that the major source of ATP turnover comes from aerobic metabolism, rather than adenosine monophosphate (AMP) deamination, towards the end of the repeated swims.

In previous studies we have shown that 0.17 g/kg body weight BCAA, twice of the dosage in the present study, could delay the development of central fatigue and prevent the decline in perceptual-motor functions [22,23]. Although the dosage can be tolerated, the large number of capsules may discourage some athletes from following this supplement regime. The lower dosage of 0.085 g/kg body weight in this study still resulted in an average plasma BCAA concentration of 0.8 mM prior to exercise, which was similar or slightly lower to the previous results [21–23]. The combined dosage of citrulline and arginine is 0.1 g/kg body weight, which is in line with previous studies using acute citrulline supplementations [22,23,39].

We specifically use the acute supplementation protocol for the purpose of creating higher plasma BCAA concentration during exercise to compete with free tryptophan in cross blood brain barrier. The longer-term of BCAA supplementation would not lead to significantly higher plasma concentrations compared to the acute regimen. In addition, it has been shown that acute supplements of citrulline-malate or other nitrate sources 40 min to 2.5 h prior to exercise could improve performance and lead to lower steady-state VO_2 at the same intensity [40–42], while several days of beetroot juice supplementation did not elicit greater improvements [40]. Therefore, the acute supplementation regime used in this study would be sufficient for the ergogenic effect.

HIIT has been reported to elicit similar, or even better effects on performance compared to high-volume lower-intensity training in competitive swimmers [6,43]. The shorter duration and lower training volume make HIIT ideal for most athletes and general populations who are living on a tight schedule. Most HIIT in swimming includes repeated sprints of 10 to 30 s duration with resting intervals of 2 to 5 min [44]. Our study adopted this protocol so that the results can be applied to practical training situations. It appears that our participants can maintain speed throughout the present protocol with 3-min recovery between each lap, whereas those with shorter (60 s) recovery periods resulted in significant decrements in performance toward the end [45]. Plasma lactate values in the present study are in line with others using similar protocols in swimmers [37,46], and similar to that after a 100 m swimming competition [47].

Swimming velocity is the product of stroke rate, the number of stroke cycle per min, and stroke length, the distance travelled with each stroke cycle [48]. Elite swimmers usually increase stroke rate, while ignoring stroke length, in order to increase or maintain speed when fatigue is accumulating [49]. In the present study, the participants in the AA trial swam significantly faster in the first and second lap with a trend of higher stroke rate. It is possible that the supplements provided higher central drive in the early stages of 8 × 50 m swim. In the seventh lap, the stroke rate is similar between the two trials. Therefore, the faster speed may result from the longer stroke length, indicating higher muscle strength or technical efficiency in the AA trial during the lap. In the third and fourth laps, the participants swam the fastest, while using the lowest stroke count. This suggests that the participants swam with the highest efficiency in the middle of the 8 × 50 m test.

One of the limitations to this study is the low energy provided in the standard meals during the two days prior to the trial. The meal boxes provided were similar to the participants' usual diet, both in terms of energy content and food choice. The participants may have had relatively low muscle glycogen levels on the morning of the trials. However, the breakfast before the trial, containing 1.0 g/kg body weight carbohydrate, would ensure the euglycemic state throughout the test. Another limitation is the relatively short exercise time. Whether the ergogenic effect shown in this study can be extended to the entire training period requires further investigation. The lack of data in oxygen consumption and phosphocreatine concentration in the muscle also prevented us from gaining further understanding of the mechanisms.

5. Conclusions

The present results showed that BCAA, arginine, and citrulline allowed young participants to swim faster in a 8 × 50 m high-intensity interval protocol while feeling the same level of effort. Future research may focus on the modifications in training load to utilize the enhanced physiological and psychological mechanisms associated with these supplements.

Author Contributions: Conceptualization, C.-K.C. and C.-L.W.; Methodology, C.-K.C., H.-J.W. and C.-F.H.; Data acquisition and analysis, C.-F.H. and T.-S.T.; Interpretation of data, C.-K.C., T.-S.T. and C.-F.H.; Funding acquisition, C.-K.C.; Manuscript preparation: C.-K.C. and C.-L.W.

Funding: This research was funded by the Ministry of Science and Technology, Taiwan, grant number MOST 105-2410-H-028-005.

Acknowledgments: The authors thank Yu-Fang Huang for her technical assistance.

Conflicts of Interest: The authors declare no conflict of interest.

References

1. Buchheit, M.; Laursen, P.B. High-intensity interval training, solutions to the programming puzzle: Part I: Cardiopulmonary emphasis. *Sports Med.* **2013**, *43*, 313–338. [CrossRef] [PubMed]

2. Buchheit, M.; Laursen, P.B. High-intensity interval training, solutions to the programming puzzle. Part II: Anaerobic energy, neuromuscular load and practical applications. *Sports Med.* **2013**, *43*, 927–954. [CrossRef] [PubMed]

3. Altenburg, T.M.; Degens, H.; van Mechelen, W.; Sargeant, A.J.; de Haan, A. Recruitment of single muscle fibers during submaximal cycling exercise. *J. Appl. Physiol. (1985)* **2007**, *103*, 1752–1756. [CrossRef] [PubMed]

4. Garcia-Hermoso, A.; Cerrillo-Urbina, A.J.; Herrera-Valenzuela, T.; Cristi-Montero, C.; Saavedra, J.M.; Martinez-Vizcaino, V. Is high-intensity interval training more effective on improving cardiometabolic risk and aerobic capacity than other forms of exercise in overweight and obese youth? A meta-analysis. *Obes. Rev.* **2016**, *17*, 531–540. [CrossRef]

5. Midgley, A.W.; McNaughton, L.R.; Carroll, S. Reproducibility of time at or near VO2max during intermittent treadmill running. *Int. J. Sports Med.* **2007**, *28*, 40–47. [CrossRef] [PubMed]

6. Faude, O.; Meyer, T.; Scharhag, J.; Weins, F.; Urhausen, A.; Kindermann, W. Volume vs. intensity in the training of competitive swimmers. *Int. J. Sports Med.* **2008**, *29*, 906–912. [CrossRef] [PubMed]

7. Sperlich, B.; Zinner, C.; Heilemann, I.; Kjendlie, P.L.; Holmberg, H.C.; Mester, J. High-intensity interval training improves VO(2peak), maximal lactate accumulation, time trial and competition performance in 9-11-year-old swimmers. *Eur. J. Appl. Physiol.* **2010**, *110*, 1029–1036. [CrossRef] [PubMed]

8. Spencer, M.; Bishop, D.; Dawson, B.; Goodman, C. Physiological and metabolic responses of repeated-sprint activities:specific to field-based team sports. *Sports Med.* **2005**, *35*, 1025–1044. [CrossRef] [PubMed]

9. Girard, O.; Mendez-Villanueva, A.; Bishop, D. Repeated-sprint ability—Part I: Factors contributing to fatigue. *Sports Med.* **2011**, *41*, 673–694. [CrossRef]

10. Sidhu, S.K.; Bentley, D.J.; Carroll, T.J. Locomotor exercise induces long-lasting impairments in the capacity of the human motor cortex to voluntarily activate knee extensor muscles. *J. Appl. Physiol. (1985)* **2009**, *106*, 556–565. [CrossRef]

11. Davis, J.M.; Bailey, S.P. Possible mechanisms of central nervous system fatigue during exercise. *Med. Sci. Sports Exerc.* **1997**, *29*, 45–57. [CrossRef] [PubMed]

12. Newsholme, E.A.; Blomstrand, E. Branched chain amino acids and central fatigue. *J. Nutr.* **2006**, *136*, 274S–276S. [CrossRef] [PubMed]

13. Soares, D.D.; Coimbra, C.C.; Marubayashi, U. Tryptophan-induced central fatigue in exercising rats is related to serotonin content in preoptic area. *Neurosci. Lett.* **2007**, *415*, 274–278. [CrossRef] [PubMed]

14. Yamamoto, T.; Azechi, H.; Board, M. Essential role of excessive tryptophan and its neurometabolites in fatigue. *Can. J. Neurol. Sci.* **2012**, *39*, 40–47. [CrossRef] [PubMed]

15. Pardridge, W.M. Blood-brain barrier carrier-mediated transport and brain metabolism of amino acids. *Neurochem. Res.* **1998**, *23*, 635–644. [CrossRef] [PubMed]

16. Fernstrom, J.D. Branched-chain amino acids and brain function. *J. Nutr.* **2005**, *135*, 1539S–1546S. [CrossRef] [PubMed]

17. Brown, M.D.; Srinivasan, M.; Hogikyan, R.V.; Dengel, D.R.; Glickman, S.G.; Galecki, A.; Supiano, M.A. Nitric oxide biomarkers increase during exercise-induced vasodilation in the forearm. *Int. J. Sports Med.* **2000**, *21*, 83–89. [CrossRef] [PubMed]

18. Bailey, S.J.; Vanhatalo, A.; Winyard, P.G.; Jones, A.M. The nitrate-nitrite-nitric oxide pathway: Its role in human exercise physiology. *Eur. J. Sport Sci.* **2012**, *12*, 309–320. [CrossRef]

19. Yavuz, H.U.; Turnagol, H.; Demirel, A.H. Pre-exercise arginine supplementation increases time to exhaustion in elite male wrestlers. *Biol. Sport* **2014**, *31*, 187–191. [CrossRef] [PubMed]

20. Bailey, S.J.; Blackwell, J.R.; Lord, T.; Vanhatalo, A.; Winyard, P.G.; Jones, A.M. l-Citrulline supplementation improves O2 uptake kinetics and high-intensity exercise performance in humans. *J. Appl. Physiol. (1985)* **2015**, *119*, 385–395. [CrossRef]

21. Chang, C.K.; Chang Chien, K.M.; Chang, J.H.; Huang, M.H.; Liang, Y.C.; Liu, T.H. Branched-chain amino acids and arginine improve performance in two consecutive days of simulated handball games in male and female athletes: A randomized trial. *PLoS ONE* **2015**, *10*, e0121866. [CrossRef] [PubMed]

22. Yang, C.C.; Wu, C.L.; Chen, I.F.; Chang, C.K. Prevention of perceptual-motor decline by branched-chain amino acids, arginine, citrulline after tennis match. *Scand. J. Med. Sci. Sports* **2017**, *29*, 935–944. [CrossRef] [PubMed]

23. Chen, I.F.; Wu, H.J.; Chen, C.Y.; Chou, K.M.; Chang, C.K. Branched-chain amino acids, arginine, citrulline alleviate central fatigue after 3 simulated matches in taekwondo athletes: A randomized controlled trial. *J. Int. Soc. Sports Nutr.* **2016**, *13*, 28. [CrossRef] [PubMed]

24. Suzuki, T.; Morita, M.; Hayashi, T.; Kamimura, A. The effects on plasma L-arginine levels of combined oral L-citrulline and L-arginine supplementation in healthy males. *Biosci. Biotechnol. Biochem.* **2016**, 1–4. [CrossRef] [PubMed]

25. Naderi, A.; Earnest, C.P.; Lowery, R.P.; Wilson, J.M.; Willems, M.E. Co-ingestion of Nutritional Ergogenic Aids and High-Intensity Exercise Performance. *Sports Med.* **2016**, *46*, 1407–1418. [CrossRef] [PubMed]

26. Borg, G.A. Psychophysical bases of perceived exertion. *Med. Sci. Sports Exerc.* **1982**, *14*, 377–381. [CrossRef] [PubMed]

27. Green, L.C.; Wagner, D.A.; Glogowski, J.; Skipper, P.L.; Wishnok, J.S.; Tannenbaum, S.R. Analysis of nitrate, nitrite, and [15N] nitrate in biological fluids. *Anal. Biochem.* **1982**, *126*, 131–138. [CrossRef]

28. Costill, D.L.; Fink, W.J. Plasma volume changes following exercise and thermal dehydration. *J. Appl. Physiol.* **1974**, *37*, 521–525. [CrossRef]

29. Atkinson, G. Analysis of repeated measurements in physical therapy research: Multiple comparisons amongst level means and multi-factorial designs. *Phys. Ther. Sport* **2002**, *3*, 191–203. [CrossRef]

30. Tan, X.R.; Low, I.C.C.; Stephenson, M.C.; Soong, T.W.; Lee, J.K.W. Neural Basis of Exertional Fatigue in the Heat: A Review of Magnetic Resonance Imaging Methods. *Scand. J. Med. Sci. Sports* **2017**. [CrossRef]

31. Benwell, N.M.; Mastaglia, F.L.; Thickbroom, G.W. Changes in the functional MR signal in motor and non-motor areas during intermittent fatiguing hand exercise. *Exp. Brain Res.* **2007**, *182*, 93–97. [CrossRef] [PubMed]

32. Blomstrand, E.; Hassmen, P.; Ek, S.; Ekblom, B.; Newsholme, E.A. Influence of ingesting a solution of branched-chain amino acids on perceived exertion during exercise. *Acta Physiol. Scand.* **1997**, *159*, 41–49. [CrossRef]

33. Cheng, I.S.; Wang, Y.W.; Chen, I.F.; Hsu, G.S.; Hsueh, C.F.; Chang, C.K. The supplementation of branched-chain amino acids, arginine, and citrulline improves endurance exercise performance in two consecutive days. *J. Sports Sci. Med.* **2016**, *15*, 509–515. [PubMed]

34. Perez-Guisado, J.; Jakeman, P.M. Citrulline malate enhances athletic anaerobic performance and relieves muscle soreness. *J. Strength Cond. Res.* **2010**, *24*, 1215–1222. [CrossRef] [PubMed]

35. Cuenca, E.; Jodra, P.; Perez-Lopez, A.; Gonzalez-Rodriguez, L.G.; Fernandes da Silva, S.; Veiga-Herreros, P.; Dominguez, R. Effects of Beetroot Juice Supplementation on Performance and Fatigue in a 30-s All-Out Sprint Exercise: A Randomized, Double-Blind Cross-Over Study. *Nutrients* **2018**, *10*. [CrossRef]

36. Dominguez, R.; Mate-Munoz, J.L.; Cuenca, E.; Garcia-Fernandez, P.; Mata-Ordonez, F.; Lozano-Estevan, M.C.; Veiga-Herreros, P.; da Silva, S.F.; Garnacho-Castano, M.V. Effects of beetroot juice supplementation on intermittent high-intensity exercise efforts. *J. Int. Soc. Sports Nutr.* **2018**, *15*, 2. [CrossRef]

37. Peyrebrune, M.C.; Toubekis, A.G.; Lakomy, H.K.; Nevill, M.E. Estimating the energy contribution during single and repeated sprint swimming. *Scand. J. Med. Sci. Sports* **2014**, *24*, 369–376. [CrossRef]

38. Peyrebrune, M.C.; Nevill, M.E.; Donaldson, F.J.; Cosford, D.J. The effects of oral creatine supplementation on performance in single and repeated sprint swimming. *J. Sports Sci.* **1998**, *16*, 271–279. [CrossRef]
39. Cunniffe, B.; Papageorgiou, M.; O'Brien, B.; Davies, N.A.; Grimble, G.K.; Cardinale, M. Acute Citrulline-Malate Supplementation and High-Intensity Cycling Performance. *J. Strength Cond. Res.* **2016**, *30*, 2638–2647. [CrossRef]
40. Vanhatalo, A.; Bailey, S.J.; Blackwell, J.R.; DiMenna, F.J.; Pavey, T.G.; Wilkerson, D.P.; Benjamin, N.; Winyard, P.G.; Jones, A.M. Acute and chronic effects of dietary nitrate supplementation on blood pressure and the physiological responses to moderate-intensity and incremental exercise. *Am. J. Physiol. Regul. Integr. Comp. Physiol.* **2010**, *299*, R1121–R1131. [CrossRef]
41. Glenn, J.M.; Gray, M.; Jensen, A.; Stone, M.S.; Vincenzo, J.L. Acute citrulline-malate supplementation improves maximal strength and anaerobic power in female, masters athletes tennis players. *Eur. J. Sport Sci.* **2016**, *16*, 1095–1103. [CrossRef] [PubMed]
42. Larsen, F.J.; Weitzberg, E.; Lundberg, J.O.; Ekblom, B. Dietary nitrate reduces maximal oxygen consumption while maintaining work performance in maximal exercise. *Free Radic. Biol. Med.* **2010**, *48*, 342–347. [CrossRef] [PubMed]
43. Kilen, A.; Larsson, T.H.; Jorgensen, M.; Johansen, L.; Jorgensen, S.; Nordsborg, N.B. Effects of 12 weeks high-intensity & reduced-volume training in elite athletes. *PLoS ONE* **2014**, *9*, e95025. [CrossRef]
44. Maglischo, E.W. *Swimming Fastest*; Human Kinetics: Champaign, IL, USA, 2003.
45. Theodorou, A.S.; Havenetidis, K.; Zanker, C.L.; O'Hara, J.P.; King, R.F.; Hood, C.; Paradisis, G.; Cooke, C.B. Effects of acute creatine loading with or without carbohydrate on repeated bouts of maximal swimming in high-performance swimmers. *J. Strength Cond. Res.* **2005**, *19*, 265–269. [CrossRef] [PubMed]
46. Toubekis, A.G.; Adam, G.V.; Douda, H.T.; Antoniou, P.D.; Douroundos, I.I.; Tokmakidis, S.P. Repeated sprint swimming performance after low- or high-intensity active and passive recoveries. *J. Strength Cond. Res.* **2011**, *25*, 109–116. [CrossRef] [PubMed]
47. Avlonitou, E. Maximal lactate values following competitive performance varying according to age, sex and swimming style. *J. Sports Med. Phys. Fitness* **1996**, *36*, 24–30.
48. Craig, A.B., Jr.; Pendergast, D.R. Relationships of stroke rate, distance per stroke, and velocity in competitive swimming. *Med. Sci. Sports* **1979**, *11*, 278–283. [CrossRef]
49. Craig, A.B., Jr.; Skehan, P.L.; Pawelczyk, J.A.; Boomer, W.L. Velocity, stroke rate, and distance per stroke during elite swimming competition. *Med. Sci. Sports Exerc.* **1985**, *17*, 625–634. [CrossRef]

© 2018 by the authors. Licensee MDPI, Basel, Switzerland. This article is an open access article distributed under the terms and conditions of the Creative Commons Attribution (CC BY) license (http://creativecommons.org/licenses/by/4.0/).

Article

Evaluation of Dietary Supplement Use in Wheelchair Rugby Athletes

Robyn F. Madden [1], Jane Shearer [1], David Legg [2] and Jill A. Parnell [2,*]

[1]   Department of Kinesiology, University of Calgary, Calgary, AB T2N 1N4, Canada;
      rmadd656@mtroyal.ca (R.F.M.); jshearer@ucalgary.ca (J.S.)
[2]   Department of Health and Physical Education, Mount Royal University, Calgary, AB T3E 6K6, Canada;
      dlegg@mtroyal.ca
*   Correspondence: jparnell@mtroyal.ca; Tel.: +1-403-440-8672

Received: 15 November 2018; Accepted: 6 December 2018; Published: 11 December 2018

Abstract: Wheelchair rugby is a rapidly growing Paralympic sport; however, research remains predominantly in the realms of physiology and biomechanics. Currently, there is little investigation into nutrition and dietary supplement use among wheelchair rugby athletes (WRA). The aim of this study was to assess the types of dietary supplements (DS) used, the prevalence of usage, and the reasons for use among WRA. The secondary aim was to report utilized and preferred sources of nutritional information among this population. A valid, reliable Dietary Supplement Questionnaire was used to report supplement use and reasons for use. Male ($n = 33$) and female ($n = 9$) WRA were recruited at a national tournament and through emailing coaches of various Canadian teams. Dietary supplement usage was prevalent as 90.9% of males and 77.8% of females reported usage within the past three months with the most regularly used supplements being vitamin D (26.2%), electrolytes (19.5%), and protein powder (19.5%). The most common reason for usage was performance. The top sources of nutrition information were dietitian/nutritionist and the internet. Further investigation into DS use is needed to help create nutritional guidelines that are accessible to WRA and athletes with disabilities in general.

Keywords: wheelchair rugby; Paralympic athlete; quadriplegic athletes; dietary supplements

---

1. Introduction

Originally named "Murderball", wheelchair rugby is the most rapidly growing wheelchair sport worldwide [1]. Invented in Canada during the 1970s [2], the sport provides a Paralympic alternative to wheelchair basketball and track and field events, since the latter sports were dominated by more mobile paraplegic athletes versus those with quadriplegia, which wheelchair rugby is designed for [1]. The sport is comprised of intense physicality, aggression, and full body contact [1,3] while requiring efficient wheelchair manoeuvrability skills [4]. Wheelchair rugby differs from other wheelchair sports in that players must be tetraplegic also known as quadriplegic [5]. Furthermore, participants are classified based on impairment level, ranging from 0.5 (most impaired) to 3.5 (least impaired) [6] and the classification points of a team on the court cannot exceed eight points [2]. Wheelchair rugby is considered an intermittent sport and is played in eight minute quarters [6]. It is recognized by the International Paralympic Committee as a summer sport [7] and has held medal status in the Paralympic Games since 2000 [2]. Athletes who compete in wheelchair rugby typically have a spinal cord injury, whereas, athletes in Paralympic sport, in general, can include other disability sports and disability types.

Despite the growth in popularity of Paralympic sport [8], and wheelchair rugby in particular, research has predominantly occurred within the physiological and biomechanical domains [9]. As a result, few studies have focused on performance nutrition and behaviours of athletes with disabilities [10] not to mention those with a spinal cord injury (SCI). This is a noteworthy omission, as sport nutrition in this demographic is complex and nutritional plans should be individually-tailored [11] and consider both health and performance. Furthermore, most athletes with physical impairments have lower overall energy requirements in comparison to their able-bodied (AB) counterparts [12], which may lead to micronutrient deficiencies [13]. Depending upon the micronutrient, deficiencies could result in increased risk of illness and injury [14], compromised immune systems [15], and fatigue [16]; all of which can jeopardize sports performance.

In an attempt to offset dietary deficiencies, dietary supplements (DS) are often incorporated into an athlete's nutritional plan. For the purpose of this study, a DS can be defined as "a food, food component, nutrient, or non-food compound that is purposefully ingested in addition to the habitually-consumed diet with the aim of achieving a specific health and/or performance benefit" [17]. Currently, DS use in wheelchair rugby athletes (WRA) has not been explored and few studies have reported the types and prevalence of use in other athletes with physical impairments [8,10,18,19] in comparison to the plethora of literature available in AB athletes [20–28]. To our knowledge, only one study has reported reasons of use in athletes with an impairment [10]. In addition, it is important to know the reasons athletes use dietary supplements (i.e., medical, performance, etc.) to ensure they are not over or under-supplementing. Furthermore, it is vital to know which sources of information athletes are utilizing for DS information, with prior research suggesting that dietitians are the main source for athletes with physical impairments [8,10]. Lastly, understanding *preferred* sources of information can help ensure accurate nutritional information is accessible to these athletes.

Presently, there are no nutritional guidelines tailored for populations with physical impairments [29]; and in particular those with quadriplegia. As a result, these athletes are forced to default to AB recommendations as a proxy. Although athletes with physical impairments can follow these recommendations, notable physiological differences including less muscle mass, limited sweating response, and impaired bowel function may affect the amounts needed [29]. Given the lack of sports nutrition investigation in WRA, the primary purpose of this study was to assess the types of dietary supplements used, the prevalence of usage, and the reasons for use among this population. The secondary objective was to report preferred sources of nutritional information for wheelchair rugby athletes.

2. Materials and Methods

2.1. Participants

Male and female athletes, 18 years and older, who met the criteria for an athlete with a physical disability, as set out by the International Paralympic Committee (IPC) [30], were recruited from various teams during the 2018 Canadian Wheelchair Rugby National Championships tournament or online via emailing various coaches of Canadian teams. Athletes who complete in wheelchair rugby are defined as those with impaired muscle power, impaired passive range of movement, limb deficiency, hypertonia, ataxia, or athetosis [7] and thus were eligible to participate in the study. Exclusion criteria included athletes who did not speak English or who were classified as having an intellectual impairment, as the questionnaires were not validated for these populations. The one exception to these exclusions was if English was an athlete's second language. In this case, they were eligible to participate, if they confirmed that they understood the question or had a translator present. The study was approved by the University of Calgary Conjoint Faculties Research Ethics Board (Ethics ID: REB18-0236). Athlete written consent was provided prior to completing the Dietary Supplement Questionnaire.

2.2. Dietary Supplement Questionnaire

A modified version of a dietary supplement questionnaire [8,25,26] (Supplementary Files: S1 Dietary Supplement Questionnaire) was used to assess supplement intakes, reasons for use, and sources of supplement information. First, it identified the athlete's age, gender, weight, height, sporting event, classification (0.5–3.5), current training phase, top level of competition, and hours of training per week. The questionnaire asked athletes if they took dietary supplements (i.e., yes or no) and to rank their overall diet as "not very healthy", "pretty healthy (average)", and "very healthy". Athletes were also encouraged to describe the nature of their impairment (injury or medical condition). Second, the survey requested quantified supplement frequencies, reasons for using each individual supplement, and sources of information athletes used to receive information about supplements. Athletes were also asked about their preferred sources of nutritional information. Supplement usage was assessed using a list of 28 supplement classes including vitamins, minerals, fortified and sport beverages, electrolytes (sport or electrolyte drinks or supplements), protein powders, sport (carbohydrate-dense or protein-dense) bars and gels, buffers (i.e., sodium bicarbonate), fatty acids, plant extracts, and probiotics. If a supplement was not listed, space was provided to add it. Athletes indicated how often they used each type of supplement, using three options: regularly (at least 2 times per week), at times (you've tried it or only use it at certain times like during competition or when sick), and never (you have not tried it or you don't know what this is) [8]. If an athlete indicated usage of a supplement, they were encouraged to provide the brand name and dosage beside it.

2.3. Procedures

Members of the research team attended the tournament and provided athletes with either a hard or electronic copy of the Dietary Supplement Questionnaire to fill out in a face-to-face manner. Two participants were recruited through emailing their coaches and filled out the online version of the questionnaire outside of the tournament. Members of the research team acted as scribes if a participant did not have the necessary dexterity to write or type.

2.4. Statistical Analysis

Data from the question 'do you take any dietary supplements' was categorized as yes or no. Dietary supplement use data were categorized into groups based on gender, age ($\leq$30 years and 31 years and older), and sport classification (0.5–1.5 and 2.0–3.5). The age groups were based on dietary reference intake (DRI) values [31]. Sport classification groups were based on low pointers (0.5–1.5) and high pointers (2.0–3.5), which are characterized by differences in functional abilities and roles during a game [32]. Responses from the dietary supplement questionnaire were quantified as the percentage of respondents who consumed the supplement "Regularly", "At Times", or "Never". Differences between gender, age, and classification groups were determined by a Pearson's chi square test. The most common dietary supplements used overall were based on the total number of "Regularly" and "At Times" responses. The most common dietary supplements used by gender were based on the total number of "Regularly" scores. The reasons of use data were condensed into four categories: Performance, Medical/Health, Dietary, and Recommendation, and were quantified as the percentage of respondents who consumed the supplement. Performance reasons included 'increase energy', 'increase endurance', 'improve exercise recovery', 'increase or maintain muscle mass, strength and/or power', and 'enhance overall athletic performance'. Medical/Health reasons include 'medical', 'stay healthy', 'enhance immune system', and 'weight loss or weight gain' Dietary reasons include 'to improve your diet', 'enjoy the taste', 'convenient when hungry or thirsty', 'food allergy/sensitivity/intolerance', and 'special dietary needs'. Recommendation reasons include 'someone told you to', and 'because others do'. If a participant put more than one reason, those reasons were condensed into the four categories. Sources of information by gender were analyzed using a chi square test. All analyses were performed using SPSS Statistics version 24 (IBM Corporation, Armonk, NY, USA).

3. Results

3.1. Participant Characteristics

A total of 42 athletes agreed to participate in the study and completed the questionnaires. Response rate to the individual questions ranged from 93% to 100%. Descriptive characteristics of the participants are outlined in Table 1.

Table 1. Descriptive Characteristics.

Descriptive Characteristics	All	Males	Females
Participants	42	33 (78.6%)	9 (21.4%)
Age, years	36.3 (9.5)	36.8 (9.2)	34.3 (10.6)
Weight, kg	n/a	74.5 (14.7)	58.5 (8.1)
Height, m	n/a	1.79 (0.08)	1.68 (0.07)
BMI, kg/m^2	22.8 (4.1)	23.3 (4.1)	20.5 (3.3)
Classification			
0.5–1.5	22 (52.4%)	17 (51.5%)	5 (55.6%)
2.0–3.5	20 (47.6%)	16 (48.5%)	4 (44.4%)
Level of Competition			
Provincial	5 (11.9%)	4 (12.1%)	1 (11.1%)
National	24 (57.1%)	18 (54.6%)	6 (66.7%)
International	13 (31.0%)	11 (33.3%)	2 (22.2%)

Descriptive characteristics are mean (standard deviation). BMI, body mass index, n/a, not applicable.

Most of the athletes used at least one supplement over the course of the past three months (males 90.9%; females 77.8%, $p = 0.281$). When athletes were asked to rank their diet, 15.2% of males and 11.1% of females ranked their diet as "not very healthy", 48.5% of males and 55.6% of females ranked their diet as "average", and 36.4% of males and 33.3% of females thought their diet was "very healthy". There were no differences between genders.

3.2. Dietary Supplement Use

Athlete use of dietary supplements is presented as those who consume the supplement regularly, at times, or never over the past three months (Figure 1).

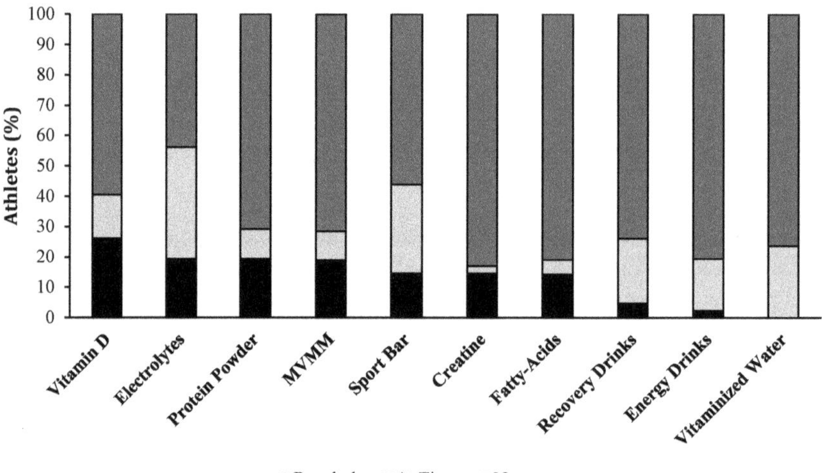

Figure 1. Dietary supplements commonly used by wheelchair rugby athletes. Data is presented in order of the most regularly used supplements. MVMM, multivitamin multi-mineral. Total numbers can be seen in Table 2.

Total supplement use and use by gender are provided in Table 2. The most common supplements based on 'regular' and 'at times' were electrolytes, sport bars, vitamin D, protein powder, and MVMM. Males most regularly used vitamin D, protein powder, and electrolytes, whereas females most regularly used MVMM and vitamin D. It is important to note that females did not 'regularly' use many dietary supplements; however, they were prominent in consuming supplements 'at times', including electrolytes, sport bars, and recovery drinks. Significant differences between genders were found in magnesium ($p = 0.019$) and probiotic pills ($p = 0.022$). Further, when the statistics were conducted by age group, athletes aged 31 years or older regularly used protein or sports bars 16.1% versus 10.0% in younger athletes. Conversely, younger athletes used bars occasionally 60.0% of the time versus 19.4% in the older group ($p = 0.048$). No other significant differences were seen in age group and no differences were seen in classification.

Table 2. Supplement use by Gender.

Supplement	Total n (%)	Male Regularly n (%)	Female Regularly n (%)	Male at Times n (%)	Female at Times n (%)	p
MVMM	12 (28.6)	6 (18.2)	2 (22.2)	4 (12.1)	0 (0.0)	0.544
Vitamin B	4 (9.5)	3 (9.1)	0 (0.0)	0 (0.0)	1 (11.1)	0.106
Vitamin C	6 (14.3)	2 (6.1)	1 (11.1)	3 (9.1)	0 (0.0)	0.582
Vitamin E	2 (4.8)	1 (3.0)	0 (0.0)	1 (3.0)	0 (0.0)	0.751
Vitamin D	17 (40.5)	9 (27.3)	2 (22.2)	5 (15.2)	1 (11.1)	0.883
Iron	2 (4.8)	1 (3.0)	1 (11.1)	0 (0.0)	0 (0.0)	0.313
Calcium	6 (14.3)	2 (6.1)	0 (0.0)	3 (9.1)	1 (11.1)	0.745
Magnesium	3 (7.1)	1 (3.0)	0 (0.0)	0 (0.0)	2 (22.2)	**0.019**
Vitaminized Water	10 (23.8)	0 (0.0)	0 (0.0)	10 (30.3)	0 (0.0)	0.058
Protein Powder	12 (29.3)	8 (25.0)	0 (0.0)	2 (6.3)	2 (22.2)	0.124
Protein or Sport Bars	18 (43.9)	6 (18.8)	0 (0.0)	9 (28.1)	3 (33.3)	0.371
BCAA	4 (9.5)	2 (6.1)	0 (0.0)	1 (3.0)	1 (11.1)	0.468
Glutamine	1 (2.4)	1 (3.0)	0 (0.0)	0 (0.0)	0 (0.0)	0.597
Buffers	1 (2.4)	0 (0.0)	0 (0.0)	1 (3.0)	0 (0.0)	0.597
Fatty-Acid Preparations	8 (19.0)	6 (18.2)	0 (0.0)	2 (6.1)	0 (0.0)	0.260
Sport or Electrolyte Drinks	23 (56.1)	8 (25.0)	0 (0.0)	11 (34.4)	4 (44.4)	0.246
Energy Drinks	8 (19.5)	1 (3.1)	0 (0.0)	7 (21.9)	0 (0.0)	0.247
Caffeine Pills	4 (9.5)	0 (0.0)	0 (0.0)	4 (12.1)	0 (0.0)	0.272
Pre-Workout Supplement	4 (9.8)	0 (0.0)	0 (0.0)	3 (9.4)	1 (11.1)	0.877
Creatine	7 (17.1)	6 (18.8)	0 (0.0)	1 (3.1)	0 (0.0)	0.305
Recovery Drinks	11 (26.2)	2 (6.1)	0 (0.0)	6 (18.2)	3 (33.3)	0.501
Plant Extracts/Herbal Supplements	6 (14.3)	2 (6.1)	0 (0.0)	3 (9.1)	1 (11.1)	0.745
Probiotic Pills	3 (7.3)	1 (3.1)	0 (0.0)	0 (0.0)	2 (22.2)	**0.022**
Sport Gels	2 (4.8)	0 (0.0)	0 (0.0)	2 (6.1)	0 (0.0)	0.449
Gummy/Bean	3 (7.1)	0 (0.0)	0 (0.0)	2 (6.1)	1 (11.1)	0.602

Intakes are presented as the number of athletes (%) who consume a dietary supplement regularly (i.e., at least twice per week) or at times (i.e., tried it or only use it at certain times such as during competition or when sick). Differences between genders were determined using a chi square test. $p < 0.05$ was considered significant. Athletes did not report using glucosamine, beta alanine, or beet root. Significant differences are bolded. MVMM, multivitamin multi-mineral. BCAA, branched chain amino acids. n/a, not applicable.

3.3. Reasons for Use

Reasons for use are presented as the number and percentage of participants who took a supplement for a specific reason (Table 3). Performance was the most frequent reason for supplementation ($n = 67$), followed by medical/health ($n = 64$), dietary ($n = 28$), and recommendation ($n = 3$).

Table 3. Supplement Reason Frequencies.

Supplement	Performance *n* (%)	Medical/Health *n* (%)	Dietary *n* (%)	Recommendation *n* (%)
MVMM *	2 (15.4)	10 (76.9)	1 (7.7)	0 (0.0)
Vitamin B	0 (0.0)	4 (100.0)	0 (0.0)	0 (0.0)
Vitamin C	0 (0.0)	6 (100.0)	0 (0.0)	0 (0.0)
Vitamin E	0 (0.0)	2 (100.0)	0 (0.0)	0 (0.0)
Vitamin D	0 (0.0)	13 (92.9)	0 (0.0)	1 (7.1)
Iron	0 (0.0)	2 (100.0)	0 (0.0)	0 (0.0)
Calcium	0 (0.0)	4 (80.0)	0 (0.0)	1 (20.0)
Magnesium	0 (0.0)	2 (66.7)	0 (0.0)	1 (33.3)
Vitaminized Water	2 (28.6)	1 (14.3)	4 (57.1)	0 (0.0)
Protein Powder	4 (36.4)	5 (45.5)	2 (18.2)	0 (0.0)
Protein or Sport Bars	8 (47.1)	1 (5.9)	8 (47.1)	0 (0.0)
BCAA	4 (100.0)	0 (0.0)	0 (0.0)	0 (0.0)
Glutamine	0 (0.0)	1 (100.0)	0 (0.0)	0 (0.0)
Fatty-Acid Preparations	1 (12.5)	3 (37.5)	4 (50.0)	0 (0.0)
Sport or Electrolyte Drinks	13 (61.9)	2 (9.5)	6 (28.6)	0 (0.0)
Energy Drinks	4 (80.0)	0 (0.0)	1 (20.0)	0 (0.0)
Caffeine Pills	4 (100.0)	0 (0.0)	0 (0.0)	0 (0.0)
Pre-Workout Supplement	4 (100.0)	0 (0.0)	0 (0.0)	0 (0.0)
Creatine	7 (100.0)	0 (0.0)	0 (0.0)	0 (0.0)
Recovery Drinks	9 (81.8)	0 (0.0)	2 (18.2)	0 (0.0)
Plant Extracts/Herbal Supplements	0 (0.0)	6 (100.0)	0 (0.0)	0 (0.0)
Probiotic Pills	0 (0.0)	2 (100.0)	0 (0.0)	0 (0.0)
Sport Gels	2 (100.0)	0 (0.0)	0 (0.0)	0 (0.0)
Gummy/Bean	3 (100.0)	0 (0.0)	0 (0.0)	0 (0.0)

Dietary supplement reason frequencies. * One person put medical and performance reasons for taking MVMM and a "yes" response was placed in both categories. Buffers was not included, as athletes who responded "yes" did not provide a reason.

3.4. Sources of Information

Sources of information are provided in Table 4. Males' top utilized sources of dietary supplement information were dietician/nutritionist (51.5%), the internet (33.3%), and sport/fitness trainer (30.3%), whereas females' sources were teammates/friends (44.4%) and the internet (33.3%). Two participants reported that they receive information from 'research' and 'self'.

Table 4. Information Sources for Dietary Supplements.

Information Source	Number of Athletes *n* (%)
Internet (Websites, Facebook)	14 (33.3)
Sport/Fitness Trainer	11 (26.2)
Health Food Store	4 (9.5)
Product Labels	5 (11.9)
Pharmacist	7 (16.7)
Family	6 (14.3)
Print Media (magazines, books)	1 (2.4)
Workshops/Classes	0 (0.0)
Naturopath/Chiropractor	2 (4.8)
Television	3 (7.1)
Coach	3 (7.1)
Teammates/Friends	12 (28.6)
Medical Physician (Doctor)	9 (21.4)
Physio/Massage Therapist	2 (4.8)
Dietician/Nutritionist	19 (45.2)

Sources for information about dietary supplements.

When the question was changed to "which way do you *prefer* to receive information about dietary supplements?" the most popular sources were individual nutrition consultation (dietitian), the internet (webpage/blogs), and coach/trainer (Table 5). Males' top preferences were individual nutrition consultation (45.5%), the internet (27.3%), and coach/trainer (24.2%). Females' top preferences were individual nutrition consultation (55.6%), the internet (33.3%), doctor/chiropractor/physiotherapist (22.2%), and coach/trainer (22.2%). No other preferred sources were listed; however, five athletes reported they were not interested in receiving information about dietary supplements.

Table 5. Preferred Information Sources for Dietary Supplements.

Preferred Information Source	Number of Athletes *n* (%)
Presentations	3 (7.1)
Family/Friends	7 (16.7)
Social media	4 (9.5)
Print media (pamphlets, books, magazines)	4 (9.5)
Coach/Trainer	10 (23.8)
E-mail	6 (14.3)
Individual nutrition consultation (Dietician)	20 (47.6)
Health food store/pharmacy	3 (7.1)
Internet (webpage, blogs)	12 (28.6)
Doctor/Chiropractor/Physiotherapist	6 (14.3)

Preferred sources for information about dietary supplements.

4. Discussion

This research provides the first analysis of dietary supplement use in Canadian wheelchair rugby athletes. Furthermore, it is novel in that it directly links the individual supplement choice to a reason for use.

4.1. Dietary Supplement Usage

The WRA demonstrated a high usage of dietary supplements with 84.4% of the participants reporting using at least one DS over the past three months. This allows us to develop a benchmark as few studies have reported DS usage among athletes with disabilities [8,10,18,19]. One study investigating usage over five impairment types revealed 58% of athletes took at least one nutritional supplement in the previous six months [10]. Usage in the current study was higher; however, these results are in alignment with another study in Canadian athletes with a physical impairment [8]. Dietary supplement use in athletes of wheelchair sports may be beneficial given their reduced energy intakes [13], which, combined with a physical impairment, may contribute to micronutrient deficiencies [13]. Another factor potentially influencing usage is the training phase or season. The majority of the participants in the current study were in a competition phase (88.1%) and a high DS usage was reported. Divergent results were found by Krempien and Barr [19], as they reported that 44% of athletes with a SCI consumed supplements at home vs. 34% when at a training camp. In able bodied athletes, Erdman et al. [33] reported that DS usage was higher during the athletes' training phase (93.1%) versus their competition phase (84.1%). Further investigation into supplement use during specific training phases may provide better insight into usage patterns as they relate to training periodization.

4.2. Dietary Supplement Types

The most regularly used dietary supplement was vitamin D. This finding coincides with the current literature, which reports frequent use of this vitamin in athletes with a physical impairment generally and those with a SCI specifically [8,34]. Maintaining sufficient levels of this micronutrient is important, as it has implications for athletic performance, including functionality of the skeletal muscles and maximal oxygen uptake [13]. Vitamin D deficiency can occur even in individuals without impairment [35] and supplementation of 1000 IU/day is often recommended in Canada [36]. Athletes with physical impairments possess a heightened deficiency risk compared to their AB counterparts due to potentially inadequate energy intakes [13]. Exposure to sunlight may not be feasible for some athletes with a SCI, as impaired thermoregulatory abilities can make skin hyper-sensitive when exposed to sunlight [14]. The coupling of vitamin D supplementation and monitoring levels in Para athletes, especially those competing in indoor sports and living in northern environments, is recommended [8,13].

Among the top supplements used were electrolyte drinks/supplements and protein powder. These findings are similar to studies that reported medication and supplement use at the 2004 Paralympic Games [18] and findings in elite, Canadian able-bodied athletes [33]. Some athletes with a SCI have compromised thermoregulation, thus have reduced sweat responses below the lesion level, but heightened sweat above the lesion [37]. This physiological factor may contribute to the regular use of electrolytes in an attempt to maintain hydration.

In sport, protein has roles in the synthesis of contractile and metabolic proteins and contributes to structural changes in tendon and bone tissue [11]. Current protein recommendations for AB athletes range from 1.2 to 2.0 g/kg and differ depending on the nature of the exercise performed (i.e., endurance vs. resistance), current phase within a periodization program, athletic goals, nutrient requirements, energy expenditure, and food choices [11]. Athletes with physical disabilities, such as a SCI, typically use less muscle mass than their AB counterparts [38]; therefore, specific protein needs for these athletes is unknown. Recent studies investigating dietary intakes in Paralympic athletes report adequate protein intake, based on AB recommendations, are being consumed from food alone [8,39].

4.3. Reasons for Use

Understanding the reasons for consuming DS is important to ensure that athletes are taking supplements for valid reasons. Findings of this study suggest that the most common reason WRA took DS was for performance. Athletes are keen to improve performance through supplementation; however, little evidence exists to suggest supplementation may actually improve performance outside of established deficiencies, at least in able bodied (AB) athletes. Exceptions include caffeine [40] and carbohydrate/electrolyte replacements [41]. The finding as to motivation for consumption is congruent with a study in elite AB athletes, which revealed the top reason for taking DS was to 'increase energy' [42].

The reasons associated with medical health were also common among participants. A noteworthy theme in medical/health reasons emerged, as some WRA reported taking cranberry pills, citing 'bladder' (infections) as their reason. According to the literature, urinary tract infections are common in those with a SCI [43]; however, the efficacy of cranberry to treat or prevent urinary tract infections requires further study [44,45].

4.4. Sources of Information

Being aware of dietary supplement information sources WRA utilize is vital to ensure they are receiving sound advice and that future nutritional guidelines are accessible. Our findings reported dietitian/nutritionist as the most utilized source, which is both congruent [8,10] and divergent [46,47] with other Paralympic studies. Both male and female athletes indicated the internet as their second most utilized and preferred source, which is concerning due to the high risk of non-scientific information and lack or absence of regulatory controls when purchasing such products online [10]. Teammates/friends, and sport/fitness trainers were also top choices for both genders; however, this also increases the risk of being given inaccurate information. Athletes and trainers should undergo basic sports nutrition training to enhance their knowledge and awareness [8].

The study is limited in that if a scribe or a translator for language was used, their answers may not be as accurate as if they had answered on their own. Furthermore, a small sample size was used, and multiple comparisons were made with a 5% level of significance, thus there is a risk of false positives. Caution is advised when interpreting the findings; however, given the paucity of research in this area, these preliminary findings are of high value.

5. Conclusions

Dietary supplement usage in WRA and their reasons for supplementation were evaluated. A variety of supplements were utilized for primarily performance and medical/health reasons. Future studies should evaluate dietary intakes and physiological levels of nutrients to determine optimal supplementation strategies in WRA.

Supplementary Materials: The following are available online at http://www.mdpi.com/2072-6643/10/12/1958/s1.

Author Contributions: R.F.M., J.S., D.L., and J.A.P. conceived and designed the experiments. R.F.M. performed the experiments; R.F.M. and J.A.P. analyzed the data; R.F.M. wrote the paper; J.S., D.L., and J.A.P. critically reviewed and revised the paper. All authors approved the final version of the manuscript.

Funding: This research was funded by the National Science and Engineering Research Council of Canada (J.S.).

Acknowledgments: The authors would like to acknowledge Nancy Gammack, Madison Fullerton, Ash Kolstad, and Jordan Parkin for their help with data collection.

Conflicts of Interest: The authors declare no conflict of interest. The funders had no role in the design of the study; in the collection, analyses, or interpretation of data; in the writing of the manuscript, or in the decision to publish the results.

References

1. Lindemann, K.; Cherney, J.L. Communicating in and through "Murderball": Masculinity and disability in wheelchair rugby. *West. J. Commun.* **2008**, *72*, 107–125. [CrossRef]
2. Sarro, K.J.; Misuta, M.S.; Burkett, B.; Malone, L.A.; Barros, R.M.L. Tracking of wheelchair rugby players in the 2008 demolition derby final. *J. Sports Sci.* **2010**, *28*, 193–200. [CrossRef] [PubMed]
3. Goodwin, D.; Johnston, K.; Gustafson, P.; Elliott, M.; Thurmeier, R.; Kuttai, H. It's okay to be a quad: Wheelchair rugby players' sense of community. *Adapt. Phys. Act. Q.* **2009**, *26*, 102–117. [CrossRef]
4. Morgulec-Adamowicz, N.; Kosmol, A.; Bogdan, M.; Molik, B.; Rutkowska, I.; Bednarczuk, G. Game efficiency of wheelchair rugby athletes at the 2008 Paralympic Games with regard to player classification. *Hum. Mov.* **2010**, *11*, 29–36. [CrossRef]
5. Black, K.E.; Huxford, J.; Perry, T.; Brown, R.C. Fluid and sodium balance of elite wheelchair rugby players. *Int. J. Sport Nutr. Exerc. Metab.* **2013**, *23*, 110–118. [CrossRef] [PubMed]
6. Rhodes, J.M.; Mason, B.S.; Malone, L.A.; Goosey-Tolfrey, V.L. Effect of team rank and player classification on activity profiles of elite wheelchair rugby players. *J. Sports Sci.* **2015**, *33*, 2070–2078. [CrossRef] [PubMed]
7. Committee, I.P. Explanatory guide to Paralympic Classification Paralympic Summer Sports. Available online: https://www.paralympic.org/sites/default/files/document/150915170806821_2015_09_15%2BExplanatory%2Bguide%2BClassification_summer%2BFINAL%2B_5.pdf (accessed on 12 September 2018).
8. Madden, R.F.; Shearer, J.; Parnell, J.A. Evaluation of dietary intakes and supplement use in paralympic athletes. *Nutrients* **2017**, *9*, 1266. [CrossRef] [PubMed]
9. Tawse, H.; Bloom, G.A.; Sabiston, C.M.; Reid, G. The role of coaches of wheelchair rugby in the development of athletes with a spinal cord injury. *Qual. Res. Sport Exerc. Heal.* **2012**, *4*, 206–225. [CrossRef]
10. Graham-Paulson, T.S.; Perret, C.; Smith, B.; Crosland, J.; Goosey-Tolfrey, V.L. Nutritional supplement habits of athletes with an impairment and their sources of information. *Int. J. Sport Nutr. Exerc. Metab.* **2015**, *25*, 387–395. [CrossRef]
11. Thomas, D.T.; Erdman, K.A.; Burke, L.M. Position of the Academy of Nutrition and Dietetics, Dietitians of Canada, and the American College of Sports Medicine: Nutrition and athletic performance. *J. Acad. Nutr. Diet.* **2016**, *116*, 501–528. [CrossRef]
12. Price, M. Energy expenditure and metabolism during exercise in persons with a spinal cord injury. *Sport. Med.* **2010**, *40*, 681–696. [CrossRef] [PubMed]
13. Scaramella, J.; Kirihennedige, N.; Broad, E. Key nutritional strategies to optimize performance in Para athletes. *Phys. Med. Rehabil. Clin. N. Am.* **2018**, *29*, 283–298. [CrossRef] [PubMed]
14. Flueck, J.L.; Perret, C. Vitamin D deficiency in individuals with a spinal cord injury: A literature review. *Spinal Cord* **2016**, *55*, 1–7. [CrossRef] [PubMed]

15. Grams, L.; Garrido, G.; Villacieros, J.; Ferro, A. Marginal micronutrient intake in high-performance male wheelchair basketball players: A dietary evaluation and the effects of nutritional advice. *PLoS ONE* **2016**, *11*, 1–13. [CrossRef] [PubMed]

16. Longo, D.L.; Camaschella, C. Iron-deficiency anemia. *N. Engl. J. Med.* **2015**, *372*, 1832–1843. [CrossRef]

17. Maughan, R.J.; Burke, L.M.; Dvorak, J.; Larson-Meyer, D.E.; Peeling, P.; Phillips, S.M.; Rawson, E.S.; Walsh, N.P.; Garthe, I.; Geyer, H.; et al. IOC consensus statement: Dietary supplements and the high-performance athlete. *Int. J. Sport Nutr. Exerc. Metab.* **2018**, *28*, 104–125. [CrossRef] [PubMed]

18. Tsitsimpikou, C.; Jamurtas, A.; Fitch, K.; Papalexis, P.; Tsarouhas, K. Medication use by athletes during the Athens 2004 Paralympic Games. *Br. J. Sports Med.* **2009**, *43*, 1062–1066. [CrossRef]

19. Krempien, J.L.; Barr, S.I. Risk of nutrient inadequacies in elite canadian athletes with spinal cord injury. *Int. J. Sport Nutr. Exerc. Metab.* **2011**, *21*, 417–425. [CrossRef]

20. Froiland, K.; Koszewski, W.; Hingst, J.; Kopecky, L. Nutritional supplement use among college athletes and their sources of information. *Int. J. Sport Nutr. Exerc. Metab.* **2004**, *14*, 104–120. [CrossRef]

21. Diehl, K.; Thiel, A.; Zipfel, S.; Mayer, J.; Schnell, A.; Schneider, S. Elite adolescent athletes' use of dietary supplements: Characteristics, opinions, and sources of supply and information. *Int. J. Sport Nutr. Exerc. Metab.* **2012**, 165–174. [CrossRef]

22. Braun, H.; Koehler, K.; Geyer, H.; Kleinert, J.; Mester, J.; Schaenzer, W. Dietary supplement use among elite young German athletes. *Int. J. Sport Nutr. Exerc. Metab.* **2009**, *19*, 97–109. [CrossRef]

23. Kristiansen, M.; Levy-Milne, R.; Barr, S.; Flint, A. Dietary supplement use by varsity athletes at a Canadian university. *Int. J. Sport Nutr. Exerc. Metab.* **2005**, *15*, 195–210. [CrossRef] [PubMed]

24. Lun, V.; Erdman, K.A.; Fung, T.S.; Reimer, R.A. Dietary supplementation practices in Canadian high-performance athletes. *Int. J. Sport Nutr. Exerc. Metab.* **2012**, *22*, 31–37. [CrossRef]

25. Wiens, K.; Erdman, K.A.; Stadnyk, M.; Parnell, J.A. Dietary supplement usage, motivation, and education in young Canadian athletes. *Int. J. Sport Nutr. Exerc. Metab.* **2014**, *24*, 613–622. [CrossRef]

26. Parnell, J.A.; Wiens, K.P.; Erdman, K.A. Dietary intakes and supplement use in pre-adolescent and adolescent Canadian athletes. *Nutrients* **2016**, *8*, 526. [CrossRef] [PubMed]

27. Heikkinen, A.; Alaranta, A.; Helenius, I.; Vasankari, T. Dietary supplementation habits and perceptions of supplement use among elite Finnish athletes. *Int. J. Sport Nutr. Exerc. Metab.* **2011**, *21*, 271–279. [CrossRef]

28. Ziegler, P.J.; Nelson, J.A.; Jonnalagadda, S.S. Use of dietary supplements by elite figure skaters. *Int. J. Sport Nutr. Exerc. Metab.* **2003**, *13*, 266–276. [CrossRef] [PubMed]

29. Ferro, A.; Garrido, G.; Villacieros, J.; Pérez, J.; Grams, L. Nutritional habits and performance in male elite wheelchair basketball players during a precompetitive period. *Adapt. Phys. Act. Q.* **2017**, *34*, 295–310. [CrossRef] [PubMed]

30. Committee, I.P. IPC Policy on Eligible Impairments in the Paralympic Movement. Available online: https://www.paralympic.org/sites/default/files/document/141113170238135_2014_10_13+Sec+ii+chapter+3_13+IPC+Policy+on+Eligible+Impairments+in+the+Paralympic+Movement.pdf (accessed on 25 July 2017).

31. Institute of Medicine Dietary Reference Intakes Tables and Application. Available online: http://www.nationalacademies.org/hmd/Activities/Nutrition/SummaryDRIs/DRI-Tables.aspx (accessed on 16 September 2018).

32. Molik, B.; Lubelska, E.; Kosmol, A.; Bogdan, M.; Yilla, A.B.; Hyla, E. An examination of the international wheelchair rugby federation classification system utilizing parameters of offensive game efficiency. *Adapt. Phys. Act. Q.* **2008**, *25*, 335–351. [CrossRef]

33. Erdman, K.A.; Fung, T.S.; Doyle-Baker, P.K.; Verhoef, M.J.; Reimer, R. A Dietary supplementation of high-performance Canadian athletes by age and gender. *Clin. J. Sport Med.* **2007**, *17*, 458–464. [CrossRef] [PubMed]

34. Doubelt, I.; de Zepetnek, J.T.; MacDonald, M.J.; Atkinson, S.A. Influences of nutrition and adiposity on bone mineral density in individuals with chronic spinal cord injury: A cross-sectional, observational study. *Bone Rep.* **2015**, *2*, 26–31. [CrossRef]

35. Flueck, J.L.; Schlaepfer, M.W.; Perret, C. Effect of 12-week vitamin D supplementation on 25[OH]D status and performance in athletes with a spinal cord injury. *Nutrients* **2016**, *8*. [CrossRef]

36. Society, C.C. Should I take a Vitamin D Supplement? Available online: http://www.cancer.ca/en/prevention-and-screening/reduce-cancer-risk/make-healthy-choices/eat-well/should-i-take-a-vitamin-d-supplement/?region=on (accessed on 9 October 2018).

37. Pritchett, R.C.; Al-Nawaiseh, A.M.; Pritchett, K.K.; Nethery, V.; Bishop, P.A.; Green, J.M. Sweat gland density and response during high-intensity exercise in athletes with spinal cord injuries. *Biol. Sport* **2015**, *32*, 249–254. [CrossRef]
38. Goosey-Tolfrey, V.L.; Crosland, J. Nutritional practices of competitive British wheelchair games players. *Adapt. Phys. Act. Q.* **2010**, *27*, 47–59. [CrossRef]
39. Gerrish, H.R.; Broad, E.; Lacroix, M.; Ogan, D.; Pritchett, R.C.; Pritchett, K. Nutrient intake of elite Canadian and American athletes with spinal cord injury. *Int. J. Exerc. Sci.* **2017**, *10*, 1018–1028.
40. Shearer, J.; Graham, T.E. Performance effects and metabolic consequences of caffeine and caffeinated energy drink consumption on glucose disposal. *Nutr. Rev.* **2014**, *72*, 121–136. [CrossRef]
41. Tarnopolsky, M.A.; Zawada, C.; Richmond, L.B.; Carter, S.; Shearer, J.; Graham, T.; Phillips, S.M. Gender differences in carbohydrate loading are related to energy intake. *J. Appl. Physiol.* **2001**, *91*, 225–230. [CrossRef]
42. Erdman, K.A.; Fung, T.S.; Reimer, R.A. Influence of performance level on dietary supplementation in elite Canadian athletes. *Med. Sci. Sports Exerc.* **2006**, *38*, 349–356. [CrossRef]
43. Brinkhof, M.W.G.; Al-Khodairy, A.; Eriks-Hoogland, I.; Fekete, C.; Hinrichs, T.; Hund-Georgiadis, M.; Meier, S.; Scheel-Sailer, A.; Schubert, M.; Reinhardt, J.D. Health conditions in people with spinal cord injury: Contemporary evidence from a population-based community survey in Switzerland. *J. Rehabil. Med.* **2016**, *48*, 197–209. [CrossRef]
44. Guay, D.R.P. Cranberry and urinary tract infections. *Drugs* **2009**, *69*, 775–807. [CrossRef]
45. Jepson, R.G.; Williams, G.; Craig, J. Cranberries for preventing urinary tract infections. *Cochrane Database Syst. Rev.* **2012**. [CrossRef]
46. Rastmanesh, R.; Taleban, F.A.; Kimiagar, M.; Mehrabi, Y.; Salehi, M. Nutritional knowledge and attitudes in athletes with physical disabilities. *J. Athl. Train.* **2007**, *42*, 99–105.
47. Eskici, G.; Ersoy, G. An evaluation of wheelchair basketball players' nutritional status and nutritional knowledge levels. *J. Sports Med. Phys. Fit.* **2016**, *56*, 259–268.

© 2018 by the authors. Licensee MDPI, Basel, Switzerland. This article is an open access article distributed under the terms and conditions of the Creative Commons Attribution (CC BY) license (http://creativecommons.org/licenses/by/4.0/).

Review

Honey Supplementation and Exercise: A Systematic Review

Samuel P. Hills, Peter Mitchell, Christine Wells and Mark Russell *

School of Social and Health Sciences, Leeds Trinity University, Horsforth, Leeds LS18 5HD, UK
* Correspondence: m.russell@leedstrinity.ac.uk; Tel.: +44-(0)113-283-7100 (ext. 649)

Received: 7 June 2019; Accepted: 2 July 2019; Published: 12 July 2019

Abstract: Honey is a natural substance formed primarily of carbohydrates (~80%) which also contains a number of other compounds purported to confer health benefits when consumed. Due to its carbohydrate composition (low glycaemic index, mostly fructose and glucose), honey may theoretically exert positive effects when consumed before, during or after exercise. This review therefore appraised research examining the effects of honey consumption in combination with exercise in humans. Online database (PubMed, MEDLINE, SPORTDiscus) searches were performed, yielding 273 results. Following duplicate removal and application of exclusion criteria, nine articles were reviewed. Large methodological differences existed in terms of exercise stimulus, population, and the nutritional interventions examined. All nine studies reported biochemical variables, with four examining the effects of honey on exercise performance, whilst five described perceptual responses. Acute supplementation around a single exercise session appeared to elicit similar performance, perceptual, and immunological responses compared with other carbohydrate sources, although some performance benefit has been observed relative to carbohydrate-free comparators. When consumed over a number of weeks, honey may dampen immunological perturbations arising from exercise and possibly improve markers of bone formation. More well-controlled research is required to better understand the role for honey in a food-first approach to exercise nutrition.

Keywords: carbohydrate; antioxidant; immune function; endurance; intermittent exercise; glucose; fructose

---

1. Introduction

Honey is defined by European Communities legislation as "the natural sweet substance produced by *Apis mellifera* bees from the nectar of plants or from secretions of living parts of plants or excretions of plant-sucking insects on the living parts of plants, which the bees collect, transform by combining with specific substances of their own, deposit, dehydrate, store and leave in honeycombs to ripen and mature" and is categorised primarily according to origin and mode of production or presentation [1]. A popular foodstuff, honey is comprised of ~80% carbohydrate, and ~19% water [2], and typically contains a wide variety of other components such as organic acids, proteins, amino acids, minerals, polyphenols, vitamins, aroma compounds, and approximately 500 enzymes [2,3]. This diverse profile has seen honey being used for a variety of different health and medicinal purposes [2–7]. For example, amongst numerous other proposed health benefits, honey is espoused to have antioxidant, antimicrobial, and anti-inflammatory properties [3–6,8]. Such effects may be at least partially attributable to honey's high osmolarity inhibiting bacterial growth, the antimicrobial effects of glucose oxidase and resultant hydrogen peroxide production, and/or the presence of antibacterial substances such as polyphenols [2]. Notably, there exists over 320 different varieties of honey, and the composition of this substance can vary substantially depending upon the variety of plant from which nectar is derived, in addition to the environmental conditions within which the plants grow [2,3,6,9,10].

The primary carbohydrates present in honey are the monosaccharides glucose (~30–35%) and fructose (~35–40%). However, ~5–10% of honey's volume may consist of up to 25 additional di- and trisaccharides in varying quantities [2,3,6,7]. In keeping with the natural variation in composition, the glycaemic index (GI) of honey (and thus potentially the postprandial insulinaemic response) also appears to differ between varieties. Depending upon botanical source, GI values ranging from ~32–85 have been reported [3], with the GI of any given honey appearing to depend largely upon its relative concentration of fructose. Indeed, a higher fructose to glucose ratio is associated with a lower GI value [3,11], and a number of studies have identified fructose-rich varieties of honey, such as Acacia honey, with "low" (i.e., values ≤ 55 [12]) or "moderate" (i.e., values of 55–69 [12]) GI ratings [3,11–13].

Carbohydrate has been recognised as an important fuel for exercise since the early 1900s [14], and it is now well established that commencing activity with high concentrations of muscle glycogen may enhance physical performance during exercise of >60–90 min in duration [15–18]. Moreover, consuming carbohydrates immediately prior to, and during, exercise can help to maintain performance throughout both prolonged endurance events [14,16,19–21] and intermittent exercise representative of team sports [14,22,23]. Indeed, carbohydrate ingestion may augment exercise capacity via a sparing of endogenous fuel stores (i.e., muscle and liver glycogen, and blood glucose concentrations), and/or by acting directly on the central nervous system. This latter suggestion is supported by increases in short-term (i.e., events lasting ≤60 min) exercise performance when carbohydrate solutions are simply swilled around the mouth [14,24].

The American College of Sports Medicine have provided broad recommendations for carbohydrate consumption during exercise, and therein suggest intakes of ~30–60 g·h^{-1} [25]. However, whilst it was traditionally believed that ~60 g·h^{-1} represented the upper limit of carbohydrate (i.e., glucose) oxidation during endurance exercise, more recent evidence suggests that simultaneously ingesting carbohydrates from multiple sources (e.g., glucose and fructose) may increase oxidation capacity by up to ~75% [15,21]. Notably, consuming a combination of glucose and fructose at a rate of 108–144 g·h^{-1} has improved performance during prolonged cycling exercise when compared with the equivalent dose of glucose alone [14,26,27]. Such developments may be important for athletes engaged in prolonged endurance events (i.e., events of ≥2.5 h in duration) in which fuel availability is likely to be a substantial performance-limiting factor [14]. Given its multiple-carbohydrate composition, there may be a theoretical basis to suggest that honey supplementation could offer a viable and natural alternative to traditional forms of exogenous carbohydrate provision.

In team sports such as soccer, players experience limited opportunities to consume carbohydrates outside of scheduled stoppages in play (i.e., half-time). For this reason, research designs based around a regular feeding pattern throughout exercise (i.e., every 15 min) may be limited in their ecological validity. Moreover, when high GI carbohydrates, including those contained within most commercially available sports drinks, are ingested before and during team sport specific exercise, including at half-time, sharp declines in blood glucose concentrations are typically observed during the early stages of the second half [28–31]. Given the likely mechanisms involved (for a review of this topic, please see [29]), it has been proposed that altering the GI of carbohydrates consumed pre-match and at half-time, may help to counteract these responses in team sports athletes [29]. Theoretically, low GI carbohydrates produce a lower insulinaemic response and a slower delivery of glucose into the systemic circulation, thus helping to maintain blood glucose concentrations throughout the second half [29,31]. In support, Stevenson et al. [31] observed better maintenance of blood glucose concentrations during the second half of simulated soccer match-play when an 8% solution of low GI isomaltulose (GI: 32) was consumed during the pre-exercise warm-up and at half-time, when compared with an equivalent volume of high GI maltodextrin (GI: 90–100). Whilst no between-trial differences were observed for any performance measure, these findings appear to suggest a potential role for low GI carbohydrate sources for athletes engaged in intermittent team sports when ecologically valid feeding patterns are used.

It is well established that a heavy schedule of prolonged and/or intense exercise can lead to immunity impairment [32,33], and an increased risk of sustaining upper respiratory tract infections [34,35]. An in-depth discussion on the relationship between nutritional strategies and immune responses to exercise is beyond the scope of this article (interested readers please see [33,34,36]), but it is noteworthy that, whilst a viable short term approach for augmentation of endurance training adaptations [15,19,37], exercising in a carbohydrate-depleted state (i.e., following days of low carbohydrate intake and/or prior glycogen depleting exercise) elicits greater elevations in circulating stress hormones and further disruption of several markers of immune function (e.g., interleukin-6; IL-6, interleukin-1 receptor antagonist; IL-1ra and interleukin-10; IL-10), compared with when carbohydrate availability is greater [34]. In addition, consuming ~30–60 g·h^{-1} of carbohydrates during exercise, particularly throughout prolonged endurance exercise, may also attenuate many of these negative responses through better maintenance of blood glucose concentrations and a concomitant blunting of stress hormone release [33,34,36,38].

In addition to carbohydrates, supplementing with antioxidants may have the potential to somewhat counter the immune disturbances experienced following exercise. Whilst evidence for the efficacy of this strategy is at best mixed (see [33]), a 60 day program of antioxidant supplementation attenuated the cytokine (i.e., tumor necrosis factor alpha; TNF-α, interleukin-1 beta; IL-1β, and IL-6) response to a 45 min cycling bout, compared with when the same exercise was completed pre-supplementation [39]. It should be noted that homeostatic disruption may represent a key driver of cellular adaptations to training, and some evidence suggests that interventions aiming to artificially reduce oxidative stress (e.g., supplementing with high doses of antioxidants), have the potential to interfere with molecular signalling [15,33,40]. Although the impact on long-term training adaptations remains unclear, athletes seeking to maintain immune function may wish to initially consider a food-first approach, which prioritises consumption of a range of antioxidant- and phytochemical-rich foods [33].

It has been suggested that honey, due to its high carbohydrate content, may be a suitable energy source for athletes or exercising populations [3]. While possible gastric tolerance issues remain to be confirmed, when consumed around exercise, honey may provide multiple transportable carbohydrates as recommended for endurance athletes, whilst the lower GI of honey compared with that of most commercially available sports drinks has potential applications for athletes engaged in intermittent sports [29,31]. Moreover, honey exhibits natural antioxidant properties that may provide an appropriate balance between controlling the immunosuppressive response to exercise, and maintaining the signalling pathways necessary for positive training adaptation. To this end, a small body of research is beginning to surface surrounding the potential application of honey as a strategy to either enhance athletic performance, improve recovery, or otherwise influence responses to exercise. With this in mind, the aim of the current review was to systematically identify and appraise the current body of research that has examined the effects of honey supplementation in combination with exercise in humans.

2. Materials and Methods

This review was undertaken in accordance with the Preferred Reporting Items for Systematic Reviews and Meta-Analyses (PRISMA) guidelines [41]. Computerised searches were run in the online databases PubMed, Medline, and SPORTDiscus during May 2019, thus articles published up until this time were considered. The search strategy incorporated the terms (honey) AND (exercis* OR soccer OR football OR rugby OR cycling OR resistance-exercis* OR sport* OR dancer OR dancing OR cyclist* OR running OR runner* OR hockey OR basketball OR handball OR swim* OR "team sport*" OR team-sport* OR endurance OR performance OR rowing OR rower* OR sprint OR jump OR power OR strength OR training OR hurling OR weightlift*), and the filters applied were: English language, humans, clinical trial, journal article, and peer-reviewed. References listed within bibliographies of the retrieved records, in addition to articles already known to the authors, were also considered for inclusion.

2.1. Study Selection

Based upon the specific aims of this review, studies identified from the original search strategy were systematically excluded according to the following criteria: (A) studies not conducted with living human participants; (B) studies which involved either no exercise stimulus, no nutritional intervention which included honey, or both; or (C) studies that were review articles.

2.2. Quality Assessment

After application of the pre-defined exclusion criteria, the remaining full text articles were assessed for methodological quality via the Physiotherapy Evidence Database (PEDro) scale. This assessment scores experimental studies out of a maximum of 10 points, based upon satisfaction of a range of criteria. Eight of these criteria evaluated a study's internal validity, and a further two criteria related to whether sufficient statistical information was presented. The PEDro scale has previously been identified as a valid and reliable indicator of methodological quality [42,43]. As per previous review papers in this field, only articles with a PEDro score of at least five out of 10 were included in order to improve the credibility of the analyses [44].

3. Results

A total of 273 records were identified through the original search strategy (including one record previously known to the authors). Following removal of 129 duplicates, 144 records were screened according to the pre-defined exclusion criteria. Of the 133 records excluded at this stage, 56 studies were not conducted with living human participants (exclusion criteria A); 71 involved either no exercise stimulus, no nutritional intervention which included honey, or both (exclusion criteria B); whilst six records were excluded on the basis that they were review articles (exclusion criteria C). When the remaining 11 full text articles were assessed for eligibility and quality, a further two records were excluded on the basis of scoring <5 out of 10 on the PEDro scale. Therefore, a total of nine articles were retained and included in this review (Figure 1).

In the nine eligible articles, outcomes were presented for 186 participants (individual study sample sizes ranging from nine to 40 participants), being mostly amateur athletes, of whom 125 were male and 61 were female. Four studies reported crossover, repeated measures designs, whilst five studies assigned participants to independent groups. Exercise modalities included team sport simulations, running, cycling, rowing, resistance exercise, and dance. Considerable methodological variation also existed with regards to the patterns and dosages of honey supplementation, with some studies feeding honey either before, during or after a single exercise session, and others investigating the effects of honey supplementation over several weeks (e.g., through a training 'block'). Articles were pooled according to three broad themes and two sub-themes (Tables 1–4), which are presented in turn below. Studies have examined the effects of honey supplementation (either acutely or over multiple weeks) on (a) biochemical markers, (b) physical and skilled performance, and (c) perceptual responses. A number of articles investigated multiple constructs and were therefore included within more than one theme.

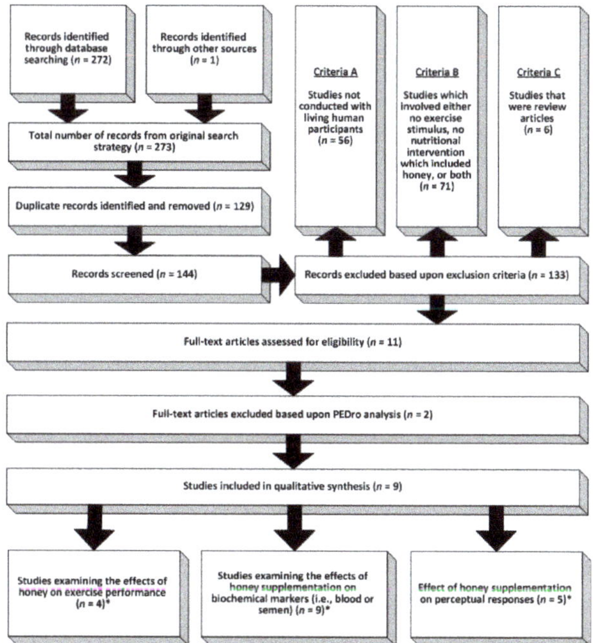

Figure 1. Selection process for articles included in the systematic review. * denotes that some studies examined multiple constructs, thus are included in more than one category.

3.1. Effect of Honey Supplementation on Biochemical Markers (i.e., Blood or Semen)

All nine eligible studies included at least one outcome variable derived from bodily fluid. Blood samples (venous or capillary) were taken in eight of these instances, whilst the remaining study derived indices of immune status from markers present within semen. As substantial methodological variation was observed, results are separated into (a) studies investigating the acute effects of honey consumption around a single exercise session (Table 1) and (b) those in which honey supplementation occurred over multiple weeks (Table 2).

3.1.1. Acute Honey Consumption around a Single Exercise Session

Five of the nine studies concerned the acute effects of honey supplementation (Table 1). In 59 males and 21 females, honey was administered in various doses, frequencies and forms (i.e., solution, gel, or powder) during rowing [45], cycling [46], and soccer specific exercise [47], as well as between running bouts in hot conditions [48], and immediately following resistance exercise [49]. Partly due to the inherent difficulty in consolidating findings from such different methodological approaches, the influence of acute honey supplementation on blood glucose concentrations and insulin responses during and following exercise remains inconclusive. In addition, consuming honey around a single exercise session appears to produce similar immune and hormonal (e.g., testosterone and cortisol concentrations) responses when compared with consumption of other carbohydrate sources.

3.1.2. Honey Supplementation over Multiple Weeks

The remaining four articles (involving a total of 76 males and 40 females) investigated biochemical changes when honey was consumed over periods ranging from 31 days to 16 weeks (Table 2). Three studies investigated immunological markers, two via blood samples, and one via semen analysis [50–52], whilst one study assessed whether honey supplementation combined with aerobic dance exercise influenced markers of bone formation and/or resorption [53]. Consuming 70 g honey prior to each training session over periods of eight to 16 weeks attenuated the negative immune response to a programme of moderate to intense cycling exercise when compared with no nutritional supplement [50,51]. Although potential benefits have been observed, the influence of daily honey supplementation is less clear with respect to markers of bone formation/resorption [53], and when consumed in smaller quantities (i.e., 3 × 10 mL of a honey and yeast product per day) over 31 days of running training [52]. Unfortunately, in each of these four studies it is unclear whether the groups used as comparators were energy or carbohydrate-matched compared with those consuming honey.

3.2. Effect of Honey Supplementation on Physical or Skilled Performance

Four eligible studies (Table 3) have assessed the influence of honey supplementation on at least one measure of physical or skilled performance in a total of 53 males. Three studies investigated the acute effects of honey consumption during team sport [47], running [48], or cycling [46] exercise, whilst one study used various cycling ergometer tests to assess the influence of consuming 70 g of honey 90 min prior to each training session on adaptations throughout a 16 week period of training [51]. Whilst benefits were observed when honey was compared to consuming no carbohydrate at all [46,48], findings have been largely inconsistent with regards to the influence of honey supplementation on exercise performance (Table 3).

3.3. Effect of Honey Supplementation on Perceptual Responses

Five articles have reported perceptual responses from a total of 59 males and 21 females, when honey was consumed before, during, or immediately after exercise (Table 4). A variety of Likert scales were employed to measure constructs relating to taste, texture, gut comfort, and perceived fatigue [45–49]. Although honey may elicit a sweeter taste compared with water [48], no differences in ratings of perceived exertion or perceptions of fatigue, either during or after exercise, were reported with honey as opposed to water or other forms of carbohydrate.

Table 1. Studies examining the effect of acute honey supplementation on biochemical responses to a single exercise session.

Study	Participants	Design	Exercise Stimulus	Nutritional Intervention	Data Collection Method and Time-Points	Outcome Variables	Main Results
Abbey and Rankin [47]	Male soccer players of NCAA Division I, post collegiate, and club standard ($n = 10$).	Randomised, single blind, crossover.	Soccer specific exercise (5×15 min blocks, plus 10 min half-time), followed by progressive 20 m shuttle run to fatigue.	6% CHO-electrolyte solution (honey, commercial sports drink, or placebo). 8.8 $mL \cdot kg^{-1}$ (0.5 $g \cdot kg$ body $mass^{-1}$) CHO consumed 30 min pre-exercise and at half-time.	Blood samples at 30 min pre-exercise (T1), immediately post-exercise (T2), 60 min post-exercise (T3).	Blood glucose, insulin, cortisol, plasma volume, IL-1ra, IL-6 and IL-10, total ORAC, and plasma ORAC.	Plasma IL-1ra was ↓ at T2 for honey vs. sports drink, and ↓ at T3 for honey vs. placebo. ↔ between trials for glucose, insulin, cortisol, plasma volume, IL-6, IL-10, total or plasma ORAC.
Ahmad et al. [48]	Male runners of recreational standard ($n = 10$).	Randomised, single blind, crossover.	60 min run at ~65% $\dot{V}O_2$ max, followed by 2 h 'rehydration phase' and 20 min treadmill running test.	6.8% CHO solution (honey) or water to recover 150% of body mass lost during run one. Fluid consumed 0 min (60% of mass loss), 30 min (50%), and 60 min (40%) after run one.	Blood samples at pre (T1), immediately post (T2), 30 min post (T3), 60 min post (T4), 90 min post (T5), and 120 min post (T6) 60 min run, and immediately post 20 min run (T7).	Blood glucose, serum insulin, haematocrit, and serum osmolality.	Serum insulin was ↑ at T3-T6 for honey vs. water. Serum osmolality was ↑ at T4 for honey vs. water. ↔ between trials for glucose or haematocrit.
Earnest et al. [46]	Amateur male cyclists ($n = 9$).	Randomised, double blind, counterbalanced, crossover.	64 km time trial on cycling ergometer.	15 g of gel (honey, dextrose, or placebo) with 250 mL H_2O consumed every 16 km (5×15 g total) plus an additional 250 mL of water every 3.2 km.	Blood samples at pre-exercise (T1), 16 km (T2), 32 km (T3), 48 km (T4), and 64 km (T5).	Blood glucose and insulin concentrations.	↔ between trials for glucose or insulin. In dextrose, glucose at T4 was ↓ vs. T1 (not the case for honey or placebo).

Table 1. *Cont.*

Study	Participants	Design	Exercise Stimulus	Nutritional Intervention	Data Collection Method and Time-Points	Outcome Variables	Main Results
Kreider et al. [49]	Resistance trained individuals ($n = 40$; males: $n = 19$, females: $n = 21$).	Randomised, four independent groups.	3 sets of 10 repetitions at approximately 70% of 1RM on chest press, seated row, shoulder press, lat pulldown, leg extension, leg curl, biceps curl, triceps extension, and leg press.	40 g of whey protein with 120 g of either sucrose, powdered honey, or maltodextrin consumed within 5 min post-exercise. Other group consumed no supplement.	Blood samples at pre-exercise (T1), post exercise (T2), and 30 min (T3), 60 min (T4), 90 min (T5), and 120 min (T6) post-feeding.	Glucose, insulin, testosterone, cortisol, testosterone: cortisol ratio, WBC, neutrophils, total neutrophils: total lymphocytes ratio, creatinine, BUN, BUN: creatinine ratio, CK, LDH, AST, and ALT.	Glucose at T4 was ↑ for honey vs sucrose, and at T3 was ↑ for honey vs. sucrose, maltodextrin, and no CHO. Insulin at T3-T6 was ↑ for honey, sucrose, and maltodextrin vs. no CHO. ↔ between groups for testosterone, cortisol, testosterone:cortisol, WBC, neutrophils, total neutrophils, total lymphocytes, LDH, AST, or ALT. BUN: creatinine at T5, was ↑ for honey and maltodextrin, and at T6 was ↑ for honey and sucrose vs. no CHO.
Łagowska et al. [45]	Trained male rowers ($n = 11$).	Randomised, crossover.	Rowing ergometer: 2 × 40 min with a 5 min break, performed at an intensity corresponding to ~75% of the onset of blood lactate accumulation.	150 mL of CHO solution (either commercial sports drink; 7.8% CHO, or 'natural' drink containing banana, fruit juice, and honey; 6.7% CHO), consumed immediately pre-exercise and every 15 min during exercise (6 × 150 mL total).	Blood samples at pre-exercise (T1), and 3 min post-exercise (T2).	Blood glucose, lactate, chemical antioxidants, urea, CK, haematocrit, leukocytes, WBC, lymphocytes, monocytes, and granulocytes.	Glucose was ↓ at T2, for natural vs. commercial drink. Glucose at T2, was ↓ for natural, but ↑ for commercial drink vs. T1. Chemical antioxidant level at T2, was ↓ for natural vs. commercial drink. ↔ between trials for lactate, urea, CK, haematocrit, leukocytes, WBC, lymphocytes, monocytes, or granulocytes.

ALT: alanine aminotransaminase, AST: aspartate aminotransaminase, BUN: blood urea nitrogen, CHO: carbohydrate, CK: creatine kinase, IL-1ra: interleukin-1 receptor antagonist, IL-6: interleukin-6, IL-10: interleukin-10, LDH: lactate dehydrogenase, NCAA: National Collegiate Athletic Association, ORAC: oxygen radical absorbance capacity, WBC: white blood cell counts, 1RM: one repetition maximum, ↑: increased/higher, ↓: decreased/lower, ↔: no difference.

Table 2. Studies examining the effect of honey supplementation over the course of multiple weeks on biochemical responses to exercise.

Study	Participants	Design	Exercise Stimulus	Nutritional Intervention	Data Collection Method and Time-Points	Outcome Variables	Main Results
Gmunder et al. [52]	Long distance runners ($n =$ 16: male: $n =$ 13, female: $n =$ 3). Standard not specified.	Randomised, double blind, two independent groups.	27 day training period, followed by 21 km running time trial.	3×10 mL per day (30 mL per day for 31 days) of either a supplement comprised of herbal yeast, malt, honey, and orange juice, or a supplement comprised of sucrose and caramel.	Blood samples at day 0 (T1), pre 21 km run at day 27 (T2), immediately post run (T3), two days post run (T4).	WBC, leukocyte counts, lymphocyte counts, Con A, CD3, CD4, CD16, CD16/8, CD19, IgA, IgE, IgM, IgG subclasses 1-4, β2M, IL-2 receptors, neopterin, plasma proteins, cortisol, adrenaline, noradrenaline.	$\leftrightarrow$ between groups for any blood marker. In honey, natural killer/suppressor cells (CD8, CD16, and CD16/8), and IgG subclass 1 at T4 were $\downarrow$, whilst Con A, and IgG subclass 2 at T4 were $\uparrow$ vs. T3 (not the case for sucrose). In sucrose, neopterin and β2M at T4 were $\downarrow$ vs. T3 (not the case for honey).
Hajizadeh et al. [51]	Amateur male cyclists ($n =$ 24)	Randomised, two independent groups.	16 week training period.	70 g honey dissolved in 250 mL distilled water. Consumed 90 min prior to each training session for 16 weeks. Other group consumed no supplement.	Blood samples at week 0 (T1), immediately (T2), 12 h (T3), and 24 h (T4) after the last training session in week 8, and immediately (T5), 12 h (T6), 24 h (T7), 7 days (T8), and 30 days (T9) after the last training session in week 16.	Lymphocyte counts, DNA damage, IL-1β, IL-6, IL-8, IL-10, TNF-a, glutathione, TAS, hydrogen peroxide, and lipid peroxide.	Hydrogen peroxide, and lipid peroxide, IL-1β, IL-6, IL-10, and TNF-a at T2-T8, were $\downarrow$ for honey vs. no supplement. Glutathione at T4-T7 was $\uparrow$ for honey vs. no supplement. T2-T7 was $\uparrow$ for honey vs. no supplement.

Table 2. *Cont.*

Study	Participants	Design	Exercise Stimulus	Nutritional Intervention	Data Collection Method and Time-Points	Outcome Variables	Main Results
Ooi et al. [53]	Females ($n = 37$).	Four (matched) independent groups.	Two groups: 3 aerobic dance sessions per week for 6 weeks. Other two groups: no exercise.	Two groups: 20 g honey with 300 mL water consumed daily for 6 weeks. Other two groups: no supplementation.	Blood samples at week 0 (T1) and after week 6 (T2).	ALP and 1CTP.	ALP at T2 for honey, and honey plus exercise was ↑ vs. T1 (not the case with no honey). ↔ between groups for ALP or 1CTP. ↔ between T1 and T2 for 1CTP.
Tartiban and Maleki [50]	Male amateur road cyclists ($n = 39$).	Randomised, double blind, two independent groups.	8 weeks intensified training period.	70 g of honey or 70 g CHO-free sweetener, with 250 mL water. Consumed 90 min prior to each training session for 8 weeks.	Semen samples at week 0 (T1), immediately (T2), 12 h (T3), and 24 h (T4) after the last training session in week 4, and immediately (T5), 12 h (T6), and 24 h (T7) after the last training session in week 8.	Semen volume, motility, morphology, concentration, and number of spermatozoa, plus IL-1β, IL-6, IL-8, TNF-α, ROS, malondialdehyde, and antioxidants superoxide dismutase, catalase, and TAC.	Semen volume at T6 was ↑ for honey vs. placebo. Semen motility, morphology and TAC at T6-T7, semen concentration at T4-T5, number of spermatozoa at T2-T3 and T5-T7, catalase at T4 and T5-T7, and superoxide dismutase at T2-T3 and T5-T7 were all ↑ for honey vs. placebo. IL-1β at T2 and T5-T7, IL-6 at It2-T7, IL-8 at T2-T3, TNF-α at T4, ROS at T4-T7, and malondialdehyde at T5-T7 were all ↓ for honey vs. placebo.

ALP: serum alkaline phosphatase, β2m: beta-2 microglobulin, CD3: cluster of differentiation 3, CD4: cluster of differentiation 4, CD16: cluster of differentiation 16, CD16/8: cluster of differentiation 16/8, CD19: cluster of differentiation 19, CHO: carbohydrate, Con A: concanavalin A, IgA: immunoglobulin A, IgE: immunoglobulin E, IgM: immunoglobulin M, IgG: immunoglobulin G, IL-1β: interleukin-1 beta, IL1ra: interleukin-1 receptor antagonist, IL-2: interleukin 2, IL-6: interleukin-6, IL-8: interleukin-8, IL-10: interleukin-10, LDH: lactate dehydrogenase, ORAC: oxygen radical absorbance capacity, ROS: reactive oxygen species, TAC: total antioxidant capacity, TAS: total antioxidant status, TNF-α: tumor necrosis factor alpha, WBC: white blood cell counts, 1CTP: serum C-terminal telopeptide of type 1 collagen, ↑: increased/higher, ↓: decreased/lower, ↔: no difference.

Table 3. Studies examining the effect of honey supplementation on exercise (physical or skilled) performance.

Study	Participants	Design	Exercise Stimulus	Nutritional Intervention	Data Collection Method and Time-Points	Outcome Variables	Main results
Abbey and Rankin [47]	Male soccer players of NCAA Division I, post collegiate, and club standard ($n = 10$).	Randomised, single blind, crossover.	Soccer specific exercise (5×15 min blocks, plus 10 min half-time), followed by progressive 20 m shuttle run to fatigue.	6% CHO-electrolyte solution (either honey, commercial sports drink, or placebo). 8.8 mL·kg^{-1} (0.5 g·kg·body mass^{-1}) CHO consumed 30 min pre-exercise and at half-time.	220 m running time trial, dribbling/agility assessment, and soccer shooting test all performed every 15 min during exercise: 220 m running time trial Progressive 20 m shuttle run to fatigue performed following 75 min of exercise.	Time taken to complete (time trial, and dribbling/agility test), number of targets hit (shooting assessment). Time to exhaustion.	↔ between trials for any performance measure.
Ahmad et al. [48]	Male runners of recreational standard ($n = 10$).	Randomised, single blind, crossover.	60 min run at ~65% $\dot{V}O_2$ max, followed by 2 h 'rehydration phase' and 20 min treadmill running test.	Either 6.8% CHO solution (honey) or water, to recover 150% of body mass lost during run one consumed at 0 min (60% of mass loss), 30 min (50%), and 60 min (40%) after run one.	20 min treadmill running test performed 120 min following completion of 60 min run	Total distance covered.	20 min running performance was ↑ for honey vs. water.
Earnest et al. [46]	Amateur male cyclists ($n = 9$).	Randomised, double blind, counterbalanced, crossover.	64 km time trial on cycling ergometer.	15 g of gel (honey, dextrose, or placebo) with 250 mL water consumed every 16 km (5×15 g total). Plus an additional 250 mL of water every 3.2 km.	64 km cycling ergometer test.	Time taken to complete 64 km, and per 16 km. Mean power over 64 km, and per 16 km.	↔ between trials for time taken to complete, or mean power over 64 km. In honey and dextrose trials, mean power over 48–64 km was ↑ vs. 0–16 km, 16–32 km, and 32–48 km (not the case for placebo). In placebo, time taken for 48–64 km and 32–48 km was ↑ vs. 0–16 km, (not the case for honey or dextrose).

Table 3. *Cont.*

Study	Participants	Design	Exercise Stimulus	Nutritional Intervention	Data Collection Method and Time-Points	Outcome Variables	Main results
Hajizadeh et al. [51]	Amateur male road cyclists ($n = 24$).	Randomised, two independent groups.	16 week training period.	70 g honey dissolved in 250 mL distilled water. Consumed 90 min prior to each training session for 16 weeks. Other group consumed no supplement.	5 km, and 40 km cycling ergometer tests, and power assessment at week 0 (T1) and week 16 (T2).	Time taken to complete (5 km and 40 km tests), peak power output (power assessment).	↔ between groups for time taken to complete 5 km or 40 km, or peak power output at T2. For all measures, performance in both groups was ↑ at T2 vs. T1

CHO: carbohydrate, NCAA: National Collegiate Athletic Association, ↑: increased/higher, ↓: decreased/lower, ↔: no difference.

Table 4. Studies examining the effect of honey supplementation on perceptual responses around exercise.

Study	Participants	Design	Exercise Stimulus	Nutritional Intervention	Data collection Method and Time-Points	Outcome Variables	Main Results
Abbey and Rankin [47]	Male soccer players of NCAA Division I, post-collegiate, and club standard ($n = 10$).	Randomised, single blind crossover.	Soccer-specific exercise (5×15 min blocks, plus 10 min half-time), followed by progressive 20 m shuttle run to fatigue.	6% CHO-electrolyte solution (honey, commercial sports drink, or placebo). 8.8 mL·kg^{-1} (0.5 g·kg·body mass^{-1}) CHO consumed 30 min pre-exercise and at half-time.	Every 15 min during exercise, scale not reported.	RPE.	↔ between trials for RPE.
Ahmad et al. [48]	Male runners of recreational standard ($n = 10$).	Randomised, single blind crossover.	60 min run at ~65% $\dot{V}O_2$ max, followed by 2 h 'rehydration phase' and 20 min treadmill running test on.	6.8% CHO solution (honey) or H_2O to recover 150% of body mass lost during run one. Fluid consumed 0 min (60% of mass loss), 30 min (50%), and 60 min (40%) after run one.	Likert scale (1–5), 0 min (T1), 30 min (T2), and 60 min (T3) after run one.	Perceptions of thirst, sweetness, nausea, fullness and stomach upset.	Perceptions of sweetness at T1-T3 were ↑ for honey vs. for honey vs. water. ↔ between trials for perceptions of thirst nausea, fullness, or stomach upset between trials.

Table 4. *Cont.*

Study	Participants	Design	Exercise Stimulus	Nutritional Intervention	Data collection Method and Time-Points	Outcome Variables	Main Results
Earnest et al. [48]	Amateur male cyclists ($n = 9$).	Randomised, double blind, counterbalanced, crossover.	64 km time trial on cycling ergometer.	15 g of gel (honey, dextrose, or placebo) with 250 mL water consumed every 16 km. Plus an additional 250 mL of water every 3.2 km.	Likert scale (6–20) at pre-exercise, 16 km, 32 km, 48 km, and 64 km.	RPE.	↔ between trials for RPE.
Kreider et al. [49]	Resistance trained individuals ($n = 40$; males: $n = 19$, females: $n = 21$).	Randomised, four independent groups.	3 sets of 10 repetitions at approximately 70% of 1 RM on chest press, seated row, shoulder press, lat pulldown, leg extension, leg curl, biceps curl, triceps extension, and leg press.	40 g of whey protein with 120 g of either sucrose, powdered honey, or maltodextrin. Consumed within 5 min post-exercise. Other group consumed no supplement.	Likert scale (1–10) at pre-exercise (T1), post exercise (T2), and 30 min (T3), 60 min (t4), 90 min (T5), and 120 min (T6) post-feeding.	Severity of perceived hypoglycaemia, dizziness, fatigue, headache, and stomach upset.	↔ between groups for perceptions of hypoglycaemia, dizziness, fatigue, headache, or stomach upset.
Łagowska et al. [45]	Trained male rowers ($n = 11$).	Randomised, crossover.	Rowing ergometer: 2 × 40 min with a 5 min break, performed at an intensity corresponding to ~75% of the onset of blood lactate accumulation.	150 mL of CHO solution (either commercial sports drink; 7.8% CHO, or 'natural' drink containing banana, fruit juice, and honey; 6.7% CHO), consumed immediately pre-exercise and every 15 min during exercise (6 × 150 mL total).	Likert scale (1–5) at post-exercise:	Perceptions of taste, smell, thirst quenching ability, beverage consistency, and refreshment.	Satisfaction with beverage consistency was ↓ for natural vs. commercial drink. ↔ between trials for perceptions of taste, smell, thirst quenching ability, or refreshment.

CHO: carbohydrate, NCAA: National Collegiate Athletic Association, RPE: rating of perceived exertion, 1RM: one repetition maximum, ↑: increased/higher, ↓: decreased/lower, ↔: no difference.

4. Discussion

This review aimed to systematically evaluate the current body of research that has assessed the influence of honey supplementation on a number of physiological, performance, and perceptual responses to exercise. Whilst nine eligible articles were identified, substantial methodological variation exists between studies, thus making it difficult to draw firm conclusions with regards to the potential efficacy of honey supplementation for exercising populations. As is the case with many supplements, it seems likely that a number of factors, including but not limited to, the dose and timing of honey ingestion, individual responsiveness, and the type, duration, and/or intensity of exercise, modulate the responses observed [44].

4.1. Effect of Honey Supplementation on Biochemical Markers (i.e., Blood or Semen)

4.1.1. Acute Honey Consumption around a Single Exercise Session

One of the primary aims for athletes consuming carbohydrates before and during exercise is to maintain fuel availability (i.e., preserve glycogen stores and increase blood glucose concentrations) to help attenuate declines in performance as exercise progresses. With regards to honey consumption prior to and/or during an exercise bout, blood glucose responses have been largely inconsistent but notably the effects on glycogen degradation remain to be examined (Table 1).

Only one study has reported a statistically significant difference in blood glucose concentrations as a result of consuming a honey-containing supplement. Lagowska et al. [45] observed lower post-exercise blood glucose concentrations when a "natural" carbohydrate beverage containing honey, fruit juice, and banana was consumed during 80 min (2 × 40 min, separated by 5 min) of rowing exercise, compared with a commercially available carbohydrate solution. Whilst the reasons for these findings remain unclear, the differential responses may be attributable to the amount of carbohydrate consumed. Although both conditions entailed consumption of six boluses of 150 mL of solution (i.e., one every 15 min during exercise), the "natural" drink contained 6.7% carbohydrates, whereas the comparator was a 7.8% solution. It seems likely that differences in the absolute quantity of carbohydrate consumed, irrespective of source, may at least partly explain the differential blood glucose responses in favour of the higher amount. Unfortunately, because the precise quantity of each constituent within the two drinks was not reported, further speculation is rendered difficult.

Blood glucose concentration maintenance may be of particular interest to athletes engaged in sports which require high levels of technical skill (e.g., team sport athletes). Whist a definitive link between hypoglycaemia and reductions in sport-specific physical or skilled performance has not yet been established in exercising participants, the role of blood glucose in brain function is clear. Indeed, the brain is one of the few human organs that relies heavily on blood glucose to maintain optimal functioning [54]. Notably, exercise studies have indicated that the rate of cerebral glucose uptake begins to decline when blood glucose concentrations fall below ~3.6 mmol·L^{-1} [55], a concentration which has previously been observed in team sport players ~15–30 min following half-time [28]. Moreover, the link between blood glucose concentrations and cognitive functioning is highlighted by increased fine motor speed, psycho-motor speed, and visual discrimination speed accompanying increased glycaemia following soccer match-play in the heat [56]. As cognitive processes are likely to be vital, not only to the skilled actions involved in team sports but also to tactical and/or strategic decision making, nutritional strategies that help maintain or increase blood glucose concentrations could be of benefit to team sports players during the latter stages of a match [29]. Given its typical carbohydrate composition (i.e., containing primarily low GI fructose), there exists a theoretical basis to suggest honey as a potentially worthwhile intervention in this context [29].

When high GI carbohydrates are consumed during team sport specific exercise, a temporary lowering of blood glucose concentrations may be experienced during the early stages of the second half [28–31]. To date, only one study has combined honey supplementation with simulated soccer match-play, and no differences in insulin or blood glucose concentrations were observed immediately or one hour following exercise cessation when either a honey solution, a carbohydrate-matched commercially available sports drink, or an energy-free placebo were consumed [47]. However, because blood samples were not taken regularly throughout exercise, whether or not honey influenced transient changes in blood glucose responses as seen previously, remains unclear. Given the role of blood glucose concentrations in the maintenance of brain function, future research into the potential influence of honey supplementation on blood glucose concentrations, in addition to physical, skilled, and cognitive performance throughout intermittent exercise, would be worthwhile.

Other studies involving honey consumption prior to and/or during exercise have also reported no significant treatment effects for blood glucose concentrations [46,48]. Notwithstanding, blood glucose concentrations were maintained throughout a 64 km cycling time-trial when 15 g of either honey or dextrose were consumed every 16 km [46]. In contrast, in the placebo condition, blood glucose concentrations had declined after 48 km relative to the initial 16 km [46]. Although no differences between honey and dextrose were observed, it is unclear whether the honey and dextrose trials were carbohydrate- or energy-matched. Indeed, whilst the authors state that 15 g of gel was consumed in each condition, it should be considered that whereas dextrose is ~100% carbohydrate, honey in its natural form may be only ~80% carbohydrate or less [2]. The lack of information regarding whether carbohydrates were consumed in equivalent doses makes interpretation of the results difficult, but if discrepancies existed, the differences in absolute carbohydrate intake may have influenced the results in addition to the source of carbohydrates alone.

Alongside providing amino acids for muscle repair and fluid for rehydration, rapid replenishment of glycogen stores is of primary importance for facilitating recovery following exercise [17,57]. Moreover, co-ingesting carbohydrates and protein during the post-exercise period may help to promote an anabolic environment, offset the acute immunosuppressive effects of intense exercise, and facilitate glycogen restoration [17,19,33,57]. Krieder et al. [19] investigated the effects of consuming 40 g whey protein alongside 120 g of either sucrose, powdered honey, or maltodextrin within five minutes of completing a bout of resistance exercise. When compared with individuals taking no supplement, all three carbohydrate groups returned higher insulin concentrations at 30, 60, 90 and 120 min post-feeding. Although this may be expected given that carbohydrates and whey protein are insulinogenic [58], it is interesting to note that blood glucose concentrations in the honey group were higher after 30 min post-feeding compared with those in the sucrose group, and higher than all three other conditions at 60 min following consumption. In keeping with the lower GI of honey compared with the other sources of carbohydrate assessed, these patterns appear to suggest a more prolonged appearance of carbohydrate in the bloodstream when honey was consumed. Whilst no treatment effect was observed for testosterone or cortisol, these blood glucose and insulin responses may suggest potential for the use of honey when post-exercise recovery is required.

When honey is consumed prior to, and/or during, a single exercise session, limited effects have been observed with regards to immunological markers (Table 1). Although the plasma IL-1ra response to team sport specific exercise was dampened following honey supplementation (i.e., a 6% solution providing a total of 1 g·kg^{-1} body mass) compared with when a commercial sports drink or carbohydrate-free placebo were consumed, similar IL-6 responses were observed [47]. The authors thus proposed that the antioxidant and/or polyphenol content of the honey supplement may have somewhat interrupted the intracellular production or release of IL-1ra in response to exercise induced elevations in IL-6. Conversely, previous carbohydrate research showed that ingesting a 6.4% glucose and maltodextrin solution before, and at 15 min intervals during, simulated soccer match-play attenuated a number of post-exercise immune disturbances (i.e., increases in plasma cortisol and IL-6, and inhibition of bacterially stimulated neutrophil degranulation), compared with a placebo [59]. Although comparisons are complicated by inconsistencies in the specific methods employed (i.e., length of exercise bout, pattern, dose, and type of carbohydrate provision, and the time-frame of measurement), it appears plausible that differences in blood glucose responses may at least partially explain the divergent findings. In support, increases in stress hormone concentrations during exercise may be attenuated when blood glucose is maintained via the feeding of exogenous carbohydrates [33,34,36,38]. That said, Abbey and Rankin [47] observed no differences in blood glucose concentrations between the honey, commercial beverage, and placebo conditions.

Although the evidence remains limited (Table 1), other studies have also shown that honey exhibits similar effects on immunological responses compared with other carbohydrate sources, at least where acute consumption around a single exercise session is concerned [45,49]. More well-controlled research is therefore required to determine whether honey has any ameliorating influence on immune markers, compared with that provided by other sources of carbohydrate.

4.1.2. Honey Supplementation over Multiple Weeks

In contrast to findings from acute honey supplementation studies, the literature available to date suggests that honey, consumed over the course of multiple weeks, may dampen the inflammatory response to periods of repeated exercise [50,51]. Indeed, ingesting 70 g of unprocessed honey ~90 min prior to each training session over 16 weeks of moderate-to-intense cycling training resulted in reduced plasma cytokine (i.e., IL-1β, IL-6, IL-10, and TNF-α) concentrations immediately, 12 h, and 24 h after the last training session in the eighth week, and immediately, 12 h, 24 h, and seven days after the last training session in week 16, compared with when no supplement was consumed [51]. Moreover, glutathione levels and total antioxidant status were higher with honey supplementation. Very similar patterns of cytokine and antioxidant responses were reported when Tartiban et al. [50] implemented the same supplementation and exercise strategy over an eight week period. This latter study, albeit investigating semen as opposed to blood markers, also noted reductions in indices of oxidative stress (i.e., reactive oxygen species and malondialdehyde) when honey was consumed throughout the eight week period. Whilst oxidative stress may play an important role in cellular adaptations to training [15,33,40], these studies in amateur male cyclists appear to demonstrate a potential application for honey supplementation during periods in which reductions in exercise-induced immune disturbances are desired. However, whilst it was stated that all participants were advised to maintain their normal diets for the duration of both investigations, and that diet diaries were completed to ensure compliance, neither articles provided detailed information outlining participants' overall energy, or macronutrient intake and/or distribution. As carbohydrate availability may influence immunological responses to exercise [34], it remains unclear whether the reported results stem from the inherent properties of honey itself, or simply reflect the outcome of an increased carbohydrate or energy intake in the honey group. In addition, more research, using ecologically valid and robust methods, is required to determine whether attenuation of acute immunological perturbations results in decreased incidences of infection.

In contrast to the above studies in cyclists [50,51], Gmunder et al. [52] reported no significant between-group differences in immunological responses to a 21 km run completed after 27 days of treatment when participants consumed 30 mL per day of either a supplement containing (a) herbal yeast, malt, honey, and orange juice, or (b) sucrose and caramel. Whilst definitive conclusions are difficult to draw, it seems plausible that the absolute quantity of honey consumed by runners in this study was insufficient to confer any antioxidant effect. Indeed, the primary health benefits of honey consumption have been demonstrated with intakes >50 g [3]. Considering the paucity of well-controlled studies published to date, further work is required to investigate the immunological effects of acute and longer term honey supplementation within exercising populations. In particular, research examining the effects of different variables which may influence the efficacy of supplementation; such as dose, duration and/or form of honey supplementation, different exercise modalities and/or intensities, different levels of athlete, and/or the role played by an individual's starting nutritional status, would be of particular interest [3,44]. From a practical perspective, the relationship between potential reductions in immunological perturbations as a result of honey supplementation and (a) the risk of developing an infection, (b) recovery and subsequent exercise performance, and (c) long-term training adaptations, must represent a research priority to help guide practitioners and athletes.

Ooi et al. [53] studied 37 females over the course of six weeks with an interest in markers of bone formation (i.e., serum alkaline phosphatase) and resorption (i.e., serum C-terminal telopeptide of type 1 collagen). Participants were assigned to one of four groups, whereby two groups performed aerobic dance exercises three times per week (with one group also consuming 20 g honey per day), whilst the remaining two groups remained sedentary (again with one group also consuming 20 g honey per day). Although there were no differences for either outcome variable when groups were directly compared, the two groups who consumed honey experienced significant increases in serum alkaline phosphatase, tentatively indicating greater bone formation over the six week period. In contrast, both non-supplemented groups maintained similar levels at week six compared with week zero [53]. Honey supplementation has previously shown potential to enhance markers of bone structure in rodent studies [60], and it could be the case that certain components such as vitamin K, and minerals such as calcium, phosphorus, iron, and magnesium found in honey contribute to improved bone health, especially when combined with an exercise programme.

4.2. Effect of Honey Supplementation on Physical or Skilled Performance

It is recommended that consuming carbohydrates prior to, and during, exercise may enhance indices of physical (i.e., time to exhaustion) and skilled (i.e., soccer passing and shooting accuracy) performance [14,16,19–23,44]. Moreover, combining different forms of carbohydrates may increase oxidation rates and allow for worthwhile ingestion of a greater overall total volume of exogenous energy during prolonged endurance exercise [14,26,27]. Given that honey contains multiple sources of carbohydrates (i.e., primarily fructose and glucose), this natural substance seems intuitively able to offer potential as a "food-first" approach to carbohydrate supplementation. However, studies investigating the influence of honey on physical or skilled performance have reported mixed results (Table 3), with only one investigation reporting a clear performance benefit [48].

When amateur runners used a 6.8% Acacia honey solution to restore ~150% of body mass losses following a 60 min run in the heat, improved running performance (i.e., distance covered) versus ingestion of an equivalent volume of water occurred during the 20 min treadmill test that followed 120 min later [48]. Given the established role of carbohydrate–electrolyte beverages in fuelling for, and recovering from, exercise, such findings are not unexpected. However, because no carbohydrate-matched alternative to honey was assessed, it cannot be determined whether the positive outcomes are linked the unique properties of honey, or a result of carbohydrate consumption (i.e., as opposed to water) per se.

Whilst no significant between-condition differences were observed for overall time taken to complete a 64 km time-trial, ingesting 15 g of either honey or dextrose in gel form every 16 km enabled cyclists to maintain average 16 km time throughout the duration of exercise, and to increase average power output during the final 16 km compared with the preceding 16 km segments [46]. In contrast, when an energy-free placebo was consumed, significant declines in performance were observed over 48–64 km compared with the opening 16 km of exercise. Although this investigation identified no differences between honey and dextrose for any performance measure, the rate of carbohydrate consumption may have been a factor. Indeed, because carbohydrates consumed from a single source (i.e., glucose) may be oxidised at up to ~60 g·h^{-1}, the full benefits of consuming multiple transportable carbohydrates (i.e., in terms of increased oxidation rates) may not be realised until absolute carbohydrate intake exceeds this level [14]. Moreover, a clear dose-response relationship exists between carbohydrate intake and endurance performance [14,61,62]. As noted above in relation to blood glucose responses, it is unclear whether the honey and dextrose trials were carbohydrate- or energy-matched. Due to reporting being vague on this matter, there is the potential that discrepancies in absolute carbohydrate intake may have influenced the results in addition to the source of carbohydrates alone.

With regards to intermittent exercise, carbohydrate supplementation has previously demonstrated benefits in terms of maintaining physical and skilled performance during the latter stages of simulated team sport match-play, whether consumed frequently (i.e., every 15 min) during exercise [28,30] or in more ecologically valid feeding patterns (i.e., before exercise and at half-time) [23]. However, ingesting a honey solution containing 6% carbohydrates demonstrated no benefit to any measure of physical or skilled performance assessed either during, or immediately following, 75 min of soccer specific exercise, when compared with a carbohydrate-matched commercially available sports drink and an energy-free placebo [46]. Whilst it may appear surprising that neither carbohydrate intervention influenced performance, it has been suggested that when limited opportunities exist to ingest carbohydrates, consuming solutions containing upwards of 10% carbohydrates may enable ergogenic rates of energy intake (e.g., >50 g·h^{-1}) to be achieved whilst minimising abdominal discomfort [29]. For example, Harper et al. [23] observed significant improvements in dribbling speed and self-paced exercise performance during the latter stages of simulated soccer match-play when a 12% carbohydrate–electrolyte solution was delivered in 250 mL boluses prior to the beginning of each half, compared with either water or an electrolyte placebo. Future research into the effects of consuming higher concentrations of honey (i.e., ≥10% carbohydrate) on physical, skilled, and cognitive performance during team sport specific exercise would be beneficial.

Whilst short periods of deliberately training with low endogenous and exogenous carbohydrate availability may promote positive training adaptations through increases in mitochondrial enzyme activity, increases in lipid oxidation, and potential improvements in exercise capacity [15,19,37], consuming a diet rich in carbohydrates may help athletes to fuel and recover from training and thus itself promote beneficial responses via an increase in training intensity [19]. In the only eligible study to have investigated the performance effects of honey supplementation over a period of multiple days, amateur male road cyclists who supplemented with 70 g of honey 90 min prior to each training session demonstrated no additional improvements in 5 km or 40 km cycling time-trial performance, or peak power output over a 16 week period compared with those taking no supplement [51]. Unfortunately, detailed dietary analysis was not provided, thus it is not possible to comment upon the adequacy of, or differences in, overall carbohydrate intake in either group over the 16 week duration. Well-controlled research to determine whether longer-term honey consumption translates into favourable performance adaptations will be difficult to conduct. Whilst longer-term consumption may have a number of other benefits for health and wellbeing [3,5,7], research into the performance effects of honey supplementation on an acute level (i.e., immediately prior to and/or during exercise) may have a more sound theoretical underpinning.

4.3. Effect of Honey Supplementation on Perceptual Responses

Honey has demonstrated similar effects on subjective ratings of effort, compared with when other sources of carbohydrates, energy free placebos, or water have been consumed (Table 4). Such findings reflect previous reports in relation to intermittent exercise, whereby nutritional interventions failed to influence perceived exertion [30,31]. Moreover, whilst possible reductions in ratings of perceived exertion have been observed during prolonged endurance cycling when a mixture of glucose and fructose were consumed versus glucose alone, these responses occurred at substantially higher rates of carbohydrate ingestion (i.e., 90 g·h^{-1}) than those provided by the studies presented in this review [63]. Indeed, the only perceptual responses affected by honey consumption relate to flavour and texture as opposed to exertion or fatigue. Specifically, a honey solution elicited a sweeter taste compared with plain water [48], and a natural beverage based upon honey, banana, and fruit juice provided a less satisfying consistency than a commercially available sports drink [45]. Within tolerable limits, sweeter tasting beverages may promote increases in voluntary fluid intake during exercise [64], and such findings may suggest a potential application of honey to encourage a greater rate of carbohydrate consumption in individuals suffering from flavour fatigue.

The fact that a honey solution (6% carbohydrate) produced similar perceptions of thirst, nausea, fullness, and stomach upset compared with plain water [48], may be an important observation. One of the traditional concerns with recommending low GI carbohydrates around exercise has surrounded the risk of gastric distress when the time-frame for appearance of exogenous carbohydrate is prolonged [65].

Although Ahmad et al. [48] only assessed perceptual responses immediately after each carbohydrate feeding, and not during the 20 min exercise bout that followed, previous research incorporating soccer match simulation has shown similar abdominal discomfort values when an 8% isomaltulose solution was consumed, compared with an equivalent volume of high GI maltodextrin [31]. Taken together, these studies may provide food for thought for practitioners and athletes who may previously have been deterred from considering lower GI carbohydrates before and during exercise.

5. Conclusions and Future Research Recommendations

Due to the potential health benefits and to offset the risks posed by supplement contamination, many athletes, practitioners and researchers espouse a "food-first" approach to sports nutrition. As honey is a natural substance comprised of ~80% carbohydrate (primarily fructose and glucose), and is known to possess antioxidant, antimicrobial, and anti-inflammatory properties, there exists a theoretical basis for its use as a nutritional supplement in exercising populations. This review summarised the available literature which has investigated the effects of honey supplementation when combined with exercise over a number of different time-frames, dosages, and modalities. Whilst the large methodological differences within the studies represented a substantial limitation, information is presented which may inform worthwhile future research.

Compared with other forms of carbohydrate, honey ingestion has had a similar effect on exercise performance, perceptions of fatigue, blood glucose concentrations, and immunological responses when consumed immediately prior to and/or during exercise, although some positive influence has been observed. When routinely consumed over multiple weeks, honey may attenuate many of the immune perturbations typically associated with a programme of moderate-to-intense exercise. Unfortunately, the research designs employed and the level of detail with which the methods have been reported make it difficult to establish whether the observed responses are attributable to the intrinsic properties of honey itself, or reflect other factors such as discrepancies in carbohydrate intake between conditions. Similarly, the same limitations mean that honey may have had certain effects (either positive or negative) which could have been masked by "noise" from external influences. Despite the lack of conclusive evidence, theory supports the use of honey, particularly as a potential ergogenic aid when consumed around exercise. Future research should take a robust approach to assessing whether honey may offer benefits to physical, skilled, or cognitive performance during different modalities, durations, and intensities of exercise, when directly compared to equivalent volumes of carbohydrates delivered in traditional forms. From these authors' perspective, the application to skilled performance during team sports may be of particular interest. Moreover, studies directly comparing the responses of male and female athletes would be a valuable addition to the knowledge base.

Author Contributions: S.P.H., P.M., C.W., and M.R. contributed to the structure and writing of the article. All authors were also involved in reviewing the available literature, and highlighting opportunities for future research.

Funding: This research received no external funding.

Conflicts of Interest: The authors declare no conflict of interest.

References

1. Council, E. Council Directive 2001/110/EC of 20 December 2001 relating to honey. *Off. J. Eur. Commun. L* **2002**, *10*, 47–52.
2. Schneider, M.; Coyle, S.; Warnock, M.; Gow, I.; Fyfe, L. Anti-microbial activity and composition of manuka and portobello honey. *Phytother. Res.* **2013**, *27*, 1162–1168. [CrossRef] [PubMed]
3. Bogdanov, S.; Jurendic, T.; Sieber, R.; Gallmann, P. Honey for nutrition and health: A review. *J. Am. Coll. Nutr.* **2008**, *27*, 677–689. [CrossRef] [PubMed]
4. Manyi-Loh, C.E.; Clarke, A.M.; Ndip, N. An overview of honey: Therapeutic properties and contribution in nutrition and human health. *Afr. J. Microbiol. Res.* **2011**, *5*, 844–852.

5. Erejuwa, O.O.; Sulaiman, S.A.; Ab Wahab, M.S. Honey: A novel antioxidant. *Molecules* **2012**, *17*, 4400–4423. [CrossRef]

6. Ahmed, S.; Othman, N.H. Review of the medicinal effects of tualang honey and a comparison with manuka honey. *Malays. J. Med. Sci.* **2013**, *20*, 6–13.

7. Samarghandian, S.; Farkhondeh, T.; Samini, F. Honey and health: A review of recent clinical research. *Pharmacogn. Res.* **2017**, *9*, 121.

8. Meo, S.A.; Al-Asiri, S.A.; Mahesar, A.L.; Ansari, M.J. Role of honey in modern medicine. *Saudi J. Biol. Sci.* **2017**, *24*, 975–978. [CrossRef]

9. Conti, M.E.; Stripeikis, J.; Campanella, L.; Cucina, D.; Tudino, M.B. Characterization of Italian honeys (Marche Region) on the basis of their mineral content and some typical quality parameters. *Chem. Cent. J.* **2007**, *1*, 14. [CrossRef]

10. Batista, B.; Da Silva, L.; Rocha, B.; Rodrigues, J.; Berretta-Silva, A.; Bonates, T.; Gomes, V.; Barbosa, R.; Barbosa, F. Multi-element determination in Brazilian honey samples by inductively coupled plasma mass spectrometry and estimation of geographic origin with data mining techniques. *Food Res. Int.* **2012**, *49*, 209–215. [CrossRef]

11. Deibert, P.; König, D.; Kloock, B.; Groenefeld, M.; Berg, A. Glycaemic and insulinaemic properties of some German honey varieties. *Eur. J. Clin. Nutr.* **2010**, *64*, 762–764. [CrossRef] [PubMed]

12. Atkinson, F.S.; Foster-Powell, K.; Brand-Miller, J.C. International tables of glycemic index and glycemic load values: 2008. *Diabetes Care* **2008**, *31*, 2281–2283. [CrossRef] [PubMed]

13. Robert, S.D.; Ismail, A. Two varieties of honey that are available in Malaysia gave intermediate glycemic index values when tested among healthy individuals. *Biomed. Pap. Med. Fac. Univ. Palacky Olomouc Czech. Repub.* **2009**, *153*, 145–147. [CrossRef] [PubMed]

14. Jeukendrup, A. A step towards personalized sports nutrition: Carbohydrate intake during exercise. *Sports Med.* **2014**, *44*, 25–33. [CrossRef]

15. Close, G.L.; Hamilton, D.L.; Philp, A.; Burke, L.M.; Morton, J.P. New strategies in sport nutrition to increase exercise performance. *Free Radic. Biol. Med.* **2016**, *98*, 144–158. [CrossRef] [PubMed]

16. Coyle, E.F.; Coggan, A.R.; Hemmert, M.; Ivy, J.L. Muscle glycogen utilization during prolonged strenuous exercise when fed carbohydrate. *J. Appl. Physiol.* **1986**, *61*, 165–172. [CrossRef] [PubMed]

17. Burke, L.M.; van Loon, L.J.; Hawley, J.A. Postexercise muscle glycogen resynthesis in humans. *J. Appl. Physiol.* **2016**, *122*, 1055–1067. [CrossRef] [PubMed]

18. Hawley, J.A.; Schabort, E.J.; Noakes, T.D.; Dennis, S.C. Carbohydrate-loading and exercise performance. *Sports Med.* **1997**, *24*, 73–81. [CrossRef]

19. Burke, L.M.; Hawley, J.A.; Wong, S.H.; Jeukendrup, A.E. Carbohydrates for training and competition. *J. Sports Sci.* **2011**, *29*, S17–S27. [CrossRef]

20. Jeukendrup, A.E. Carbohydrate intake during exercise and performance. *Nutrition* **2004**, *20*, 669–677. [CrossRef]

21. Jeukendrup, A.E. Carbohydrate feeding during exercise. *Eur. J. Sports Sci.* **2008**, *8*, 77–86. [CrossRef]

22. Nicholas, C.W.; Williams, C.; Lakomy, H.K.; Phillips, G.; Nowitz, A. Influence of ingesting a carbohydrate-electrolyte solution on endurance capacity during intermittent, high-intensity shuttle running. *J. Sports Sci.* **1995**, *13*, 283–290. [CrossRef] [PubMed]

23. Harper, L.D.; Stevenson, E.J.; Rollo, I.; Russell, M. The influence of a 12% carbohydrate-electrolyte beverage on self-paced soccer-specific exercise performance. *J. Sci. Med. Sport* **2017**, *20*, 1123–1129. [CrossRef] [PubMed]

24. Burke, L.M.; Maughan, R.J. The Governor has a sweet tooth–mouth sensing of nutrients to enhance sports performance. *Eur. J. Sports Sci.* **2015**, *15*, 29–40. [CrossRef] [PubMed]

25. Rodriguez, N.R.; Di, N.M.; Langley, S. American College of Sports Medicine position stand. Nutrition and athletic performance. *Med. Sci. Sport Exerc.* **2009**, *41*, 709–731.

26. Currell, K.; Jeukendrup, A. Superior endurance performance with ingestion of multiple transportable carbohydrates. *Med. Sci. Sport Exerc.* **2008**, *40*, 275. [CrossRef] [PubMed]

27. Triplett, D.; Doyle, J.A.; Rupp, J.C.; Benardot, D. An isocaloric glucose-fructose beverage's effect on simulated 100-km cycling performance compared with a glucose-only beverage. *Int. J. Sport Nutr. Exerc. Metab.* **2010**, *20*, 122–131. [CrossRef] [PubMed]

28. Kingsley, M.; Penas-Ruiz, C.; Terry, C.; Russell, M. Effects of carbohydrate-hydration strategies on glucose metabolism, sprint performance and hydration during a soccer match simulation in recreational players. *J. Sci. Med. Sport* **2014**, *17*, 239–243. [CrossRef] [PubMed]

29. Hills, S.; Russell, M. Carbohydrates for soccer: A focus on skilled actions and half-time practices. *Nutrients* **2017**, *10*, 22. [CrossRef] [PubMed]

30. Russell, M.; Benton, D.; Kingsley, M. Influence of carbohydrate supplementation on skill performance during a soccer match simulation. *J. Sci. Med. Sport* **2012**, *15*, 348–354. [CrossRef] [PubMed]

31. Stevenson, E.J.; Watson, A.; Theis, S.; Holz, A.; Harper, L.D.; Russell, M. A comparison of isomaltulose versus maltodextrin ingestion during soccer-specific exercise. *Eur. J. Appl. Physiol.* **2017**, *117*, 2321–2333. [CrossRef] [PubMed]

32. Walsh, N.P.; Gleeson, M.; Shephard, R.J.; Gleeson, M.; Woods, J.A.; Bishop, N.; Fleshner, M.; Green, C.; Pedersen, B.K.; Hoffman-Goete, L.; et al. Position statement part one: Immune function and exercise. *Exerc. Immunol. Rev.* **2011**, *17*, 6–63. [PubMed]

33. Peake, J.M.; Neubauer, O.; Walsh, N.P.; Simpson, R.J. Recovery of the immune system after exercise. *J. Appl. Physiol.* **2016**, *122*, 1077–1087. [CrossRef] [PubMed]

34. Gleeson, M.; Nieman, D.C.; Pedersen, B.K. Exercise, nutrition and immune function. *J. Sports Sci.* **2004**, *22*, 115–125. [CrossRef] [PubMed]

35. Nieman, D.C.; Johanssen, L.M.; Lee, J.W.; Arabatzis, K. Infectious episodes in runners before and after the Los Angeles Marathon. *J. Sports Med. Phys. Fitness* **1990**, *30*, 316–328. [PubMed]

36. Walsh, N.P.; Gleeson, M.; Pyne, D.B.; Nieman, D.C.; Dhabhar, F.S.; Shephard, R.J.; Oliver, S.J.; Bermon, S.; Kajeniene, A. Position statement part two: Maintaining immune health. *Exerc. Immunol. Rev.* **2011**, *17*, 64–103. [PubMed]

37. Bartlett, J.D.; Hawley, J.A.; Morton, J.P. Carbohydrate availability and exercise training adaptation: Too much of a good thing? *Eur. J. Sport Sci.* **2015**, *15*, 3–12. [CrossRef] [PubMed]

38. Nieman, D.C.; Nehlsen-Cannarella, S.L.; Fagoaga, O.R.; Henson, D.A.; Utter, A.; Davis, J.M., Williams, F.; Butterworth, D.E. Influence of mode and carbohydrate on the cytokine response to heavy exertion. *Med. Sci. Sport Exerc.* **1998**, *30*, 671–678. [CrossRef]

39. Vassilakopoulos, T.; Karatza, M.-H.; Katsaounou, P.; Kollintza, A.; Zakynthinos, S.; Roussos, C. Antioxidants attenuate the plasma cytokine response to exercise in humans. *J. Appl. Physiol.* **2003**, *94*, 1025–1032. [CrossRef]

40. Morrison, D.; Hughes, J.; Della Gatta, P.A.; Mason, S.; Lamon, S.; Russell, A.P.; Wadley, G.D. Vitamin C and E supplementation prevents some of the cellular adaptations to endurance-training in humans. *Free Radic. Biol. Med.* **2015**, *89*, 852–862. [CrossRef]

41. Liberati, A.; Altman, D.G.; Tetzlaff, J.; Mulrow, C.; Gøtzsche, P.C.; Ioannidis, J.P.; Clarke, M.; Devereaux, P.J.; Kleijnen, J.; Moher, D. The PRISMA statement for reporting systematic reviews and meta-analyses of studies that evaluate health care interventions: Explanation and elaboration. *PLoS Med.* **2009**, *6*, e1000100. [CrossRef] [PubMed]

42. Maher, C.G.; Sherrington, C.; Herbert, R.D.; Moseley, A.M.; Elkins, M. Reliability of the PEDro scale for rating quality of randomized controlled trials. *Phys. Ther.* **2003**, *83*, 713–721. [PubMed]

43. De Morton, N.A. The PEDro scale is a valid measure of the methodological quality of clinical trials: A demographic study. *Aust. J. Physiother.* **2009**, *55*, 129–133. [CrossRef]

44. Russell, M.; Kingsley, M. The efficacy of acute nutritional interventions on soccer skill performance. *Sports Med.* **2014**, *44*, 957–970. [CrossRef] [PubMed]

45. Łagowska, K.; Podgórski, T.; Celińska, E.; Wiertel, Ł.; Kryściak, J. A comparison of the effectiveness of commercial and natural carbohydrate–electrolyte drinks. *Sci. Sports* **2017**, *32*, 160–164. [CrossRef]

46. Earnest, C.P.; Lancaster, S.L.; Rasmussen, C.J.; Kerksick, C.M.; Lucia, A.; Greenwood, M.C.; Almada, A.L.; Cowan, P.A.; Kreider, R.B. Low vs. high glycemic index carbohydrate gel ingestion during simulated 64-km cycling time trial performance. *J. Strength Cond. Res.* **2004**, *18*, 466–472. [PubMed]

47. Abbey, E.L.; Rankin, J.W. Effect of ingesting a honey-sweetened beverage on soccer performance and exercise-induced cytokine response. *Int. J. Sport Nutr. Exerc. Metab.* **2009**, *19*, 659–672. [CrossRef]

48. Ahmad, N.S.; Foong Kiew, O.; Ismail, M.S.; Mohamed, M. Effects of post-exercise honey drink ingestion on blood glucose and subsequent running performance in the heat. *Asian J. Sports Med.* **2015**, *6*, e24044. [CrossRef]

49. Kreider, R.B.; Earnest, C.P.; Lundberg, J.; Rasmussen, C.; Greenwood, M.; Cowan, P.; Almada, A.L. Effects of ingesting protein with various forms of carbohydrate following resistance-exercise on substrate availability and markers of anabolism, catabolism, and immunity. *J. Int. Soc. Sports Nutr.* **2007**, *4*, 18. [CrossRef]

50. Tartibian, B.; Maleki, B.H. The effects of honey supplementation on seminal plasma cytokines, oxidative stress biomarkers, and antioxidants during 8 weeks of intensive cycling training. *J. Androl.* **2012**, *33*, 449–461. [CrossRef]

51. Hajizadeh Maleki, B.; Tartibian, B.; Mooren, F.C.; Kruger, K.; FitzGerald, L.Z.; Chehrazi, M. A randomized controlled trial examining the effects of 16 weeks of moderate-to-intensive cycling and honey supplementation on lymphocyte oxidative DNA damage and cytokine changes in male road cyclists. *Cytokine* **2016**, *88*, 222–231. [CrossRef] [PubMed]

52. Gmünder, F.K.; Joller, P.W.; Joller-Jemelka, H.I.; Bechler, B.; Cogoli, M.; Ziegler, W.H.; Müller, J.; Aeppli, R.E.; Cogoli, A. Effect of a herbal yeast food supplement and long-distance running on immunological parameters. *Br. J. Sport Med.* **1990**, *24*, 103–112. [CrossRef] [PubMed]

53. Ooi, F.K. Effects of combined aerobic dance exercise and honey supplementation on bone turnover markers in young females. *Asian J. Exerc. Sports Sci.* **2011**, *8*, 53–63.

54. Schönfeld, P.; Reiser, G. Why does brain metabolism not favor burning of fatty acids to provide energy?-Reflections on disadvantages of the use of free fatty acids as fuel for brain. *J. Cereb. Blood Flow Metab.* **2013**, *33*, 1493–1499. [CrossRef] [PubMed]

55. Boyle, P.J.; Nagy, R.J.; O'Connor, A.M.; Kempers, S.F.; Yeo, R.A.; Qualls, C. Adaptation in brain glucose uptake following recurrent hypoglycemia. *Proc. Natl. Acad. Sci. USA* **1994**, *91*, 9352–9356. [CrossRef] [PubMed]

56. Bandelow, S.; Maughan, R.; Shirreffs, S.; Ozgünen, K.; Kurdak, S.; Ersöz, G.; Binnet, M.; Dvorak, J. The effects of exercise, heat, cooling and rehydration strategies on cognitive function in football players. *Scand. J. Med. Sci. Sports* **2010**, *20*, 148–160. [CrossRef] [PubMed]

57. Burke, L.M.; Kiens, B.; Ivy, J.L. Carbohydrates and fat for training and recovery. *J. Sports Sci.* **2004**, *22*, 15–30. [CrossRef]

58. Salehi, A.; Gunnerud, U.; Muhammed, S.J.; Östman, E.; Holst, J.J.; Björck, I.; Rorsman, P. The insulinogenic effect of whey protein is partially mediated by a direct effect of amino acids and GIP on β-cells. *Nutr. Metab.* **2012**, *9*, 48. [CrossRef]

59. Bishop, N.C.; Gleeson, M.; Nicholas, C.W.; Ali, A. Influence of carbohydrate supplementation on plasma cytokine and neutrophil degranulation responses to high intensity intermittent exercise. *Int. J. Sport Nutr. Exerc. Metab.* **2002**, *12*, 145–156. [CrossRef]

60. Zaid, S.S.M.; Sulaiman, S.A.; Othman, N.H.; Soelaiman, I.-N.; Shuid, A.N.; Mohamad, N.; Muhamad, N. Protective effects of Tualang honey on bone structure in experimental postmenopausal rats. *Clinics* **2012**, *67*, 779–784. [CrossRef]

61. Smith, J.W.; Zachwieja, J.J.; Péronnet, F.; Passe, D.H.; Massicotte, D.; Lavoie, C.; Pascoe, D.D. Fuel selection and cycling endurance performance with ingestion of [^{13}C] glucose: Evidence for a carbohydrate dose response. *J. Appl. Physiol.* **2010**, *108*, 1520–1529. [CrossRef] [PubMed]

62. Smith, J.W.; Pascoe, D.D.; Passe, D.H.; Ruby, B.C.; Stewart, L.K.; Baker, L.B.; Zachwieja, J.J. Curvilinear dose-response relationship of carbohydrate (0–120 gh^{-1}) and performance. *Med. Sci. Sports Exerc.* **2013**, *45*, 336–341. [CrossRef] [PubMed]

63. Jeukendrup, A.E.; Moseley, L.; Mainwaring, G.I.; Samuels, S.; Perry, S.; Mann, C.H. Exogenous carbohydrate oxidation during ultraendurance exercise. *J. Appl. Physiol.* **2006**, *100*, 1134–1141. [CrossRef] [PubMed]

64. Passe, D.; Horn, M.; Murray, R. Impact of beverage acceptability on fluid intake during exercise. *Appetite* **2000**, *35*, 219–229. [CrossRef] [PubMed]

65. Oosthuyse, T.; Carstens, M.; Millen, A.M. Ingesting isomaltulose versus fructose-maltodextrin during prolonged moderate-heavy exercise increases fat oxidation but impairs gastrointestinal comfort and cycling performance. *Int. J. Sport Nutr. Exerc. Metab.* **2015**, *25*, 427–438. [CrossRef] [PubMed]

© 2019 by the authors. Licensee MDPI, Basel, Switzerland. This article is an open access article distributed under the terms and conditions of the Creative Commons Attribution (CC BY) license (http://creativecommons.org/licenses/by/4.0/).

Review

Practical Hydration Solutions for Sports

Luke N. Belval [1,*], Yuri Hosokawa [2], Douglas J. Casa [1], William M. Adams [3],
Lawrence E. Armstrong [4], Lindsay B. Baker [5], Louise Burke [6], Samuel Cheuvront [7],
George Chiampas [8], José González-Alonso [9], Robert A. Huggins [1], Stavros A. Kavouras [10],
Elaine C. Lee [4], Brendon P. McDermott [11], Kevin Miller [12], Zachary Schlader [13], Stacy Sims [14],
Rebecca L. Stearns [1], Chris Troyanos [15] and Jonathan Wingo [16]

[1]   Korey Stringer Institute, Department of Kinesiology, University of Connecticut, Storrs, CT 06269, USA
[2]   Faculty of Sport Sciences, Waseda University, Saitama 359-1192, Japan
[3]   Department of Kinesiology, University of North Carolina at Greensboro, Greensboro, NC 27402, USA
[4]   Department of Kinesiology, University of Connecticut, Storrs, CT 06269, USA
[5]   Gatorade Sports Science Institute, Barrington, IL 60010, USA
[6]   Sports Nutrition, Australian Institute of Sport, Canberra, ACT 2617, Australia
[7]   Sports Science Synergy, LLC, Franklin, MA 02038, USA
[8]   U.S. Soccer, Chicago, IL 60616, USA
[9]   Centre for Human Performance, Exercise and Rehabilitation, Brunel University London,
      Uxbridge UB8 3PH, UK
[10] Hydration Science Lab, College of Health Solutions, Arizona State University, Phoenix, AZ 85004, USA
[11] Department of Health, Human Performance and Recreation, University of Arkansas,
      Fayetteville, AR 72701, USA
[12] Department of Rehabilitation and Medical Sciences, Central Michigan University,
      Mount Pleasant, MI 48859, USA
[13] Department of Exercise and Nutrition Sciences, University at Buffalo, Buffalo, NY 14214, USA
[14] Faculty of Health, Sport and Human Performance, University of Waikato, Hamilton 3216, New Zealand
[15] International Institute of Race Medicine, Plymouth, MA 02360, USA
[16] Department of Kinesiology, University of Alabama, Tuscaloosa, AL 35487, USA
*   Correspondence: luke.belval@uconn.edu; Tel.: +860-486-5336

Received: 15 May 2019; Accepted: 3 July 2019; Published: 9 July 2019

Abstract: Personalized hydration strategies play a key role in optimizing the performance and safety of athletes during sporting activities. Clinicians should be aware of the many physiological, behavioral, logistical and psychological issues that determine both the athlete's fluid needs during sport and his/her opportunity to address them; these are often specific to the environment, the event and the individual athlete. In this paper we address the major considerations for assessing hydration status in athletes and practical solutions to overcome obstacles of a given sport. Based on these solutions, practitioners can better advise athletes to develop practices that optimize hydration for their sports.

Keywords: fluid replacement; athletics; exercise

---

1. Introduction

Maintaining euhydration, the state of preserving body water within its optimal homeostatic range, is essential to sustain life. Water contributes 50–70% of total body mass and is compartmentalized within both intracellular (65%) and extracellular (35%) spaces [1]. Euhydration is typically maintained over the course of day-to-day life via behavioral and biological controls [2]. However, exercise can cause an acute disruption to fluid balance, challenging the athlete's goal of optimal performance and safety during exercise, especially in hot environmental conditions. The process of incurring a fluid deficit is known as dehydration, while the outcome is defined as hypohydration. The loss of body water during exercise exacerbates physiological and perceptual strain [3–10] and it is well established that these changes can impair endurance performance, particularly in hot environments and may increase the risk of exertional heat illness [11–17].

While the sensation of thirst, a centrally mediated response to body water deficits, is useful in dictating the need for fluid intake during daily life, thirst is relatively insensitive in acutely tracking hydration status during exercise [18,19]. Maintaining an optimal state of hydration during exercise becomes more complicated depending on the sport, type of activity and availability of fluid. Optimal hydration is dependent on many factors but can generally be defined during exercise as avoiding losses greater than 2–3% of body mass while also avoiding overhydration [15]. Furthermore, during exercise, it is not uncommon for individuals to involuntarily dehydrate, in which they consume less fluid than their fluid needs. Excessive fluid intake can also be problematic, with hyponatremia developing in severe cases of overhydration [15]. Inappropriate management of fluid intake resulting in hypohydration, or hyperhydration, can be detrimental for performance and in some circumstances, increases health risk.

Current consensus recommends that good hydration practices include: (1) beginning exercise in a state of euhydration, (2) preventing excessive hypohydration during exercise, and (3) replacing remaining losses following exercise prior to the next exercise bout [15,20,21]. These practices attenuate the adverse effects of acute dehydration on physical activity and health [15]. However, it is acknowledged that fluid needs are individualistic and rely on factors such as personal sweat rate, exercise mode, exercise intensity, environmental conditions and exercise duration (Figure 1) [14,15,22–25]. Furthermore, characteristics and rules unique to each sport environment in which it is played, event uniform and equipment, and the availability of fluid during both training and competition may greatly influence the ability to optimize hydration during activity. The Korey Stringer Institute and Gatorade convened a meeting to address these issues as they relate to athletes. The purpose of these proceedings is to discuss practical strategies to assess and tailor hydration for sports. This manuscript will specifically focus on the factors that underlie fluid needs and provide guidance to clinicians and practitioners on how to plan for these needs in the context of a given activity.

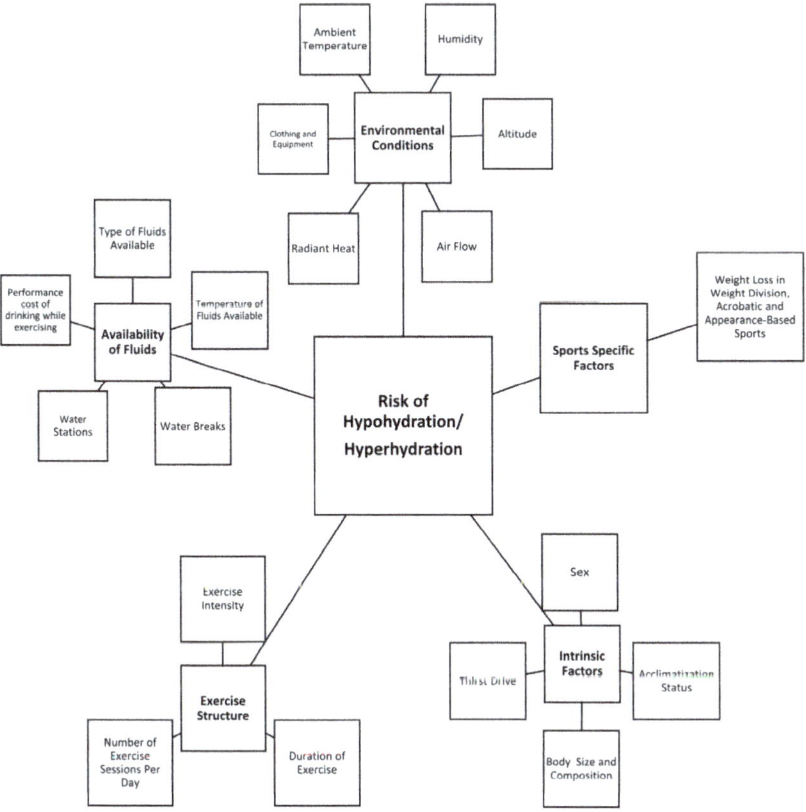

Figure 1. Factors that contribute to the risk of hypohydration or hyperhydration during exercise.

2. Hydration Assessment

Hydration assessment can be utilized to indicate one's current hydration state, but if taken serially, it can also be used to track changes in hydration and indicate fluid needs (i.e., during a bout of physical activity). While various methods of hydration assessment exist, there is no single method that can serve as a criterion measure to assess hydration status in all settings (i.e., day-to-day life and exercise, etc.). Plasma osmolality, changes in plasma volume, and the volume, osmolality and specific gravity of urine are the most commonly published metrics to assess changes in hydration status in clinical settings [26]. The use of these measures in field applications is often impractical, due to the methods or equipment needed to acquire the measure (e.g., a needle stick to draw blood or providing a urine sample) as well as the sensitivity of the measure.

In field applications, the careful assessment of changes in body mass over a bout of physical activity provides a reasonably accurate assessment of body water deficits incurred during the session, since sweat loss and fluid intake during the session underpin the major changes in body mass and body water content. This is true for most sporting activities conducted over a duration of <2–3 h; however, during very prolonged and strenuous exercise (e.g., ultra-endurance races), other factors that cause mass changes, metabolic water production and the liberation of stored water become numerically important and undermine the utility of this assessment [15,27,28]. A comparison of body mass pre- and post-exercise will help guide the athlete in understanding whether their hydration strategy during activity was effective in achieving acceptable fluid balance as well as knowing the volume of fluids that are needed following exercise to return to baseline hydration levels prior to the next exercise session. The methodology for assessing sweat losses is to assess the athlete's body mass before and after exercise with care to avoid or account for substantial amounts of fluid trapped in hair and clothes. Accounting for any fluid consumed or urine excreted, the difference between the masses can be used to calculate the amount of sweat lost as well as the residual fluid deficit that should be addressed in post-exercise recovery plans [29].

A useful paradigm for tracking daily changes in hydration status in sporting situations is to consider a combination of assessments to track daily changes. The monitoring of daily changes in body mass, coupled with urine color and thirst sensation status provides adequate sensitivity for most athletic situations [30]. Cheuvront and Kenefick established useful criteria for these variables as body mass changes greater than 1.1%, a conscious desire for water (thirst), and dark-colored urine (>5 a.u. on an 8-a.u. scale [31]) indicating varying degrees of fluid inadequacy [32]. Two of these factors combined suggest daily fluid intake is likely inadequate, while all three factors indicate that daily fluid intake is very likely inadequate. It should be noted that this assessment technique is based on first morning values and requires baseline body mass values to provide the most useful information to athletes.

Practical Solutions:

(1) Carefully monitor acute changes in body mass over an exercise bout to determine sweat rate, adequacy of fluid replacement and fluid needs for recovery for that session. Consider how well this can be used to evaluate general hydration strategies in similar situations.
(2) Use changes in body mass, urine color and thirst upon awakening to track daily changes in hydration status.

3. Exercise Structure

Body fluid loss during sport or exercise largely results from sweating. Net fluid balance is modulated to a certain extent by drinking. Rate of sweating is primarily a function of metabolic heat production [33], but can be modified by environment, clothing, acclimatization and hydration status [7,34]. As the primary mechanism of heat dissipation in many environments, the evaporation of sweat is vital for regulating body temperature, even during exercise in temperate weather. However, this heat dissipation is accompanied by typical fluid losses of ~0.5–1.9 L/h [35].

Exercise intensity is the main factor that determines metabolic heat production, meaning that the rate of fluid losses from sweat for a given exercise session can be partially explained by the intensity of the exercise [36]. Total fluid losses are a result of the sweat rate of a given exercise intensity and the total duration of that activity [27]. In most circumstances there is an inverse relationship between the exercise intensity of a session and the duration of that session. However, given the wide variability in individual sweat rates, the unique interplay between intensity, duration and sweat rate must be considered in unison. For example, a runner with a 2 L/h sweat rate who completes a marathon in 2 h will accumulate the same fluid losses as a runner with a 1 L/h sweat rate that completes the race in 4 h. In Tables 1 and 2, we define exercise intensity of a range of sporting activities into three distinct categories (High, Moderate, Low), based on typical practice and competition structures on the principles described above but comprehensive plans should consider individual athletes.

Table 1. Team Sport Factors That Influence Hypohydration.

Sport	Availability of Fluid		Environment		Intensity		Hypohydration Risk	
	Training	Competition	Training	Competition	Training	Competition	Training	Competition
Basketball	High	High	Low	Low	Mod	Mod	Low	Low
Ice Hockey	High	High	Low	Low	Mod	High	Mod	Mod
Football	High	High	Mod	Mod	Mod	High	Mod	Mod
Baseball	High	High	Mod	Mod	Low	Low	Low	Low
Softball	High	High	Mod	Mod	Low	Low	Low	Low
Volleyball	High	High	Low	Low	Low	Low	Low	Low
Soccer	Mod	Low	Mod	Mod	Mod	High	Mod	High
Lacrosse	High	High	Mod	Mod	Mod	Mod	Mod	Mod
Rugby	High	Low	Mod	Mod	Mod	High	Mod	High

Availability of Fluid: High, the dynamics of the sport allow for multiple opportunities to consume fluid; Moderate (Mod), Fluid is only available during breaks in training, competition, or carried by the athlete; Low, Fluid is limited or not available due to time restrictions, rules or dynamics of the sport, and ability to carry. Environment: High, environmental conditions that are of great risk for hypohydration; Mod, the environment is variable ranging from cool to hot conditions that may pose risk for hypohydration; Low, the environmental conditions are not a threat to hypohydration. Intensity: High, exercise intensity in the sport is increased and likely to result in large sweat losses and hypohydration; Mod, exercise intensity in the sport varies from moderate to high and may result in large sweat losses and hypohydration; Low, exercise intensity in the sport is low and less likely to result in large sweat losses and hypohydration. Hypohydration Risk: High, the risk for hypohydration in the sport is high based on reported sweat losses, the availability of fluid, environmental conditions, and the intensity of exercise; Mod, the risk for hypohydration in the sport is moderate based on reported sweat losses, the availability of fluid, environmental conditions, and the intensity of exercise; Low, the risk for hypohydration in the sport is low based on reported sweat losses, the availability of fluid, environmental conditions, and the intensity of exercise. Note: These assessments are representative of typical situations encountered in these sports. Site-specific factors (Figure 1) may differ from those presented here.

Table 2. Individual Sport Factors That Influence Hypohydration.

Sport	Availability of Fluid		Environment		Intensity		Hypohydration Risk	
	Training	Competition	Training	Competition	Training	Competition	Training	Competition
Tennis	High	Mod	Mod	Mod	High	High	Mod	Mod
Wrestling	High	High	Mod	Mod	High	High	High	Low
Gymnastics	High	High	Low	Low	Mod	Low	Low	Low
Running (<1 h)	Low	High	Mod	Mod	High	High	Low	Low
Running (1–2 h)	Low	High	Mod	Mod	Mod	Mod	Mod	Mod
Running (>2 h)	Low	High	Mod	Mod	Low	Mod	Mod	Mod
Cycling (<1 h)	High	High	Mod	Mod	High	High	Low	Low
Cycling (>2 h)	Mod	Mod	Mod	Mod	Mod	Mod	Low	High
Swimming	High	High	Low	Low	High	High	Low	Low
Triathlon (<2 h)								
Swim	Low	Low	Low	Low	Mod	Mod	Low	Low
Bike	Mod	High	Mod	Mod	Mod	Mod	Low	Low
Run	Low	High	Mod	Mod	Mod	Mod	Low	Low
Triathlon (2–5 h)								
Swim	Low	Low	Low	Low	Mod	Mod	Low	Low
Bike	Mod	High	Mod	Mod	Mod	Mod	Low	Low
Run	Low	High	Mod	Mod	Mod	Mod	Low	Low
Triathlon (5–9 h)								
Swim	Low	Low	Low	Low	Mod	Mod	Low	Low
Bike	Mod	High	Mod	Mod	Mod	Mod	Mod	Mod
Run	Low	High	Mod	Mod	Mod	Mod	Mod	Mod
Triathlon (>9 h)								
Swim	Low	Low	Low	Low	Mod	Mod	Low	Low
Bike	Mod	High	Mod	Mod	Mod	Mod	Mod	Mod
Run	Low	High	Mod	Mod	Mod	Mod	Mod	Mod

Availability of Fluid: High, the dynamics of the sport allow for multiple opportunities to consume fluid; Moderate (Mod), Fluid is only available during breaks in training, competition, or carried by the athlete; Low, Fluid is limited or not available due to time restrictions, rules or dynamics of the sport, and ability to carry. Environment: High, environmental conditions that are of great risk for hypohydration; Mod, the environment is variable ranging from cool to hot conditions that may pose risk for hypohydration; Low, the environmental conditions are not a threat to hypohydration. Intensity: High, exercise intensity in the sport is increased and likely to result in large sweat losses and hypohydration; Mod, exercise intensity in the sport varies from moderate to high and may result in large sweat losses and hypohydration; Low, exercise intensity in the sport is low and less likely to result in large sweat losses and hypohydration. Hypohydration Risk: High, the risk for hypohydration in the sport is high based on reported sweat losses, the availability of fluid, environmental conditions, and the intensity of exercise; Mod, the risk for hypohydration in the sport is moderate based on reported sweat losses, the availability of fluid, environmental conditions, and the intensity of exercise; Low, the risk for hypohydration in the sport is low based on reported sweat losses, the availability of fluid, environmental conditions, and the intensity of exercise. Note: These assessments are representative of typical situations encountered in these sports. Site-specific factors (Figure 1) may differ from those presented here.

For sports like cycling and running, the influence of exercise intensity and duration on fluid needs is very easy to determine given the consistent nature of exercise. Figure 2 demonstrates the relationship between duration and target fluid replacement for steady-state exercise. However, as shown in Table 1, some of the most common sports involve supramaximal exercise in short bursts with longer breaks. In this case, individuals should regard an overall average of exercise intensity rather than the maximal effort during the exercise bout when determining optimal fluid balance.

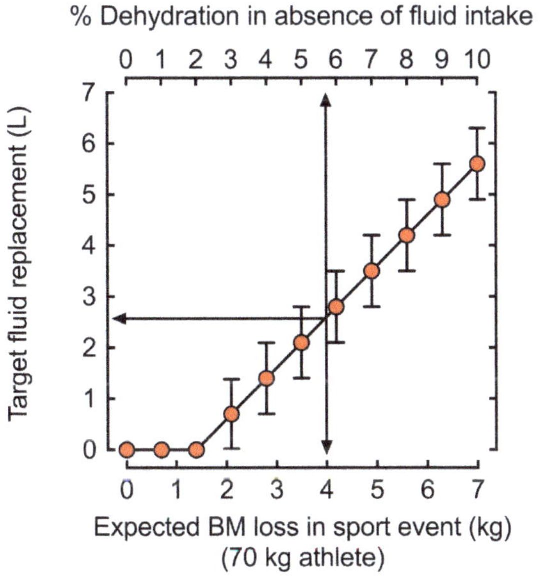

Figure 2. Target fluid replacement estimates to prevent >2 ± 1 % body mass (BM) loss as water (i.e., dehydration). The 70 kg athlete in the example would need to drink a volume of fluid equal to 2.6 ± 0.7 L to prevent >2 ± 1% dehydration when losing 4 L of body water, such as during a marathon (42.1 km). During shorter distances such as 5 or 10 km when fluid losses are unlikely to reach or exceed 2% dehydration, the same athlete would not need to ingest fluids during competition as fluid losses accumulate to <2% dehydration.

It should also be noted that exercise intensity influences gastric emptying rate [37]. Individuals striving to closely match sweat losses with fluid consumption can be challenged by maximal gastric emptying rates. However, when vigorous exercise is conducted (>70% VO_{2max}), gastric emptying decreases predictably, most likely based on decreased splanchnic perfusion [37].

Practical Solutions:

(1) Increased fluid intake is necessary with prolonged or intense exercise due to increased sweat production.
(2) During vigorous exercise (>70% VO_{2max}) understand that gastric emptying may limit fluid absorption. Athletes can train their gut to improve gastrointestinal comfort or adopt strategies to increase fluid intake before and after exercise.

4. Environment

At a fixed exercise intensity, the ambient environment further modulates sweat rate. The magnitude of evaporative, radiant, convective and conductive heat exchange between the body and the environment is a function of the gradients between the environment and the skin which provides the main physiological interface for heat exchange. A number of factors contribute to sweat rate including ambient and radiant temperature, humidity, clothing, and air velocity [34], all of which differ depending on the sport or activity (Tables 1 and 2). Therefore, individuals exercising in hot-humid environments with direct sunlight and minimal airflow will produce near maximal sweat rates and be at the greatest risk of hypohydration. Wet-bulb globe thermometry (WBGT) accounts for these environmental factors and can help inform fluid replacement decisions [38].

All clothing provides insulation and presents a barrier to heat loss, resulting in increased sweat rates to provide similar cooling to an unclothed situation [34]. Thus, sports/activities with specific clothing requirements, such as American football [39,40], are at greater risk of body fluid loss compared to similar activities in which clothing is minimal. Synthetic wicking materials can increase sweat fluid losses compared to cotton garments [41], potentially decreasing the thermal load but increasing the risk of dehydration. When the effects of hot, humid environmental conditions are combined with clothing and equipment, individuals can achieve near maximal sweat rates which can create a significant fluid deficit rapidly [40].

Other Environmental Considerations:

Exercising in the cold, or at high altitudes merits special considerations when determining the fluid needs of athletes. Athletes must also be vigilant and mindful of their fluid needs during exercise in the cold. Exercise in the cold can still produce copious sweating, especially when heavy clothing is worn, while also diminishing thirst sensitivity and reducing *ad libitum* fluid consumption, thus potentially leading to impaired fluid replacement and hypohydration [42]. If possible, athletes should know their individual fluid replacement needs, based upon sweat rate measurement, during exercise in hot and cold environments to ensure they can develop a plan for competing while optimally hydrated.

Athletes unaccustomed to exercising in higher altitudes may require additional fluids. Very high altitude (4900–7600 m) exposure tends to increase water and electrolyte losses, decrease plasma volume and total body water content [43]. In both cold air and high altitudes, respiratory water losses may increase and require additional fluid consumption due to low air water vapor pressures [44]. Therefore, athletes should acclimate to altitude over several days and maintain euhydration prior to competition to ensure optimal athletic performance.

Tables 1 and 2 summarizes typical environmental conditions found among a range of sports into three distinct categories, and how these conditions contribute to the considerations around an individualized fluid plan. Of course, there are large regional differences in environmental conditions experienced for sports at the same time of year [45]. Local measurements utilizing WBGT allow for the greatest characterization of the environmental demands placed on athletes during exercise in the heat [46].

Practical Solutions:

(1) Measure local environmental conditions to determine the risk of high sweat rates resulting in large fluid losses.
(2) Increase fluid-replacement during exercise in hot and humid environments to account for increased sweat losses.
(3) Account for clothing or equipment requirements when evaluating fluid needs.
(4) Modify fluid intake when exercising in cold or altitude according to an estimation of fluid losses noting that thirst may be less reliable as a guide to dehydration under these conditions.

5. Fluid Availability

Fluid availability refers to the factors that dictate an athlete's ability to replace fluid losses during activity. In many instances, the characteristics of the sport have a strong influence on the ability to drink during competition and in many scenarios prevent an athlete from "drinking to thirst" [21].

Meanwhile, training activities are often easily modified to allow for some degree of fluid replacement. In many instances, water-breaks during training can be determined on the basis of work-to-rest ratios set by environmental conditions with free access to fluid throughout the break [47]. Characteristics such as flavor and temperature affect the palatability of fluids and may increase voluntary intake when they are matched to the cultural preferences of the athletes and the prevailing conditions (e.g., cool drinks in a warm environment) [48,49].

In endurance sports which provide competitors with feed zones/water stations (e.g., running events), the number of water stations on a course and the frequency with which a competitor reaches them can influence drinking behavior. The International Association of Athletics Federations recommends that stations are placed approximately every 5 km, however, many races include more frequent stations which may influence athletes' drinking strategies and behaviors [50]. Slower athletes with lower sweat rates who compete in such events, particularly over prolonged distances or duration, are often able to drink in volumes that exceed their true fluid losses and are at particular risk of developing hyponatremia [51]. Athletes taking part in these events should be educated on the importance of fluid balance and the prevention of hyperhydration. It should also be noted that in many events lasting up to 45 min, the risk of dehydration is low due to the limited duration across which sweat losses can accumulate. For faster and/or more competitive athletes, extra elements related to drinking while performing continuous exercise must be taken into consideration. This includes considerations around gastrointestinal comfort when fluid consumed during higher-intensity and "gut joggling" activities (e.g., high-speed running vs. the more "gliding" movements of cross-country skiing or cycling). Furthermore, the time lost in slowing down or moving out of an aerodynamic position to obtain or consume a drink must be factored into the overall race performance. This creates different factors in the cost:benefit analysis of an individual's fluid intake plan.

The official rules and competition characteristics of "stop-start" sports such as team and racket sports create other influences, and often unique scenarios, around fluid availability. In some examples (e.g., soccer, rugby), governing rules limit the availability of fluids for athletes during competition. Soccer, for example, includes two 45-min halves (with a continuous running clock) in which fluid availability is extremely limited to players. At the other end of the spectrum are sports such as baseball, basketball and tennis with frequent rest breaks within playing time (e.g., time outs, change of ends or player rotations) during which fluids can be consumed. An athlete's drinking strategy for a competition represents a unique instance for their particular sport based on their ability to rehydrate within the rules [52]. We support recent governing body rule changes and referee decisions to add breaks to competitions (Major League Soccer, FIFA soccer matches, US Open Tennis) to facilitate safe participation by the athletes. These changes likely augment athletic performance and safety simultaneously. Individuals should understand their sport and its fluid needs/fluid availability characteristics to prepare and practice optimal fluid plans for competition. Where rule changes, or alterations are allowed, individuals and teams should attempt to ask for these alterations (e.g., extended rest periods, additional breaks) in advance to formulate an appropriate drinking strategy. In all sports, athletes should aim to practice and fine-tune their personal drinking strategy for race/competition conditions. This will help individuals to confirm its feasibility, understand their personal responses and develop any necessary behavioral practices within the expected rules of competition.

In Tables 1 and 2, we define fluid availability into three distinct categories, based on particular sport variations. Sports were categorized as having high fluid availability if there are multiple opportunities for fluid consumption, rather than only during breaks. Low fluid availability was used to describe those activities involving governing rules, time constraints, or an inability to carry personal fluids during competition. Accessibility to fluid consumption during competition represents a major variable to be used in preparation of an optimal fluid replacement strategy.

Practical Solutions:

(1)   During training, ensure that there is ample access to fluids that are palatable to athletes.

(2) Investigate or understand the opportunities for fluid intake during that are specific to a sport or event, and any other practical issues that determine fluid intake.

(3) Consider the risks of hyperhydration as well as hypohydration for any sporting event or individual athlete, and prepare appropriate practice and education strategies.

(4) Develop personalized fluid intake plans that incorporate fluid availability characteristics of the sport or event. Where there is a likelihood of hypohydration, be proactive and creative in making use of existing opportunities for fluid intake within sport rules and characteristics and be prepared to request for changes when there is a likelihood of a serious mismatch between fluid losses and the opportunity to address these.

(5) Practice intended competition drinking plans ahead of time to determine their suitability and allow time for readjustment.

6. Intrinsic Factors

A number of intrinsic factors modulate the individual variances that are observed in fluid losses. One of the greatest considerations for an individual's sweat rate is his or her body size. Larger individuals typically have higher sweat losses, with football linemen exhibiting some of the highest recorded sweat rates [14]. Therefore, required absolute drink volumes will be higher for these athletes. An individual's thirst drive also dictates how much they desire to drink during exercise, but this may not match their actual fluid needs. Indeed, multiple authors report that athletes voluntarily dehydrate during exercise due to discrepancies in fluid losses and drinking behavior [53,54]. Case studies of individuals who have developed hyponatremia due to excessive drinking during exercise also note that they reported thirst as an underlying contributor to their fluid intake [55].

Heat acclimatization contributes to variations in an individual's sweating rate responses. Individuals who are heat acclimatized exhibit greater sweat rates which can pose a greater risk of hypohydration [56]. Although the increased sweat provides extra heat dissipation, it also requires extra fluid intake.

Women may be at greater risk for exercise-induced hyponatremia. This risk has been attributed to their lower body weight and size, excess water ingestion, and longer racing times relative to men [57]. The greater incidence of hyponatremia in women is unlikely due to their greater levels of estradiol in plasma and tissue. Although female sex-hormones can also influence neural and hormonal control of thirst, fluid intake, sodium appetite and sodium regulation [58,59], there is no evidence that anything beyond stature and drinking behavior significantly impact their risk.

Practical Solutions:

(1) Consider body size, acclimatization status and thirst drive when developing hydration plans for individual athletes.

7. Sport-Specific Factors

Weight Division, Acrobatic and Appearance-Based Sports

The culture and normal behaviors surrounding specific sports can greatly affect the hydration practices of its athletes. The three most prominent examples of the cultural effects of sports on hydration practices are weight division sports, acrobatic sports and appearance-based sports. In weight division sports (e.g., combat sports, horse racing, lightweight rowing, etc.), the practice of deliberately dehydrating to manipulate body mass to meet lighter competition weight classifications is common [60]. In many cases, athletes not only sacrifice their performance through these practices but also endanger their health and well-being. In a similar fashion, sports where body-image and appearance are emphasized (e.g., cheerleading, body building and gymnastics), dangerous practices, such as extreme fluid restriction may be used by athletes to cheat "unofficial" weight checks that are self-instigated or expected within their training environment. Finally, "acrobatic" feats such as gymnastics, jumping and climbing are aided by a high power to weight ratio, but should not rely on severe hypohydration to achieve this. Excessive use of dehydration to manage body mass goals should be corrected to avoid long-term health complications [61].

Practical Solutions:

(1) Promote healthy weight management strategies in weight division, acrobatic and appearance-based sports that minimize manipulation of body water.

8. Conclusions

Based on the factors in the above sections, along with published literature on typical fluid balance observations in various sports [62,63], we assigned risks of hypohydration to the sports in Tables 1 and 2. These determinations can be used as a general guideline for sports that pose large risks for fluid imbalances that may limit sport performance. The factors for individual situations or geographical locations may vary and should be considered based on the principles mentioned above to tailor the necessary fluid replacement accommodations. An example of using this paradigm to develop a hydration plan can be found in Table 3.

Table 3. Establishing a Hydration Plan.

Guiding Question	Steps to Correct	Implementation Example
Are athletes in a state of optimal hydration?	• Assess hydration status	• Have scales available before and after practice to assess fluid deficits • Measure fluid needs via sweat rate
Is the exercise prolonged or intense?	• Increase availability of palatable fluids	• Have more breaks during longer practices or more intense exercise • Allow longer duration breaks
Is the exercise being performed in environmental conditions that lead to greater fluid losses?	• Establish breaks based upon environmental conditions	• Modify practice schedules utilizing WBGT to establish work-to-rest ratios that allow for adequate fluid intake
Is fluid available throughout the entire duration of exercise?	• Fluid is made readily available for athletes • If fluid is restricted (e.g., running races, soccer matches etc.), maximize opportunities for rehydration	• Provide free access to fluids during practice • Ensure athletes utilize breaks to rehydrate when opportunities are limited
Are there individuals with intrinsic risk factors?	• Identify individuals with high sweat rates or other limits to optimal hydration • Identify individuals whose thirst drive is not matched to their fluid losses during exercise • Counsel and monitor these athletes	• Test sweat rates of individuals who have issues with hydration • Develop individual hydration plans for high-risk athletes
Are there sport-specific factors that need to be considered?	• Counsel athletes on health and performance risks of utilizing dehydration for weight loss	• Assess hydration status alongside weight measurements to promote healthy weight management

WBGT: Wet-bulb globe temperature.

In this paper we present a paradigm that can be used by clinicians and practitioners to develop hydration strategies for sports based on fluid availability, environment and exercise intensity. These tools are provided to inform hydration education and practices in a dynamic and individualized manner so that athletes can adapt to different circumstances and optimize performance.

Author Contributions: All authors attended the meeting and contributed to drafting and revision of the manuscript. All authors approved the final version of the paper.

Funding: The meeting preceding this manuscript was funded by PepsiCo (Gatorade).

Conflicts of Interest: Douglas Casa is the Chief Executive Officer of the Korey Stringer Institute, a 503.c not for profit which subsists on the donations from our corporate partners who include Gatorade, CamelBak, NFL, NATA, Mission, Eagle Pharmaceuticals and Kestrel. He has also been a recipient of grant funds from the following entities to study hydration related products: GE Healthcare, Halo Wearables, Nix Inc., CamelBak. William Adams has consulted with the following entities regarding hydration and exercise performance or the development of hydration assessment devices: BSX Athletics; Samsung Oak Holdings, Inc; Nobo, Inc; Clif Bar &Company; The Gatorade Company, Inc. Lawrence Armstrong is a hydration consultant to Danone Nutricia Research, France and the Drinking Water Research Foundation, Alexandria VA, USA. Lindsay Baker is employed by the Gatorade Sports Science Institute, a division of PepsiCo, Inc. The views expressed in this article are those of the authors and do not necessarily reflect the position or policy of PepsiCo, Inc. Louise Burke was a member of the Gatorade Sports Science Institute's Expert Panel from 2014–2015 for which her workplace received an honorarium. Jose Gonzalez-Alonso was a member of the Gatorade Sports Science Institute's Expert Panel in 2017 for which his workplace received an honorarium. Stavros Kavouras has served as scientific consultant for Quest Diagnostics, Standard Process and Danone Research and has active grants with Danone Research. The funders had no role in the in the writing of the manuscript.

References

1. Sawka, M.; Pandolf, K.B. Effects of body water loss on physiological function and exercise performance. In *Perspectives in Exercise Science and Sports Medicine*; Gisolfi, C., Lamb, D.R., Eds.; Benchmark Press: Carmel, CA, USA, 1990; Volume 3, pp. 1–38.

2. Sawka, M.N.; Wenger, C.B.; Pandolf, K.B. Thermoregulatory Responses to Acute Exercise-Heat Stress and Heat Acclimation. In *Comprehensive Physiology*; Terjung, R., Ed.; John Wiley & Sons: Hoboken, NJ, USA, 2011; pp. 97–151.

3. Montain, S.J.; Coyle, E.F. Influence of graded dehydration on hyperthermia and cardiovascular drift during exercise. *J. Appl. Physiol.* **1992**, *73*, 1340–1350. [CrossRef] [PubMed]

4. Montain, S.J.; Sawka, M.N.; Latzka, W.A.; Valeri, C.R. Thermal and cardiovascular strain from hypohydration: Influence of exercise intensity. *Int. J. Sports Med.* **1998**, *19*, 87–91. [CrossRef] [PubMed]

5. Sawka, M.N.; Young, A.J.; Francesconi, R.P.; Muza, S.R.; Pandolf, K.B. Thermoregulatory and blood responses during exercise at graded hypohydration levels. *J. Appl. Physiol.* **1985**, *59*, 1394–1401. [CrossRef] [PubMed]

6. Adams, W.M.; Ferraro, E.M.; Huggins, R.A.; Casa, D.J. Influence of body mass loss on changes in heart rate during exercise in the heat: A systematic review. *J. Strength Cond. Res.* **2014**, *28*, 2380–2389. [CrossRef] [PubMed]

7. Kenefick, R.W.; Cheuvront, S.N. Physiological adjustments to hypohydration: Impact on thermoregulation. *Auton. Neurosci.* **2016**, *196*, 47–51. [CrossRef] [PubMed]

8. Trangmar, S.J.; Gonzalez-Alonso, J. New Insights into the Impact of Dehydration on Blood Flow and Metabolism during Exercise. *Exerc. Sport Sci. Rev.* **2017**, *45*, 146–153. [CrossRef] [PubMed]

9. Gonzalez-Alonso, J.; Crandall, C.G.; Johnson, J.M. The cardiovascular challenge of exercising in the heat. *J. Physiol.* **2008**, *586*, 45–53. [CrossRef]

10. Trangmar, S.J.; Gonzalez-Alonso, J. Heat, Hydration and the Human Brain, Heart and Skeletal Muscles. *Sports Med.* **2019**, *49*, 69–85. [CrossRef]

11. Cheuvront, S.N.; Carter, R., 3rd; Sawka, M.N. Fluid balance and endurance exercise performance. *Curr. Sports Med. Rep.* **2003**, *2*, 202–208. [CrossRef]

12. Cheuvront, S.N.; Kenefick, R.W.; Montain, S.J.; Sawka, M.N. Mechanisms of aerobic performance impairment with heat stress and dehydration. *J. Appl. Physiol.* **2010**, *109*, 1989–1995. [CrossRef]

13. Barr, S.I. Effects of dehydration on exercise performance. *Can. J. Appl. Physiol.* **1999**, *24*, 164–172. [CrossRef] [PubMed]

14. Sawka, M.N.; Burke, L.M.; Eichner, E.R.; Maughan, R.J.; Montain, S.J.; Stachenfeld, N.S. American College of Sports Medicine position stand. Exercise and fluid replacement. *Med. Sci. Sports Exerc.* **2007**, *39*, 377–390. [PubMed]

15. McDermott, B.P.; Anderson, S.A.; Armstrong, L.E.; Casa, D.J.; Cheuvront, S.N.; Cooper, L.; Kenney, W.L.; O'Connor, F.G.; Roberts, W.O. National Athletic Trainers' Association Position Statement: Fluid Replacement for the Physically Active. *J. Athl. Train.* **2017**, *52*, 877–895. [CrossRef] [PubMed]

16. Judelson, D.A.; Maresh, C.M.; Anderson, J.M.; Armstrong, L.E.; Casa, D.J.; Kraemer, W.J.; Volek, J.S. Hydration and muscular performance: Does fluid balance affect strength, power and high-intensity endurance? *Sports Med.* **2007**, *37*, 907–921. [CrossRef] [PubMed]

17. Coyle, E.F. Physiological determinants of endurance exercise performance. *J. Sci. Med. Sport* **1999**, *2*, 181–189. [CrossRef]

18. Adams, J.D.; Sekiguchi, Y.; Suh, H.G.; Seal, A.D.; Sprong, C.A.; Kirkland, T.W.; Kavouras, S.A. Dehydration Impairs Cycling Performance, Independently of Thirst: A Blinded Study. *Med. Sci. Sports Exerc.* **2018**, *50*, 1697–1703. [CrossRef] [PubMed]

19. Kenefick, R.W. Drinking Strategies: Planned Drinking versus Drinking to Thirst. *Sports Med.* **2018**, *48*, 31–37. [CrossRef] [PubMed]

20. Bergeron, M.F.; Bahr, R.; Bartsch, P.; Bourdon, L.; Calbet, J.A.L.; Carlsen, K.H.; Castagna, O.; González-Alonso, J.; Lundby, C.; Maughan, R.J.; et al. International Olympic Committee consensus statement on thermoregulatory and altitude challenges for high-level athletes. *Br. J. Sports Med.* **2012**, *46*, 770–779. [CrossRef] [PubMed]

21. Racinais, S.; Alonso, J.M.; Coutts, A.J.; Flouris, A.D.; Girard, O.; Gonzalez-Alonso, J.; Hausswirth, C.; Jay, O.; Lee, J.K.; Mitchell, N.; et al. Consensus recommendations on training and competing in the heat. *Br. J. Sports Med.* **2015**, *49*, 1164–1173. [CrossRef]

22. Carter, J.E.; Gisolfi, C.V. Fluid replacement during and after exercise in the heat. *Med. Sci. Sports Exerc.* **1989**, *21*, 532–539. [CrossRef]

23. Coyle, E.F.; Hamilton, M. Fluid replacement during exercise: Effects on physiological homeostasis and performance. In *Fluid Homeostasis During Exercise*; Gisolfi, C., Lamb, D.R., Eds.; Benchmark Press: Carmel, CA, USA, 1990; Volume 3, pp. 281–308.

24. Mora-Rodriguez, R.; Hamouti, N.; Del Coso, J.; Ortega, J.F. Fluid ingestion is more effective in preventing hyperthermia in aerobically trained than untrained individuals during exercise in the heat. *Appl. Physiol. Nutr. Metab.* **2013**, *38*, 73–80. [CrossRef] [PubMed]

25. Wittels, P.; Gunga, H.C.; Kirsch, K.; Kanduth, B.; Gunther, T.; Vormann, J.; Rocker, L. Fluid regulation during prolonged physical strain with water and food deprivation in healthy, trained men. *Wien. Klin. Wochenschr.* **1996**, *108*, 788–794. [PubMed]

26. Armstrong, L.E. Hydration assessment techniques. *Nutr. Rev.* **2005**, *63*, S40–S54. [CrossRef] [PubMed]

27. Cheuvront, S.N.; Kenefick, R.W. CORP: Improving the status quo for measuring whole body sweat losses. *J. Appl. Physiol.* **2017**, *123*, 632–636. [CrossRef] [PubMed]

28. Baker, L.B.; Lang, J.A.; Kenney, W.L. Change in body mass accurately and reliably predicts change in body water after endurance exercise. *Eur. J. Appl. Physiol.* **2009**, *105*, 959–967. [CrossRef] [PubMed]

29. Armstrong, L.E.; Casa, D.J. Methods to Evaluate Electrolyte and Water Turnover of Athletes. *Athl. Train. Sports Health Care* **2009**, *1*, 1–11. [CrossRef]

30. Cheuvront, S.N.; Kenefick, R.W. Am I Drinking Enough? Yes, No, and Maybe. *J. Am. Coll. Nutr.* **2016**, *35*, 185–192. [CrossRef] [PubMed]

31. Armstrong, L.E.; Maresh, C.M.; Castellani, J.W.; Bergeron, M.F.; Kenefick, R.W.; LaGasse, K.E.; Riebe, D. Urinary indices of hydration status. *Int. J. Sport Nutr.* **1994**, *4*, 265–279. [CrossRef]

32. Cheuvront, S.N.; Sawka, M.N. Hydration Assessment of Athletes. *Sports Sci. Exch.* **2005**, *18*, 1–5.

33. Stolwijk, J.A.; Saltin, B.; Gagge, A.P. Physiological factors associated with sweating during exercise. *Aerosp. Med.* **1968**, *39*, 1101–1105.

34. Parsons, K. *Human Thermal Environments: The Effects of Hot, Moderate and Cold Environments on Human Health, Comfort and Performance*, 3rd ed.; CRC Press: Boca Raton, FL, USA, 2014.

35. Baker, L.B.; Barnes, K.A.; Anderson, M.L.; Passe, D.H.; Stofan, J.R. Normative data for regional sweat sodium concentration and whole-body sweating rate in athletes. *J. Sports Sci.* **2016**, *34*, 358–368. [CrossRef] [PubMed]

36. Nadel, E.R. Control of sweating rate while exercising in the heat. *Med. Sci. Sports* **1979**, *11*, 31–35. [PubMed]

37. Horner, K.M.; Schubert, M.M.; Desbrow, B.; Byrne, N.M.; King, N.A. Acute exercise and gastric emptying: A meta-analysis and implications for appetite control. *Sports Med.* **2015**, *45*, 659–678. [CrossRef] [PubMed]

38. Budd, G.M. Wet-bulb globe temperature (WBGT)—Its history and its limitations. *J. Sci. Med. Sport* **2008**, *11*, 20–32. [CrossRef] [PubMed]

39. Kulka, T.J.; Kenney, W.L. Heat balance limits in football uniforms how different uniform ensembles alter the equation. *Physician Sportsmed.* **2002**, *30*, 29–39. [CrossRef]

40. Armstrong, L.E.; Johnson, E.C.; Casa, D.J.; Ganio, M.S.; McDermott, B.P.; Yamamoto, L.M.; Lopez, R.M.; Emmanuel, H. The American football uniform: Uncompensable heat stress and hyperthermic exhaustion. *J. Athl. Train.* **2010**, *45*, 117–127. [CrossRef] [PubMed]

41. Hooper, D.R.; Cook, B.M.; Comstock, B.A.; Szivak, T.K.; Flanagan, S.D.; Looney, D.P.; DuPont, W.H.; Kraemer, W.J. Synthetic garments enhance comfort, thermoregulatory response, and athletic performance compared with traditional cotton garments. *J. Strength Cond. Res.* **2015**, *29*, 700–707. [CrossRef]

42. O'Brien, C.; Freund, B.J.; Sawka, M.N.; McKay, J.; Hesslink, R.L.; Jones, T.E. Hydration assessment during cold-weather military field training exercises. *Arct. Med. Res.* **1996**, *55*, 20–26.

43. Fusch, C.; Gfrorer, W.; Koch, C.; Thomas, A.; Grunert, A.; Moeller, H. Water turnover and body composition during long-term exposure to high altitude (4900–7600 m). *J. Appl. Physiol.* **1996**, *80*, 1118–1125. [CrossRef]

44. Mitchell, J.W.; Nadel, E.R.; Stolwijk, J.A. Respiratory weight losses during exercise. *J. Appl. Physiol.* **1972**, *32*, 474–476. [CrossRef]

45. Grundstein, A.; Williams, C.; Phan, M.; Cooper, E. Regional heat safety thresholds for athletics in the contiguous United States. *Appl. Geogr.* **2015**, *56*, 55–60. [CrossRef]

46. Cheuvront, S.N.; Caruso, E.M.; Heavens, K.R.; Karis, A.J.; Santee, W.R.; Troyanos, C.; D'Hemecourt, P. Effect of WBGT Index Measurement Location on Heat Stress Category Classification. *Med. Sci. Sports Exerc.* **2015**, *47*, 1958–1964. [CrossRef] [PubMed]

47. Casa, D.J.; DeMartini, J.K.; Bergeron, M.F.; Csillan, D.; Eichner, E.R.; Lopez, R.M.; Ferrara, M.S.; Miller, K.C.; O'Connor, F.; Sawka, M.N.; et al. National Athletic Trainers' Association Position Statement: Exertional Heat Illnesses. *J. Athl. Train.* **2015**, *50*, 986–1000. [CrossRef] [PubMed]

48. Minehan, M.R.; Riley, M.D.; Burke, L.M. Effect of flavor and awareness of kilojoule content of drinks on preference and fluid balance in team sports. *Int. J. Sport Nutr. Exerc. Metab.* **2002**, *12*, 81–92. [CrossRef] [PubMed]

49. Burdon, C.A.; Johnson, N.A.; Chapman, P.G.; O'Connor, H.T. Influence of beverage temperature on palatability and fluid ingestion during endurance exercise: A systematic review. *Int. J. Sport Nutr. Exerc. Metab.* **2012**, *22*, 199–211. [CrossRef] [PubMed]

50. Williams, J.; Tzortziou Brown, V.; Malliaras, P.; Perry, M.; Kipps, C. Hydration strategies of runners in the London Marathon. *Clin. J. Sport Med.* **2012**, *22*, 152–156. [CrossRef]

51. Chorley, J.; Cianca, J.; Divine, J. Risk factors for exercise-associated hyponatremia in non-elite marathon runners. *Clin. J. Sport Med.* **2007**, *17*, 471–477. [CrossRef]

52. Maughan, R.J.; Shirreffs, S.M. Development of individual hydration strategies for athletes. *Int. J. Sport Nutr. Exerc. Metab.* **2008**, *18*, 457–472. [CrossRef]

53. Passe, D.; Horn, M.; Stofan, J.; Horswill, C.; Murray, R. Voluntary dehydration in runners despite favorable conditions for fluid intake. *Int. J. Sport Nutr. Exerc. Metab.* **2007**, *17*, 284–295. [CrossRef]

54. Hosseinlou, A.; Khamnei, S.; Zamanlu, M. The effect of water temperature and voluntary drinking on the post rehydration sweating. *Int. J. Clin. Exp. Med.* **2013**, *6*, 683–687.

55. Hew-Butler, T.; Rosner, M.H.; Fowkes-Godek, S.; Dugas, J.P.; Hoffman, M.D.; Lewis, D.P.; Maughan, R.J.; Miller, K.C.; Montain, S.J.; Rehrer, N.J.; et al. Statement of the Third International Exercise-Associated Hyponatremia Consensus Development Conference, Carlsbad, California, 2015. *Clin. J. Sport Med.* **2015**, *25*, 303–320. [CrossRef] [PubMed]

56. Tyler, C.J.; Reeve, T.; Hodges, G.J.; Cheung, S.S. The Effects of Heat Adaptation on Physiology, Perception and Exercise Performance in the Heat: A Meta-Analysis. *Sports Med.* **2016**, *46*, 1699–1724. [CrossRef] [PubMed]

57. Almond, C.S.; Shin, A.Y.; Fortescue, E.B.; Mannix, R.C.; Wypij, D.; Binstadt, B.A.; Duncan, C.N.; Olson, D.P.; Salerno, A.E.; Newburger, J.W. Hyponatremia among runners in the Boston Marathon. *N. Engl. J. Med.* **2005**, *352*, 1550–1556. [CrossRef] [PubMed]

58. Ishunina, T.A.; Swaab, D.F. Vasopressin and oxytocin neurons of the human supraoptic and paraventricular nucleus; size changes in relation to age and sex. *J. Clin. Endocrinol. Metab.* **1999**, *84*, 4637–4644. [CrossRef] [PubMed]

59. Sar, M.; Stumpf, W. Simultaneous localization of [3 H] estradiol and neurophysin I or arginine vasopressin in hypothalamic neurons demonstrated by a combined technique of dry-mount autoradiography and immunohistochemistry. *Neurosci. Lett.* **1980**, *17*, 179–184. [CrossRef]
60. Barley, O.R.; Chapman, D.W.; Abbiss, C.R. Weight Loss Strategies in Combat Sports and Concerning Habits in Mixed Martial Arts. *Int. J. Sports Physiol. Perform.* **2018**, *13*, 933–939. [CrossRef] [PubMed]
61. Bonci, C.M.; Bonci, L.J.; Granger, L.R.; Johnson, C.L.; Malina, R.M.; Milne, L.W.; Ryan, R.R.; Vanderbunt, E.M. National athletic trainers' association position statement: Preventing, detecting, and managing disordered eating in athletes. *J. Athl Train.* **2008**, *43*, 80–108. [CrossRef] [PubMed]
62. Nuccio, R.P.; Barnes, K.A.; Carter, J.M.; Baker, L.B. Fluid Balance in Team Sport Athletes and the Effect of Hypohydration on Cognitive, Technical, and Physical Performance. *Sports Med.* **2017**, *47*, 1951–1982. [CrossRef]
63. Garth, A.K.; Burke, L.M. What do athletes drink during competitive sporting activities? *Sports Med.* **2013**, *43*, 539–564. [CrossRef]

 © 2019 by the authors. Licensee MDPI, Basel, Switzerland. This article is an open access article distributed under the terms and conditions of the Creative Commons Attribution (CC BY) license (http://creativecommons.org/licenses/by/4.0/).

Review

Dietary Intakes of Professional and Semi-Professional Team Sport Athletes Do Not Meet Sport Nutrition Recommendations—A Systematic Literature Review

Sarah L. Jenner [1,2,*], Georgina L. Buckley [3], Regina Belski [3], Brooke L. Devlin [1] and Adrienne K. Forsyth [1]

[1] Department of Rehabilitation, Nutrition and Sport, La Trobe University, Bundoora, VIC 3068, Australia; B.Devlin@latrobe.edu.au (B.L.D.); A.Forsyth@latrobe.edu.au (A.K.F.)
[2] Carlton Football Club, Ikon Park, Carlton, VIC 3053, Australia
[3] School of Health Sciences, Swinburne University of Technology, Hawthorn, VIC 3122, Australia; gbuckley@swin.edu.au (G.L.B.); rbelski@swin.edu.au (R.B.)
* Correspondence: s.jenner@latrobe.edu.au; Tel.: +61-03-9479-5953

Received: 10 May 2019; Accepted: 20 May 2019; Published: 23 May 2019

Abstract: Background: to develop sport-specific and effective dietary advice, it is important to understand the dietary intakes of team sport athletes. This systematic literature review aims to (1) assess the dietary intakes of professional and semi-professional team sport athletes and (2) to identify priority areas for dietetic intervention. Methods: an extensive search of MEDLINE, Sports DISCUS, CINAHL, Web of Science, and Scopus databases in April–May 2018 was conducted and identified 646 studies. Included studies recruited team sport, competitive (i.e., professional or semi-professional) athletes over the age of 18 years. An assessment of dietary intake in studies was required and due to the variability of data (i.e., nutrient and food group data) a meta-analysis was not undertaken. Two independent authors extracted data using a standardised process. Results: 21 ($n = 511$) studies that assessed dietary intake of team sport athletes met the inclusion criteria. Most reported that professional and semi-professional athletes' dietary intakes met or exceeded recommendations during training and competition for protein and/or fat, but not energy and carbohydrate. Limitations in articles include small sample sizes, heterogeneity of data and existence of underreporting. Conclusions: this review highlights the need for sport-specific dietary recommendations that focus on energy and carbohydrate intake. Further exploration of factors influencing athletes' dietary intakes including why athletes' dietary intakes do not meet energy and/or carbohydrate recommendations is required.

Keywords: sports nutrition; carbohydrate intake; energy; nutritional recommendations

---

1. Introduction

1.1. Nutrition for Team Sport Athletes

Team sports can be defined as field- and court-based sports with intermittent and high-intensity game patterns [1]. Match patterns will vary markedly among different sports and for specific positions within a sport. Team sports are divided into three categories: (1) endurance-based sports including football (i.e., soccer), Australian football, hockey (2) strength and power sports such as rugby union and league, American football and (3) batting sports such as baseball, softball and cricket [1]. Athletes participating in professional team sports are supported by high-performance and medical staff, which aim to monitor and optimise fitness, body composition and performance outcomes. The physiological demands of team sports differ and can include a range of performance modes including: running moderate to long distances, high intensity bouts of movement, variable activity patterns and small bouts of rest periods [1–3]. The variable nature of team sport exercise requires use both anaerobic and aerobic systems to fuel performance [1]. Therefore, each team sport and position within the sport, depending on the nature of training and competition, will have unique energy demands and nutrient requirements. To optimise performance and enhance recovery, international sporting committees (i.e., International Olympic Committee (IOC), American College of Sports Medicine (ACSM), International Society of Sports Nutrition (ISSN)) have provided nutrition recommendations to support dietitians working with athletes to meet their individual nutrition needs [4–9].

Dietitians must consider a range of sport-specific factors including the rules, arena size, timing of competition, frequency of matches and length of seasons (including macrocycles: preseason, competition season, off-season) when assessing an athlete's nutrition requirements and goals. Additionally, the physique characteristics and position-specific tasks of the sport will further influence the nutritional requirements of athletes. For example, the sport of rugby union will require forwards to be heavier and stronger in comparison to backs who need to be leaner and faster [10]. Due to the sport-specific factors, physique and position differences, dietary advice for team sport athletes should be individualised.

Recommendations that support athletes to consume sufficient energy and the correct balance of macronutrients and micronutrients, with appropriate timing to enhance performance and recovery, will enable athletes to train and perform optimally [11]. An earlier review by Holway and Spriet [1] found that athletes competing in team sports commonly do not meet recommended dietary intake needs [1,12]. Those that fail to consume energy and/or maintain a diet that encompasses the appropriate balance of macronutrients may find that this impedes on training adaptations and recovery [11,12]. Deficiencies in energy can have implications for an athlete's performance including a loss of fat free mass, disturbances to immune function, decreased bone mineral density, increased susceptibility to injury and increased prevalence of symptoms of overtraining [11].

1.2. Objectives

In the past decade, thousands of new research papers have been published in sports nutrition and 17 new consensus statements and recommendation papers have been released by authoritative organisations such as the IOC, ACSM, ISSN [4–9,11,13–22]. There have also been a large number of published studies on the dietary intake of professional and semi-professional team sport athletes during this time [2,3,10,23–39]. With new sports nutrition recommendations [5–8,11,13–21] and updated literature reporting the dietary intake of team sport athletes, it is now timely to review the literature to determine whether team sport athletes consume diets that align with the sports nutrition recommendations [2,3,10,23–39]. This paper aims to review the literature on dietary intakes of professional and semi-professional team sport athletes systematically with the aim of identifying priority areas for dietetic intervention.

2. Materials and Methods

2.1. Protocol Registration

All methods and search strategies were aligned with Preferred Reporting Items for Systematic Review (PRISMA) guidelines. This review was registered with International Prospective Register of Systematic Reviews (registration number: CRD42018105168) [40,41]. A PICOS criteria (i.e., Participants, Intervention, Comparison, Outcome and Study design) for review is defined in Table 1. A systematic search using terms such as sport or team sport or dietary intake or food intake was conducted by one researcher (SJ). All keywords used in search are listed in Table 2.

Table 1. Participants, Intervention, Comparison, Outcome and Study (PICOS) criteria for inclusion and exclusion of studies.

Parameter	Description
Population	Professional and semi-professional team sport athletes
Intervention OR exposure	Baseline dietary intake
Comparison	Dietary intake in comparison to sports nutrition guidelines and recommendations.
Outcomes	Meeting/not meeting sports nutrition guidelines and recommendations
Study design	RCT (where baseline dietary intake data available), cross-sectional, longitudinal, thesis (unpublished and published)

Abbreviations: RCT Randomised Control Trial.

Table 2. Table of keywords.

Concept	Keywords
Sport OR team sport	"sport*" OR "team sport*" OR "football" OR "soccer" OR "football" OR "netball" OR "AFL" OR "Aussie rules" OR "rugby" OR "basketball" OR "grid iron" OR "American football" OR "hockey"
Dietary intake OR food intake	"nutrient requirement*" OR "dietary intake" OR "daily food intake" OR "food intake"

2.2. Search Strategy

Five electronic databases including MEDLINE (Medlars International Literature Online), Sports DISCUS, CINAHL (Cumulative Index of Nursing and Allied Health Literature), Web of Science (WOS), and Scopus (World's largest abstract and citation database of peer-reviewed literature) were searched to investigate the dietary intake of professional and semi-professional team sport athletes by one researcher (SJ). The timeframe designated for the search included studies published from 2011 to present (i.e., after the 2011 review by Holway and Spriet) [1]. An additional limit regarding age (i.e., >18 years) was included to limit results to adult athletes only. In order to identify any further relevant publications, the reference lists of the studies included were hand searched and other manual searches were conducted (i.e., Google Scholar).

2.3. Eligibility Criteria

All original research (i.e., cross-sectional, longitudinal, published and unpublished thesis) conducted in adult team sport athletes (i.e., 18 years and older) and published since 2011 was considered for inclusion (Table 3). Randomised control trials were additionally included in the review if baseline dietary intake data was available. Randomised control trials where only post intervention dietary intakes were available, conference posters, abstracts and web-based articles were not included for review. Only professional and semi-professional athletes were included in the review; amateur and recreational athletes were not included. Only English-language studies were included for this review. Included studies were required to provide baseline or habitual dietary intake data that quantified energy, macronutrients and micronutrients to allow for the specified conversions made and displayed in the data extraction table i.e., energy (MJ/day), carbohydrate (grams (g) and $g \cdot kg^{-1} \cdot day^{-1}$), protein (g and $g \cdot kg^{-1} \cdot day^{-1}$), fat (g and % total energy), calcium (mg/day) and iron (mg/day). Studies that reported dietary habits, dietary knowledge, attitudes, and education strategies where dietary intake was not quantified, were excluded from the review.

Table 3. Eligibility criteria.

Inclusion	Exclusion
1. Studies that include only team sport athletes.	1. Studies that include only individual athletes or sports.
2. Studies published since 2012.	2. Studies published up to 2011.
3. Studies that include only professional or semi-professional team sport athletes.	3. Studies that include only non-professional or amateur team sport athletes.
4. Studies that include only adult team sport athletes (i.e., >18 years of age).	4. Studies that include adolescent and child team sports athletes (i.e., <18 years of age).
5. English language studies.	5. Non-English language studies.
6. Studies that include quantitative measures of dietary intake that can be converted into units of intake per day for each nutrient.	6. Studies that include nutrition habits, attitudes, educational strategies, knowledge, where dietary intakes cannot be compared.
7. Studies that include the dietary assessment of total energy carbohydrate, protein fat, micronutrient intake (i.e., iron (mg/day), calcium (mg/day), folate etc.).	7. Studies where only supplement or antioxidant intake is assessed.
8. Studies that assess dietary intake using a validated method of assessment (i.e., 7 day food diary, 7 day weighed food diary, food records, 3 day food diaries, FFQ, diet histories etc.) and therefore estimates absolute dietary intake.	8. Studies that assess dietary intake, however methods used provide dietary assessments represented in food groups, percentage of total energy etc.
9. Only human studies that include; RCT (where baseline dietary intake data available), cross-sectional, longitudinal, thesis.	9. No animal studies, RCT (where only post intervention dietary intakes available), conference posters, reviews, abstracts and web-based articles.
10. Published and unpublished research (i.e., thesis).	

Abbreviations: FFQ Food Frequency Questionnaire, RCT Randomised Control Trial.

2.4. Study Selection

All studies were screened based on title and abstract by main author (SJ). Articles deemed eligible for full text review were screened against inclusion and exclusion criteria (Table 3) by two authors (SJ and GB). Additional reviewers (AF and RB) provided advice on eligibility if a decision for inclusion and exclusion required feedback. Selection of included studies and reasons for exclusion are reported in Chart 1.

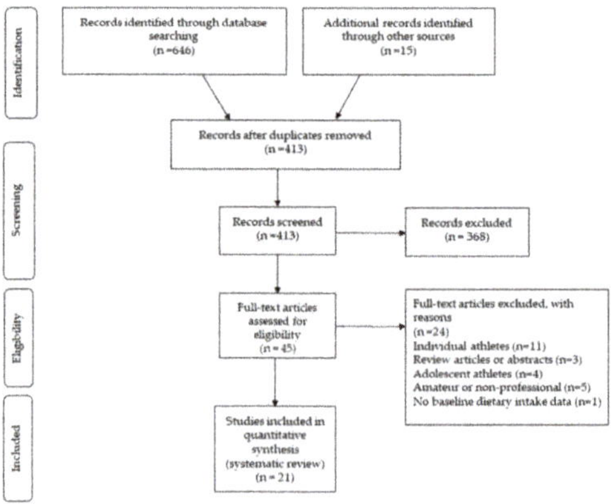

Chart 1. Study section process.

2.5. Data Collection Process

Dietary intake data were extracted from the included studies by two authors (SJ and GB). Data presented in Table 4 include participant demographics including sport, anthropometry measures (total mass (kg), body fat (%), fat free mass (kg)) and athletic level (professional or semi-professional)); details regarding the study background and methods; country of origin and survey method (i.e., 7-day food diary) and dietary intake results including total energy/calories (i.e., MJ), carbohydrate intake (i.e., g and g·kg^{-1}·day^{-1}), protein intake (i.e., g and g·kg^{-1}·day^{-1}), fat intake (i.e., g and % total energy), calcium intake (i.e., mg) and iron intake (mg). Where total energy intake was reported in calories, this was converted to MJ to enable comparison across studies.

2.6. Study Quality: Risk of Bias

The quality of studies was examined by two authors (SJ and GB) using the Academy of Nutrition and Dietetics Quality Criteria Checklist from the Academy of Nutrition and Dietetics Evidence Analysis Manual [42]. The quality criteria checklist provides an assessment based on relevance and validity criteria questions, ranking studies as either positive, neutral or negative. A third reviewer (AF) reviewed any discrepancies that occurred during the quality analysis. Studies with positive ratings needed to describe study selection adequately (including inclusion and exclusion criteria), methods of comparing groups and the study setting, and include measurements that were valid and reliable.

3. Results

3.1. Study Selection

The original search retrieved 646 studies that fit the search criteria with an additional 15 studies identified by hand searches (Chart 1). After duplicates were removed, a title and abstract exclusion was undertaken and 45 studies were retained for full text assessment. After completion of full text assessment 21 studies were included in this review for data extraction, quality assessment and analysis. All studies included in the review had a positive or neutral quality rating [41].

3.2. Study Characteristics

The majority of studies included in this review included team sport athletes competing professionally and semi-professionally in Australia (*n* = 255) [3,25,26,33,43] and Spain (*n* = 81) [28–30,36], with the remaining studies including athletes from Europe (not-specified) (*n* = 34) [10,27], England (*n* = 30) [23,34], America (*n* = 26) [32,38], Canada (*n* = 25) [39], Brazil (*n* = 19) [37], Netherlands (*n* = 14) [24], South Africa (*n* = 11) [35], United Kingdom (*n* = 10) [2], and Mexico (*n* = 6) [31,37]. The majority of studies included in this review reported the dietary intakes of professional team sport athletes [2,3,10,23,24,26–29,31,33,34,36,37], with additional studies exploring dietary intake of semi-professional team sport athletes [32,35,38,39] and studies exploring the dietary intakes of a combination of sports [25,30,43]. Studies included a range of team sport athletes with the majority of studies reporting on the dietary intakes of football athletes (*n* = 210) [23–25,28,30,31,34,37], followed by Australian football (*n* = 139) [3,26,43]. Across the remaining studies, other team sport athletes represented in this review include; rugby union (*n* = 70) [10,27,33,35], ice-hockey (*n* = 25) [39], wheelchair basketball (*n* = 17) [29], American football (*n* − 15) [32], handball (*n* − 14) [36], volleyball (*n* = 11) [38], rugby league (*n* = 10) [2].

The majority of studies included in this review used a cross-sectional study design (*n* = 14), with the remaining studies using pre-post-test [32,38], case-study [27,29], case control [30] and longitudinal [26,36] designs to assess dietary intake. Dietary intake data were collected most frequently using food diaries/records (weighed and not weighed). Studies used 7-day food diaries [2,3,23,28,33,35,39], 3-day food diaries [29,32,37,38] and 24 hr recalls [10,24,25,43]. Other studies used food diaries of different duration (i.e., 4-day, 6-day, and 8-day) [27,30,34], dietary recalls (i.e., 72-h) [26,36] and a combination of 4-day weighed food diary in addition to a food frequency questionnaire (FFQ) to assess dietary intake [31].

Table 4. Reported energy, macronutrient and micronutrients intakes of male and female professional and semi-professional team sport athletes.

| References | | | | Participant Characteristics | | | | | Dietary Intake Data | | | | | | | | |
Author, Year	Sport	Country	Level	N (Gender)	Total Mass (kg)	Lean Mass (kg)	BF%	Energy Intake Reported (Estimated MJ Were Available)	CHO (g)	CHO (g·kg^{-1}·day^{-1})	PRO (g)	PRO (g·kg^{-1}·day^{-1})	Fat (g)	Fat (% TE)	Calcium (mg·day^{-1})	Iron (mg·day^{-1})	Quality Rating
Anderson et al. [23], (2017)	Football (soccer)	England	P	n = 6 (male)	80.5 ± 8.7	11.9 ± 1.2	65.0 ± 6.7	3186 ± 585 kcal·day^{-1} (13.3)	330 ± 98	4.2 ± 1.4	205 ± 30						+
Andrews and Itsiopoulus [25] (2016)	Football (soccer)	Australia	P, SP	n = 73 (male)	P 79.6 ± 7.7 SP 75.6 ± 7.6			P 11,525 ± 1987 kJ·day^{-1} SP 10,831 ± 3842 kJ·day^{-1} (P 11.5 ± 2.0, SP 10.8 ± 3.8)	P 302.4 ± 72.3 SP 298.7 ± 148.5	P 3.5 ± 0.8 SP 3.9 ± 1.8	P 152.3 ± 27.7 SP 149.1 ± 46.8	P 1.9 ± 0.3 SP 2.0 ± 0.6	P 95.9 ±31.7 SP 85.8 ±37.8	P 30.4 ± 7.3 SP 29.5 ± 7.4			∅
Bettonviel et al. [24] (2016)	Football (soccer)	Netherlands	P	n = 14 (male)	77.0 ± 8.6			2988 ± 583 kcal·day^{-1} (12.5)	365 ± 76	4.7 ± 0.7	145 ± 24	1.9 ± 0.3	97 ± 26	29.0 ± 3.6			∅
Bilsborough et al. [26] (2016)	Australian Football	Australia	P	n = 45 (male)	86.8 ± 7.9	[7.9 ± 1.8–10.5 ± 2.7]	[71.2 ± 6.8–76.8 ± 6.4]	[11.1–13.9]	[321 ± 142–379 ± 66]	4.1 ± 1.6	[118 ± 67–215 ± 60]	1.9 ± 0.9	[61 ± 29–110 ± 40]	[19.0 ± 5.0–33.1 ±6.7]			∅
Bradley et al. [27] (2015)	Rugby Union	Europe (non-specified)	P	n = 14 (male)	F 110 ± 6.2 B 93.6 ± 5.9			F 16.6 ± 1.25 MJ·day^{-1} B 14.2 ± 1.20 MJ·day^{-1}		F 3.5 ± 0.8 B 3.4 ± 0.7		F 2.7 ± 0.5 B 2.7 ± 0.3		F 32 B 33	1733 ± 694	24 ± 9	∅
Bradley et al. [10] (2015)	Rugby Union	Europe (non-specified)	P	n = 20 (male)	F 109.3 ± 6.9 B 91.7 ± 6.6	F 13 ± 3 B 9.3 ± 2	F 94.9 ± 4.5, B 83.1 ± 5.4	F 14.8 ± 1.9 MJ·day^{-1} B 13.3 ± 1.9 MJ·day^{-1}		F 3.3 ± 0.7 B 4.14 ± 0.4		F 2.52 ± 0.3 B 2.59 ± 0.6					∅
Conejos et al [28] (2011)	Football (soccer)	Spain	P	n = 22 (male)				Fwd. 12.7 ± 2.9 MJ·day^{-1} M 14.0 ± 6.2 MJ·day^{-1} D 14.8 ± 2.6 MJ·day^{-1} G 12.2 ± 4.6 MJ·day^{-1}	Fwd. 342.5± 92.9 M 382.1 ± 187.2 D 419.1 ± 98.3 G 320.3 ± 11.9		Fwd. 138.7 ± 27.5 M 144.8 ± 56.9 D 144.5 ± 19.9, G 142.8 ± 100.1		Fwd. 120.3 ± 59.6 M 131.6 ± 62.6 D 124.5 ± 36.1 G 109.8 ± 45.3	Fwd. 34.7 M 35.6 D 32.2 G 34.9	Fwd. 1361.5 ± 549.0 M 1592.3 ± 966.1 D 1208.4 ± 457.1 G 1499.7 ± 1035.9	Fwd. 13.4 ± 3.7 M 18.9 ± 9.7 D 22.6 ± 6.8 G 15.5 ± 10.2	∅
Devlin et al. [43] (2017)	Australian Football (AF) Football (Soccer, F)	Australia	P, SP	n = 66 AF = 48, F = 18 (male)	AF (P) 87.8 ± 9.2 AF (SP) 82.9 ± 9.0 F (P) 75.6 ± 5.6	AF (P) 15.1 ± 2.4, AF (SP) 16.7 ± 2.7 F (P) 12.8 ± 1.9	AF (P) 65.4 ± 7.9 AF (SP) 61.2 ± 3.9 F (P) 56.8 ± 5.2	AF (P) 17.3 ± 4.2 MJ·day^{-1} AF (SP) 13.2 ± 2.5 MJ·day^{-1} F (P) 9.4 ± 2.3 MJ·day^{-1}	AF (P) 406 ± 132 AF (SP) 368 ± 93 F (P) 220 ± 76	AF (P) 4.6 ± 1.5 AF (SP) 4.5 ± 1.2 F (P) 2.9 ± 1.1	AF (P) 295 ± 97 AF (SP) 171 ± 52 F (P) 140 ± 35	AF (P) 3.4 ± 1.1 AF (SP) 2.1 ± 0.7 F (P) 1.9 ± 0.5		AF (P) 29 ± 6 AF (SP) 28 ± 8 F(P) 33 ± 9			∅
Grams et al. [29] (2016)	Wheelchair basketball	Spain	P	n = 17 (male)	75.5 ± 13.5			2673 ± 485 kcal·day^{-1} (11.2)		3.9 [Range: 1.8–8.1]			1.7 ± 0.6	33.7 ± 5.5			+
Gravina et al. [30] (2012)	Football (soccer)	Spain	P, SP	n = 28 (female)	61 ± 8.4	16.7 ± 3.2		2271 ± 578 MJ·day^{-1} (9.5)						33 ± 7			+
Hidalgo et al. [31] (2015)	Football (soccer)	Mexico	P	n = 6 (male)	68.3 ± 2.0		32.5 ± 1.0	3042 ± 56 kcal·day^{-1} (12.7)	364 ± 17.5	5.4 ± 0.3	145 ± 14	2.2 ± 0.2	113 ± 6.3	33 ± 0.0			∅
Jenner et al. [3] (2018)	Australian Football	Australia	P	n = 46 (male)	86.3 ± 9.4	10.8 ± 2.3	73.9 ± 9.1	9.1 ± 1.8 MJ·day^{-1}	201 ± 56	2.4 ± 0.8	150 ± 30	1.8 ± 0.4	78.9 ± 20.0	32 ± 4.5	952 ± 287		∅
Kirwan et al. [32] (2012)	American Football	America	SP	n = 15 (male)	93.8 ± 15.3		70.5 ± 7.7	3518 ± 849 kcal·day^{-1} (14.7)	353 ± 118		169 ± 52		160 ± 45				∅
MacKenzie et al. [33] (2015)	Rugby Union	Australia	P	n = 25 (male)	100.2 ± 13.3		77.0 ± 7.0	13605 ± 3639 kJ·day^{-1} (13.6)	352 ± 115	3.6 ± 1.3	211 ± 62	2.2 ± 0.7	101 ± 34	28 ± 5.0			∅

Table 4. *Cont.*

References		Participant Characteristics							Dietary Intake Data								
Author, Year	Sport	Country	Level	n (Gender)	Total Mass (kg)	Lean Mass (kg)	BF%	Energy Intake Reported (Estimated MJ Were Available)	CHO (g)	CHO (g·kg^{-1}·day^{-1})	PRO (g)	PRO (g·kg^{-1}·day^{-1})	Fat (g)	Fat (% TE)	Calcium (mg·day^{-1})	Iron (mg·day^{-1})	Quality Rating
Molina-Lopez et al. [36] (2013)	Handball	Spain	P	n = 14 (male)	86.7 ± 5.4	11.6 ± 2.5		2974.5 ± 211.1 kcal·day^{-1} (12.4)	360.9 ± 27.6	4.2 ±0.4	133.4 ±14.3	1.5 ± 0.2	118.6 ± 22.5		1251.4 ± 338.2	24.2 ± 8.5	∅
Ono et al. [34] (2011)	Football (soccer)	England	P	n = 24 (male)				[2648-4606 kcal·day^{-1}] (11.1–19.3)	505.2 ± 120		141.7 ± 22.8						∅
Potgieter, et al. [35] (2014)	Rugby Union	South Africa	SP	n = 11 (male)	95.5 ± 13.6	18.2 ± 5.7	79.7 ± 9.0	45.4 ± 9.0 kcal·kg^{-1}·day^{-1}		4.3 ± 0.4		2.4 ± 0.7		33.8 ± 4.3	1250 ± 403		∅
Raizel et al. [37] (2017)	Football (soccer)	Brazil	P	n = 19 (male)	71.8 ± 8.2	4.9 ± 1.5		40.74 ± 12.8 kcal·kg^{-1}·day^{-1}		5.4 ± 1.9	136.4 ± 57.4	1.9 ± 0.8	91.2 ± 35.9				∅
Tooley et al. [2] (2015)	Rugby League	UK	P	n = 10 (male)	97.3 ± 3.1			3731 ± 164 kcal·day^{-1} (15.6)	467 ± 24	4.9 ± 0.3	209 ± 10	2.2 ± 0.2	124 ± 10	30 ± 1.0			∅
Valliant et al. [38] (2012)	Volleyball	America	SP	n = 11 (female)	75.4 ± 13.4	24.5 ± 5.9		1756 ± 557.5 kcal·day^{-1} (7.3)	224.3 ± 64.4	3.08 ± 1.1	69.3 ± 26.8	0.9 ± 0.3	67.4 ± 27.8	33.7 ± 6.4			∅
Vermeulen et al. [39] (2017)	Ice-hockey	Canada	SP	n = 25 (female)	67.0 ± 1.7			2354 ± 74 kcal·day^{-1} (9.8)	305 ± 15	4.6 ± 0.2	91 ± 4	1.4 ± 0.1	82 ± 4	32 ± 1.0	1041 ± 64	19 ± 3	∅

Abbreviations: AF Australian Football, F Football (soccer), BF% Body fat percentage, CHO Carbohydrate, D Defenders, Fwd Forwards, GK Goal-keepers, kcal·day^{-1} Calories per day, Kcal·kg^{-1}·day^{-1} Calories per kg per day, kJ·day^{-1} Kilojoules per day, MJ·day^{-1} Mega joules per day, M Midfielders *PRO* Protein, *P* Professional, *SP* Semi-professional. Quality rating criteria: Positive (+), Neutral (∅) and Negative (−).

3.3. Dietary Intake of Team Sport Athletes

3.3.1. Energy Intake of Team Sport Athletes on Training Days

All studies included assessed the energy intake of team sport athletes. The majority of authors assessed the energy intake of athletes against recommendations advocated by IOC, ACSM, ISSN and sports specific research [1,4,12,35,44–55]. Several studies provided evidence that energy intake of team sport athletes assessed was suboptimal and did not meet recommendations [2,3,10,26,28,35,36,38]. Of the 21 studies included, five studies reported that energy intake was adequate according to the respective dietary recommendations used [23,32,33,37,39]. One study by Devlin et al. [43] reported that the Australian football athletes included met energy recommendations, however the football athletes did not. The remaining studies (*n* = 7) did not report on adequacy of energy intake (i.e., recommendations met versus not met) [24,25,27,29–31,34].

3.3.2. Energy Intake of Team Sport Athletes during Competition

The mean energy intake reported in this review ranged from 9.1–16.6 MJ/day and 7.3–9.8 MJ/day for males and females respectively. Seven studies included comparison data of dietary intake on training days and match days [2,23,24,27,31,35,39]. Two studies that explored the dietary intake of rugby union (*n* = 10) and ice hockey (*n* = 25) athletes on match days found that energy intake did not meet increased requirements for the fueling and recovery required on match days [2,39]. However, in comparison, two studies that included professional rugby union (*n* = 14) and football athletes (*n* = 6), found that energy intake was greater on match days in comparison to training days [23,27]. In particular, research by Bradley et al. [27] found that on average professional rugby union athletes increased their energy intake in preparation for game day in comparison to the first four days of training where energy intake was reduced, irrespective of energy expenditure. Anderson et al. [23] additionally found that professional football players (*n* = 6) had a greater absolute and relative energy intake on match days in comparison to training days.

3.3.3. Macronutrient Intake of Team Sport Athletes on Training Days

Overall, the macronutrient composition of the diets of team sport athletes was inadequate to meet the fuel, recovery and performance demands of their sports. All but one study assessed the dietary intake of carbohydrate and protein [30] and all but two assessed the dietary intake of fat [30,34]. Overall, a macronutrient imbalance was found in the majority of studies with most athletes reported consuming diets high in protein and fat, at the expense of carbohydrate. Team sport athletes including athletes from football (*n* = 175), Australian football (*n* = 139), rugby union (*n* = 88), volleyball (*n* = 11) and rugby league (*n* = 10) consumed diets that consistently did not meet carbohydrate recommendations [2,3,10,23–28,30,33,35–38,43]. Out of the 17 studies that provided mean intakes of carbohydrates, 15 reported low carbohydrate intake and fell below ISSN recommended intakes of 5-8 $g \cdot kg^{-1} \cdot day^{-1}$ (range: 2.4–4.9 $g \cdot kg^{-1} \cdot day^{-1}$ and 3.08–4.6 $g \cdot kg^{-1} \cdot day^{-1}$ for male and female athletes respectively). One study that explored the dietary intakes of male professional football players found that carbohydrate intakes consumed by athletes were closer to meeting recommendations for tactical or skill based sports (3–5 $g \cdot kg^{-1} \cdot day^{-1}$) [12,25].

Conversely, the majority of studies found that dietary intake of protein [2,10,24–26,28,30,31, 33,35–37,43,56] and fat [24,27,28,32,35–37] exceeded recommendations. Eight studies that included athletes competing professionally and semi-professionally in Australian football, rugby league, rugby union and football found protein intakes in excess of 2.0 $g \cdot kg^{-1} \cdot day^{-1}$ [2,10,25,27,31,33,35,43], with a study reporting that the diets of professional Australian football athletes on average contained 3.4 ± 1.1 $g \cdot kg^{-1} \cdot day^{-1}$ of protein per day [43]. Studies included in this review reported dietary intakes that were high in protein, however low in carbohydrates and/or total energy (hypocaloric) [2,3,10, 26,28,35,36,38,43]. Nine studies found that dietary intake of fat exceeded recommendations [2,27–29,31,32,35,37,39]. Three studies that included rugby union athletes (*n* = 35) and football athletes

(n = 25) found that while overall total fat intake exceeded recommendations, polyunsaturated fat intake fell below recommended intakes of 10% of total energy [31,35,37]. Six studies explored the intake of cholesterol, finding that dietary intakes of athletes exceeded recommended intakes of <300 mg/day [2,28,31,32,35,37]. Kirwan et al. [32] linked high cholesterol intakes to potential body composition goals of American football athletes (i.e., to put on mass, quickly).

3.3.4. Macronutrient Intake of Team Sport Athletes during Competition

Three of seven studies that assessed intake on competition days found that carbohydrate consumed in preparation for match day, during the match and during the recovery period post game, did not meet recommended intakes for competitions [24,31,39]. One study in female ice-hockey players found that there was no significant difference in carbohydrate, protein and fat intakes between game, training and rest days [39]. In comparison, three studies on rugby union (2) and football (1) athletes found that carbohydrate intake during and post-game day met recommendations [23,27,35]. In particular, Anderson et al. [23] found that athletes practiced a level of periodization, finding that carbohydrate consumed on the two match days were significantly greater than carbohydrate consumed on training days (p = < 0.05; 6.4 g·kg^{-1}·day^{-1} and 4.2 g·kg^{-1}·day^{-1} for match and training days respectively). In comparison, research by Bettonviel et al. [24] found that professional football players as a whole failed to meet carbohydrate recommendations on match days (5.3 ± 1.5 g·kg^{-1}·day^{-1}) and one day post-match (4.5 ± 1.0 g·kg^{-1}·day^{-1}); however, protein intakes on match day and post-match were adequate (2.0 ± 0.4 and 1.6 ± 0.3 g·kg^{-1}·day^{-1} respectively).

3.3.5. Micronutrient Intake

Six studies reported iron and/or calcium intakes [3,27,28,35–37,39]. The majority of these studies found that team sport athletes were meeting or exceeding recommended intakes of calcium and iron, when compared to the general public [27,28,35,36,39]. One study by Raizel et al. [37] found that professional football players were not meeting general (non-sport specific) recommendations (estimated average requirements) for calcium, reporting that athletes had a marginal intake of 83% of EAR.

4. Discussion

4.1. Energy Intake of Team Sport Athletes

In order for athletes to optimise training and performance, they need to consume sufficient energy for the work required and to support physiological adaptions [11,13,57]. A diet that contains insufficient energy (i.e., energy deficit) during periods of training can result in a number of performance detriments including loss of lean muscle mass and bone mineral density, increased prevalence of overtraining and injury, and may contribute to endocrine and reproductive system disturbances [11]. The mean energy intakes of professional male team sport athletes reported in the literature have decreased from those reported by a previous review [1]; however, in comparison female athletes are consuming diets of relatively similar energy density (i.e., 7.3–9.8 MJ/day) [1]. Simply looking at the difference in energy intakes of male team athletes would suggest that they are eating less and potentially not meeting energy recommendations; however, a range of factors may have influenced the dietary intake of these athletes at the time of assessment.

While not identified as an influencing factor previously [1], three studies included in this review that explored the dietary intakes of rugby union, Australian football and football athletes suggested that low energy intakes were related to the presence of team culture surrounding body composition goals (i.e., to decrease body fat and/or increase lean muscle mass) [3,10,25]. In theory, body composition goals should be individualised to the athlete, with the focus to support training adaptions; however, this is not always reflected in practice [8]. Research by Bradley et al. [10] found that during a rugby union preseason, athletes did not meet energy intake recommendations (14.8 ± 1.9 MJ/day) and it was found that the existence of body composition goals (i.e., to reduce body fat) were potentially influencing intake at the expense of fueling for training demands. Research during a preseason in Australian football, hypothesised that athletes intentionally restricted energy and carbohydrate intake surrounding body composition assessments using dual-energy X-ray absorptiometry (DXA), to meet target body composition goals (actual energy intake: 9.1 ± 1.8 MJ/day) [3]. Similarly, Andrews et al. [25] reported professional and semi-professional football athletes that were recognised as under reporters were those athletes that were identified as attempting to maintain body composition or reduce body fat at the time of recording (11.5 ± 2.0 and 10.8 ± 3.8 MJ/day). Realistic body composition goals must be promoted to prevent under fueling and support the training adaptions and recovery of athletes. Furthermore, in light of recent research regarding relative energy deficiency (RED-S) in male athletes, there is a greater need for education for coaches and support staff regarding the importance of an individualised approach when tailoring body composition goals [8].

4.2. Macronutrient Intake of Team Sport Athletes

4.2.1. Carbohydrate Intake of Team Sport Athletes

Carbohydrate intake is important for optimising performance and recovery. Team sports have varied training and physiological demands, therefore advice must be tailored to match the training demands as well as the demands of specific positions within the sport. Team sport athletes included in this review continue to consume diets that do not align with carbohydrate recommendations, intakes on average insufficient when compared to recommendations [1,11,12,52]. Research included has suggested that carbohydrate recommendations need to be better suited to the demands of the sport, individualising nutrition based on positions [10,23,28,58]. Research by Bradley et al. [27] found that professional rugby union athletes tapered carbohydrate intake across a training week; however, on average intakes fell below recommendations used by authors of 6–10 $g \cdot kg^{-1} \cdot day^{-1}$ (3.5 $g \cdot kg^{-1} \cdot day^{-1}$ and 3.4 $g \cdot kg^{-1} \cdot day^{-1}$ for forwards and backs respectively). Although these intakes did not align with recommendations, Bradley et al. [27] suggested that the tapering of intakes during a training week to match demands of training as well as enhance training adaptions (i.e., alter body fat) may be better suited to this athletic population. The idea of *"fueling for the work required"* is commonly being used by many sports nutrition professionals working in professional sport [59]. The concept of periodising intake is not simple and requires thorough knowledge of the athlete's needs, the sport and the training and competition demands. Anderson et al. [23] found that football athletes were applying the principle of carbohydrate periodization to their daily intakes; consuming significantly greater energy and carbohydrate on game days compared to training days. However, when assessing dietary intakes as a whole; carbohydrate intake the day before a competition was unlikely to maximise glycogen storage and, therefore, meet match demands. Dietitians working with professional team sport athletes need to use an individualised approach when periodising an athlete's carbohydrate intake. Further work is required to support education in this space to optimise glycogen storage and resynthesis and to support athletes' training and match day nutrition goals.

4.2.2. Protein Intake of Team Sport Athletes

Protein is important for muscle protein synthesis, supports recovery processes, promotes satiety and can aid the maintenance of body composition. It also has many other important roles in the body as enzymes, hormones, transporters and antibodies. Athletes with insufficient protein intakes have an increased risk of muscle wasting, illness and injury. ISSN recommendations state that to maintain protein balance athletes should consume 1.4–2.0 $g \cdot kg^{-1} \cdot day^{-1}$ of high-quality protein [11]. In this review team sport athletes, on average, adequately met recommendations; most athletes adopting a diet high in protein, but low in carbohydrates and/or total energy (hypocaloric) [2,3,10,26,28,35,36,38,43]. Research has suggested that high-protein diets (2.3–3.1 $g \cdot kg^{-1} \cdot day^{-1}$) may be appropriate for resistance trained athletes who are consuming hypocaloric diets with the aim to maintain lean muscle tissue while reducing fat mass [20,60]. This is supported by research by Bradley et al. [10] who found that although protein intakes of professional rugby athletes (2.5–2.6 g·kg^{-1}·day^{-1}) exceeded protein recommendations, since athletes were manipulating carbohydrate intakes during the preseason training week, protein intakes may have been appropriate at the time to optimise body composition. Given the context of the time of season this study was undertaken (i.e., preseason) and the ability for athletes to optimise training adaptions and changes to body composition without any detriment to competition performance, these intakes may be acceptable for this athletic cohort. However, in contrast, research by Potgieter et al. [35] reported that rugby union athletes' intakes in-season did not meet carbohydrate recommendations, and exceeded protein recommendations. Potgieter et al. [35] suggests that greater intakes of protein (i.e., 2.4 g·kg^{-1}·day^{-1}) may be suited in times where muscle hypertrophy is required (i.e., offseason); however, in-season where athletes are required to meet training, competition and recovery demands, protein intakes should align with recommendations and not be increased at the expense of carbohydrate intake. Similarly, research by Mackenzie et al. [33] on rugby union athletes reported that there was no compelling evidence to increase the distribution of protein for muscle protein synthesis and that an excess quantity of protein may in fact compromise lean muscle goals by promoting satiety which can result in decreased calorie intake. Taken as a whole, when working with team sport athletes greater emphasis should be placed on the distribution and timing of protein intake across a training day, instead of the total quantity [33]. In addition, it should be highlighted that the need to meet a body composition goal should not come at the expense of meeting nutrient requirements for performance and recovery.

4.2.3. Fat Intake of Team Sport Athletes

Four of the studies that reported athletes' diets exceeding recommended fat intakes (i.e., >30% of total energy) included the dietary intakes of athletes competing in strength and power sports such as rugby union (2), rugby league (1) and American football (1) [2,27,32,35]. Collectively strength and power sports are characterised as high-intensity and intermittent collision sports and may require a greater total mass to protect athletes against the physical impacts of scrumming, tackling etc. [1,2,27,32,35]. A study that explored the dietary intakes of American football athletes found that athletes consumed high intakes of fat (i.e., 41% of total energy) suggesting these intakes were a result of overfeeding to increase weight [32]. In comparison, research by Bradley et al. [27] in rugby union found that athletes had fat intakes that fell slightly above recommendations (i.e., 32% and 33% for forwards and backs respectively); however, in contrast they had suboptimal total energy intake when compared to energy expenditure. Additionally Potgieter et al. [35] found that competing rugby union athletes were consuming an excess in fat (i.e., 33.8 ± 4.3%), however failed to meet recommendations for total energy and carbohydrate. Taken together, these results indicate a reduction in dietary fat intake for these groups may not be warranted in order to support athletes to meet energy requirements. However, as many athletes did not meet these enhanced energy requirements, the addition of carbohydrate may allow athletes to meet energy balance or surplus, without the need for intakes that are in excess of dietary fat.

Although most team sport athletes included in this review consumed diets that were in line with or exceeded recommendations for dietary fat intake, the composition of saturated to unsaturated fats (i.e., mono and poly unsaturated fats) was suboptimal (i.e., saturated fat >10% total energy). This is supported by Hidalgo et al. [31] who found that football players' total dietary fat intake aligned with recommendations (i.e., 30%–33% of total energy); however, saturated fat and cholesterol intake exceeded recommendations. In addition, it was found that polyunsaturated fat intake fell below recommended intakes. Hidalgo et al. [31] and Raizel et al. [37] together suggested that football athletes' high protein intake, including a large intake of animal proteins, may have contributed to their overall high cholesterol intake. Research suggests diets that are rich in saturated fat (i.e., >10%TE), cholesterol and trans fats are linked to chronic diseases such as cardio-vascular disease (i.e., heart disease and stroke) and type 2 diabetes [61]. For the long term health of athletes, dietary advice should aim to include a varied diet with a focus on total and saturated fat, not exceeding recommendations (i.e., total fat <30% total energy and saturated fat <10% total energy) [11]. Additionally, athletes may benefit from the inclusion of mono- and poly unsaturated fat-based protein foods (i.e., fatty fish, nuts and seeds) to help meet energy and protein requirements and provide additional anti-inflammatory benefits for training and recovery [22].

This review highlights the role of dietitians in providing long-term dietary strategies to increase lean muscle and total body mass, in a manner that does not adversely affect performance and/or lipid profile, which may be evident in short term high energy and high fat diets. Greater education regarding the long-term implications of intakes that are in excess of total fat and saturated fat is required in team sport environments.

4.2.4. Dietary Intake during Competition

Due to the elevated requirements for stored glycogen and glycogen resynthesis during training and competition, athletes are recommended to undertake aggressive carbohydrate feeding prior to these periods. Inconsistencies between energy and carbohydrate intakes on training and match days were observed in team sport athletes. In particular research by Bettonviel et al. [24] found that carbohydrate intakes of football athletes did not meet recommendations (6–10 g·kg^{-1}·day^{-1}) for match days; however, in comparison these athletes exceeded recommended protein intakes. Additionally, research by Potgieter et al. [35] found that while rugby union athletes were consuming adequate energy and carbohydrate prior to competition, diets were additionally high in protein and fat. Interestingly, research by Tooley et al. [2] found that rugby league athletes had greater fat intakes on the two days post competition (i.e., "recovery period"). This research suggested that this elevated intake of fat might be considered a 'reward' post competition for these athletes [2,62]. Greater education on match day nutrition strategies are required to optimise energy and carbohydrate intake prior to competition. Additionally, in many sports athletes need to compete multiple times a week therefore it is integral that recovery tactics aim to restore muscle glycogen within 24–48 h post-competition [57]. In these instances, it is essential to consume a carbohydrate and protein-rich meal shortly after the game, followed by another carbohydrate-rich meal two hours later to accelerate glycogen resynthesis [11].

4.3. Limitations

Studies in this systematic review included small numbers of participants and may not be generalisable to team sport disciplines more broadly. In addition, the heterogeneity of the included studies led to an inability to compare results across all studies and as a result a meta-analysis of data was not possible. Underreporting is an important consideration when assessing dietary intake; however, suboptimal intakes should not be attributed solely to underreporting and dietary assessment should encompass a range of influencing factors (i.e., body composition, appetite, nutrition knowledge etc.) [3]. Many studies explored the existence of intentional and unintentional underreporting, thus the findings of these analyses should be interpreted with caution.

4.4. Conclusions

This systematic review found that despite the publication of high-quality research studies, expert consensus statements and recognition of the consequences of inadequate intakes, team sport athletes' total energy and carbohydrate intakes did not meet sports nutrition recommendations (i.e., IOC, ISSN, ASCM and sports specific research) for energy and carbohydrate. In contrast, many athletes met or exceeded recommendations for protein and/or fat. Further research into the development of sport-specific recommendations for energy and macronutrients in particular carbohydrate would be beneficial to further optimise distribution throughout a training week. Furthermore, nutrition in team sport environments requires a knowledge base of the physiological demands of training and competition, and therefore sports dietitians should work collaboratively with sports science teams when tailoring nutrition advice to meet energy and macronutrient needs. Future research is required to explore the factors that influence athletes' dietary intakes.

Author Contributions: S.L.J., R.B., and A.K.F. contributed to the design of the systematic review and the interpretation of data. S.L.J., performed the literature search. S.L.J. and G.L.B. performed the data extraction and quality assessment process using the using the Academy of Nutrition and Dietetics "Quality Criteria Checklist" from the Academy of Nutrition and Dietetics Evidence Analysis Manual. S.L.J., B.L.D., R.B, and A.K.F. contributed to the writing of the manuscript. All authors read, revised and approved the final manuscript.

Funding: This research received no external funding.

Acknowledgments: Sarah Jenner is currently undertaking her PhD studies and receives a La Trobe University scholarship. Open Access fees was funded by author's research funds.

Conflicts of Interest: The authors declare no conflict of interest.

References

1. Holway, F.E.; Spriet, L.L. Sport-specific nutrition: Practical strategies for team sports. *J. Sports Sci.* **2011**, *29*, S115–S125. [CrossRef] [PubMed]

2. Tooley, E.; Bitcon, M.; Briggs, M.A.; West, D.J.; Russell, M. Estimates of energy intake and expenditure in professional rugby league players. *Int. J. Sports Sci. Coach.* **2015**, *10*, 551–560. [CrossRef]

3. Jenner, S.L.; Trakman, G.; Coutts, A.; Kempton, T.; Ryan, S.; Forsyth, A.; Belski, R. Dietary intake of professional Australian football athletes surrounding body composition assessment. *J. Int. Soc. Sports Nutr.* **2018**, *15*, 43. [CrossRef] [PubMed]

4. Rodriguez, N.R.; Dimarco, N.M.; Langley, S. Position of the American Dietetic Association, Dietitians of Canada, and the American College of Sports Medicine: Nutrition and athletic performance. *J. Am. Diet. Assoc.* **2009**, *109*, 509–527. [CrossRef] [PubMed]

5. Potgieter, S. Sport nutrition: A review of the latest guidelines for exercise and sport nutrition from the American college of sport nutrition, the International Olympic Committee and the international society for sports nutrition. *S. Afr. J. Clin. Nutr.* **2013**, *26*, 6–16. [CrossRef]

6. Kerksick, C.M.; Arent, S.; Schoenfeld, B.J.; Stout, J.R.; Campbell, B.; Wilborn, C.D.; Taylor, L.; Kalman, D.; Smith-Ryan, A.E.; Kreider, R.B.; et al. International society of sports nutrition position stand: Nutrient timing. *J. Int. Soc. Sports Nutr.* **2017**, *14*, 33. [CrossRef] [PubMed]

7. Aragon, A.A.; Schoenfeld, B.; Wildman, R.; Kleiner, S.; Vandusseldorp, T.; Taylor, L.; Earnest, C.P.; Arciero, P.J.; Wilborn, C.; Kalman, D.S.; et al. International society of sports nutrition position stand: diets and body composition. *J. Int. Soc. Sports Nutr.* **2017**, *14*, 16. [CrossRef]

8. Mountjoy, M.; Sundgot-Borgen, J.; Burke, L.; Ackerman, K.E.; Blauwet, C.; Constantini, N.; Lebrun, C.; Lundy, B.; Melin, A.; Meyer, N.; et al. International Olympic Committee (IOC) consensus statement on relative energy deficiency in sport (RED-S): 2018 update. *Int. J. Sport Nutr. Exerc. Metab.* **2018**, *28*, 316–331. [CrossRef]

9. International Olympic Committee. IOC consensus statement on sports nutrition 2010. *Int. J. Sport Nutr. Exerc. Metab.* **2010**, *1003*. [CrossRef]

10. Bradley, W.J.; Cavanagh, B.P.; Douglas, W.; Donovan, T.F.; Morton, J.P.; Close, G.L. Quantification of training load, energy intake, and physiological adaptations during a rugby preseason: A case study from an elite european rugby union squad. *J. Strength Cond. Res.* **2015**, *29*, 534–544. [CrossRef]

11. Kerksick, C.M.; Wilborn, C.D.; Roberts, M.D.; Smith-Ryan, A.; Kleiner, S.M.; Jäger, R.; Collins, R.; Cooke, M.; Davis, J.N.; Galvan, E.; et al. ISSN exercise & sports nutrition review update: research & recommendations. *J. Int. Soc. Sports Nutr.* **2018**, *15*, 38. [CrossRef] [PubMed]

12. Burke, L.M.; Hawley, J.A.; Wong, S.H.; Jeukendrup, A.E. Carbohydrates for training and competition. *J. Sports Sci.* **2011**, *29*, S17–S27. [CrossRef]

13. Thomas, D.T.; Erdman, K.A.; Burke, L.M. American college of sports medicine joint position statement. nutrition and athletic performance. *Med. Sci. Sports Exerc.* **2016**, *48*, 543–568. [CrossRef] [PubMed]

14. Kreider, R.B.; Kalman, D.S.; Antonio, J.; Ziegenfuss, T.N.; Wildman, R.; Collins, R.; Candow, D.G.; Kleiner, S.M.; Almada, A.L.; Lopez, H.L. International Society of Sports Nutrition position stand: safety and efficacy of creatine supplementation in exercise, sport, and medicine. *J. Int. Soc. Sport Nutr.* **2017**, *14*, 18. [CrossRef] [PubMed]

15. Wilson, J.M.; Fitschen, P.J.; Campbell, B.; Wilson, G.J.; Zanchi, N.; Taylor, L.; Wilborn, C.; Kalman, D.S.; Stout, J.R.; Hoffman, J.R.; et al. International Society of Sports Nutrition Position Stand: beta-hydroxy-beta-methylbutyrate (HMB). *J. Int. Soc. Sport Nutr.* **2013**, *10*, 6. [CrossRef]

16. Campbell, B.; Wilborn, C.; La Bounty, P.; Taylor, L.; Nelson, M.; Greenwood, M.; Ziegenfuss, T.N.; Lopez, H.L.; Hoffman, J.R.; Stout, J.R.; et al. International Society of Sports Nutrition position stand: energy drinks. *J. Int. Soc. Sports Nutr.* **2013**, *10*, 1. [CrossRef]

17. Maughan, R.J.; Burke, L.M.; Dvorak, J.; Larson-Meyer, D.E.; Peeling, P.; Phillips, S.M.; Rawson, E.S.; Walsh, N.P.; Garthe, I.; Geyer, H.; et al. IOC Consensus Statement: Dietary Supplements and the High-Performance Athlete. *Int. J. Sport Nutr. Exerc. Metab.* **2018**, *28*, 104–125. [CrossRef] [PubMed]

18. Bergeron, M.F.; Bahr, R.; Bärtsch, P.; Bourdon, L.; Calbet, J.A.; Carlsen, K.H.; Castagna, O.; González-Alonso, J.; Lundby, C.; Maughan, R.J.; et al. International Olympic Committee consensus statement on thermoregulatory and altitude challenges for high-level athletes. *Br. J. Sports Med.* **2012**, *46*, 770–779. [CrossRef] [PubMed]

19. Trexler, E.T.; Smith-Ryan, A.E.; Stout, J.R.; Hoffman, J.R.; Wilborn, C.D.; Sale, C.; Kreider, R.B.; Jäger, R.; Earnest, C.P.; Bannock, L.; et al. International society of sports nutrition position stand: Beta-Alanine. *J. Int. Soc. Sports Nutr* **2015**, *12*, 30. [CrossRef] [PubMed]

20. Jäger, R.; Kerksick, C.M.; Campbell, B.I.; Cribb, P.J.; Wells, S.D.; Skwiat, T.M.; Purpura, M.; Ziegenfuss, T.N.; Ferrando, A.A.; Arent, S.M.; et al. International Society of Sports Nutrition Position Stand: Protein and exercise. *J. Int. Soc. Sports Nutr.* **2017**, *14*, 20. [CrossRef]

21. Position of the Academy of Nutrition and Dietetics. Dietitians of Canada, and the American College of Sports Medicine: Nutrition and Athletic Performance. *Can. J. Diet. Pract. Res.* **2016**, *77*, 54. [CrossRef] [PubMed]

22. Kreider, R.B.; Wilborn, C.D.; Taylor, L.; Campbell, B.; Almada, A.L.; Collins, R.; Cooke, M.; Earnest, C.P.; Greenwood, M.; Kalman, D.S.; et al. ISSN exercise & sport nutrition review: Research & recommendations. *J. Int. Soc. of Sports Nutr.* **2010**, *7*, 7. [CrossRef]

23. Anderson, L.; Orme, P.; Naughton, R.J.; Close, G.L.; Milsom, J.; Rydings, D.; O'Boyle, A.; Di Michele, R.; Louis, J.; Hambly, C.; et al. Energy Intake and Expenditure of Professional Soccer Players of the English Premier League: Evidence of Carbohydrate Periodization. *Int. J. Sport Nutr. Exerc. Metab.* **2017**, *27*, 228–238. [CrossRef] [PubMed]

24. Bettonviel, A.E.O.; Brinkmans, N.Y.J.; Russcher, K.; Wardenaar, F.C.; Witard, O.C. Nutritional Status and Daytime Pattern of Protein Intake on Match, Post-Match, Rest and Training Days in Senior Professional and Youth Elite Soccer Players. *Int. J. Sport Nutr. Exerc. Metab.* **2016**, *26*, 285–293. [CrossRef]

25. Andrews, M.C.; Itsiopoulos, C. Room for improvement in nutrition knowledge and dietary intake of male football (Soccer) players in Australia. *Int. J. Sport Nutr. Exerc. Metab.* **2016**, *26*, 55–64. [CrossRef]

26. Bilsborough, J.C.; Greenway, K.; Livingston, S.; Cordy, J.; Coutts, A.J. Changes in anthropometry, upper-body strength, and nutrient intake in professional Australian football players during a season. *Int. J. Sports Physiol. Perform.* **2016**, *11*, 290–300. [CrossRef] [PubMed]

27. Bradley, W.J.; Cavanagh, B.; Douglas, W.; Donovan, T.F.; Twist, C.; Morton, J.P.; Close, G.L. Energy intake and expenditure assessed 'in-season' in an elite European rugby union squad. *Eur. J. Sport Sci.* **2015**, *15*, 469–479. [CrossRef]

28. Conejos, C.; Giner, A.; Mañes, J.; Soriano, J.M. Energy and nutritional intakes in training days of soccer players according to their playing positions. *Arch. Med. Deporte* **2011**, *28*, 29–35.

29. Grams, L.; Garrido, G.; Villacieros, J.; Ferro, A. Marginal Micronutrient Intake in High-Performance Male Wheelchair Basketball Players: A Dietary Evaluation and the Effects of Nutritional Advice. *PLoS ONE* **2016**, *11*, e0157931. [CrossRef]

30. Gravina, L.; Ruiz, F.; Diaz, E.; Lekue, J.A.; Badiola, A.; Irazusta, J.; Gil, S.M. Influence of nutrient intake on antioxidant capacity, muscle damage and white blood cell count in female soccer players. *J. Int. Soc. Sports Nutr.* **2012**, *9*, 32. [CrossRef]

31. Hidalgo y Teran Elizondo, R.; Martin Bermudo, F.M.; Penaloza Mendez, R.; Berna Amoros, G.; Lara Padilla, E.; Berral de la Rosa, F.J. Nutritional intake and nutritional status in elite Mexican teenagers soccer players of different ages. *Nutr. Hosp.* **2015**, *32*, 1735–1743. [CrossRef]

32. Kirwan, R.D.; Kordick, L.K.; McFarland, S.; Lancaster, D.; Clark, K.; Miles, M.P. Dietary, anthropometric, blood-lipid, and performance patterns of American college football players during 8 weeks of training. *Int. J. Sport Nutr. Exerc. Metab.* **2012**, *22*, 444–451. [CrossRef]

33. MacKenzie, K.; Slater, G.; King, N.; Byrne, N. The measurement and interpretation of dietary protein distribution during a rugby preseason. *Int. J. Sport Nutr. Exerc. Metab.* **2015**, *25*, 353–358. [CrossRef]

34. Ono, M.; Kennedy, E.; Reeves, S.; Cronin, L. Nutrition and culture in professional football. A mixed method approach. *Appetite* **2012**, *58*, 98–104. [CrossRef]

35. Potgieter, S.; Visser, J.; Croukamp, I.; Markides, M.; Nascimento, J.; Scott, K. Body composition and habitual and match-day dietary intake of the FNB Maties Varsity Cup rugby players.(ORIGINAL RESEARCH). *S. Afr. J. Sports Med.* **2014**, *26*, 35–43. [CrossRef]

36. Molina-López, J.; Molina, J.M.; Chirosa, L.J.; Florea, D.; Sáez, L.; Jiménez, J.; Planells, P.; Perez de la Cruz, A.; Planells, E. Implementation of a nutrition education program in a handball team; consequences on nutritional status. *Nutr. Hosp.* **2013**, *28*, 1065–1076. [CrossRef]

37. Raizel, R.; da Mata Godois, A.; Coqueiro, A.Y.; Voltarelli, F.A.; Fett, C.A.; Tirapegui, J.; de Paula Ravagnani, F.C.; de Faria Coelho-Ravagnani, C. Pre-season dietary intake of professional soccer players. *Nutr. Health* **2017**, *23*, 215–222. [CrossRef]

38. Valliant, M.W.; Pittman Emplaincourt, H.; Wenzel, R.K.; Garner, B.H. Nutrition Education by a Registered Dietitian Improves Dietary Intake and Nutrition Knowledge of a NCAA Female Volleyball Team. *Nutrients* **2012**, *4*, 506–516. [CrossRef] [PubMed]

39. Vermeulen, T. Seven Day Dietary Intakes of Female Varsity Ice Hockey Players. Master's Thesis, The University of Guelph, Guelph, ON, Canada, 2017.

40. International Prospective Register of Systematic Reviews. Available online: https://www.crd.york.ac.uk/prospero/display_record.php?RecordID=105168 (accessed on 24 September 2018).

41. Moher, D.; Liberati, A.; Tetzlaff, J.; Altman, D.G.; PRISMA Group. Preferred Reporting Items for Systematic Reviews and Meta-Analyses: The PRISMA statement. *Ann. Intern. Med.* **2009**, *151*, 264–269. [CrossRef] [PubMed]

42. Academy of Nutrition and Dietetics. Evidence Analysis Manual. Available online: https://www.andeal.org/vault/2440/web/files/2016_April_EA_Manual.pdf (accessed on 18 September 2018).

43. Devlin, B.L.; Leveritt, M.D.; Kingsley, M.; Belski, R. Dietary Intake, Body Composition, and Nutrition Knowledge of Australian Football and Soccer Players: Implications for Sports Nutrition Professionals in Practice. *Int. J. Sport Nutr. Exerc. Metab.* **2017**, *27*, 130–138. [CrossRef]

44. Maughan, R.J.; Burke, L.M. Practical Nutritional Recommendations for the Athlete. *Nestle Nutr. Inst. Workshop Ser.* **2011**, *69*, 131–149. [CrossRef]

45. Lemon, P.W. Protein requirements of soccer. *J. Sports Sci.* **1994**, *12*, S17–S22. [CrossRef] [PubMed]

46. Hernandez, A.J.; Nahas, R.M. Modificações dietéticas, reposição hídrica, suplementos alimentares e drogas: comprovação de ação ergogênica e potenciais riscos para a saúde. *Rev. Bras. Med. Esporte* **2009**, *15*, 1–12.

47. Mettler, S.; Mitchell, N.; Tipton, K.D. Increased Protein Intake Reduces Lean Body Mass Loss during Weight Loss in Athletes. *Med. Sci. Sports Exerc.* **2010**, *42*, 326–337. [CrossRef] [PubMed]

48. Tipton, K.D.; Wolfe, R.R. Protein and amino acids for athletes. *J. Sports Sci.* **2004**, *22*, 65–79. [CrossRef]

49. Bishop, N.C.; Blannin, A.K.; Walsh, N.P.; Robson, P.J.; Gleeson, M. Nutritional Aspects of Immunosuppression in Athletes. *Sports Med.* **1999**, *28*, 151–176. [CrossRef] [PubMed]

50. American College of Sports Medicine author issuing body. *ACSM's Guidelines for Exercise Testing and Prescription*, 10th ed.; Wolters Kluwer Health: Philadelphia, PA, USA, 2018.

51. Phillips, S.M.; Van Loon, L.J. Dietary protein for athletes: From requirements to optimum adaptation. *J. Sports Sci.* **2011**, *29*, S29–S38. [CrossRef] [PubMed]

52. Nutrition for football: The FIFA/F-MARC Consensus Conference. *J. Sports Sci.* **2006**, *24*, 663–664. [CrossRef]

53. Bussau, V.A.; Fairchild, T.J.; Rao, A.; Steele, P.; Fournier, P.A. Carbohydrate loading in human muscle: An improved 1 day protocol. *Eur. J. Appli. Physiol.* **2002**, *87*, 290–295. [CrossRef]

54. Krustrup, P.; Mohr, M.; Steensberg, A.; Bencke, J.; Kjær, M.; Bangsbo, J. Muscle and Blood Metabolites during a Soccer Game: Implications for Sprint Performance. *Med. Sci. Sports Exerc.* **2006**, *38*, 1165–1174. [CrossRef]

55. Hargreaves, M. Carbohydrate and lipid requirements of soccer. *J. Sports Sci.* **1994**, *12*, S13–S16. [CrossRef] [PubMed]

56. Bradley, W.J.; Hannon, M.P.; Benford, V.; Morehen, J.C.; Twist, C.; Shepherd, S.; Cocks, M.; Impey, S.G.; Cooper, R.G.; Morton, J.P.; et al. Metabolic demands and replenishment of muscle glycogen after a rugby league match simulation protocol. *J. Sci. Med. Sport* **2017**, *20*, 878–883. [CrossRef] [PubMed]

57. Burke, L.M.; Loucks, A.B.; Broad, N. Energy and carbohydrate for training and recovery. *J. Sports Sci.* **2006**, *24*, 675–685. [CrossRef] [PubMed]

58. Collins, J.; McCall, A.; Bilsborough, J.; Maughan, R. Football nutrition: time for a new consensus? *Br. J. Sports Med.* **2017**, *51*, 1577–1578. [CrossRef] [PubMed]

59. Impey, S.G.; Hammond, K.M.; Shepherd, S.O.; Sharples, A.P.; Stewart, C.; Limb, M.; Smith, K.; Philp, A.; Jeromson, S.; Hamilton, D.L.; et al. Fuel for the work required: A practical approach to amalgamating train-low paradigms for endurance athletes. *Physiol. Rep.* **2016**, *4*, e12803. [CrossRef] [PubMed]

60. Areta, J.L.; Burke, L.M.; Camera, D.M.; West, D.W.; Crawshay, S.; Moore, D.R.; Stellingwerff, T.; Phillips, S.M.; Hawley, J.A.; Coffey, V.G. Reduced resting skeletal muscle protein synthesis is rescued by resistance exercise and protein ingestion following short-term energy deficit. *Am. J. Physiol. Endocrinol. Metab.* **2014**, *306*, E989–E997. [CrossRef]

61. National Health and Medical Research Council. *Australian Guide to Healthy Eating*; NHMRC: Canberra, Australia, 2013.

62. Lundy, B.; O'Connor, H.; Pelly, F.; Caterson, I. Anthropometric characteristics and competition dietary intakes of professional rugby league players. *Int. J. Sport Nutr. Exerc. Metab.* **2006**, *16*, 199–213. [CrossRef]

© 2019 by the authors. Licensee MDPI, Basel, Switzerland. This article is an open access article distributed under the terms and conditions of the Creative Commons Attribution (CC BY) license (http://creativecommons.org/licenses/by/4.0/).

Review

Sleep and Nutrition Interactions: Implications for Athletes

Rónán Doherty [1,2,3,*], Sharon Madigan [2], Giles Warrington [4,5] and Jason Ellis [3]

[1]   Letterkenny Institute of Technology, Port Road, Letterkenny, F92 FC93 County Donegal, Ireland
[2]   Sport Ireland Institute, National Sport Campus, Abbotstown, 15, D15 Y52H Dublin, Ireland;
      smadigan@instituteofsport.ie
[3]   Northumbria Centre for Sleep Research, Northumbria University, Newcastle NE1 8ST, UK;
      jason.ellis@northumbria.ac.uk
[4]   Health Research Institute, Schuman Building, University of Limerick, V94 T9PX County Donegal, Ireland;
      giles.warrington@ul.ie
[5]   Department of Physical Education and Sport Sciences, University of Limerick, V94 T9PX County Donegal,
      Ireland
*   Correspondence: ronan.doherty@lyit.ie; Tel.: +353-749-186-299

Received: 4 February 2019; Accepted: 10 April 2019; Published: 11 April 2019

Abstract: This narrative review explores the relationship between sleep and nutrition. Various nutritional interventions have been shown to improve sleep including high carbohydrate, high glycaemic index evening meals, melatonin, tryptophan rich protein, tart cherry juice, kiwifruit and micronutrients. Sleep disturbances and short sleep duration are behavioural risk factors for inflammation, associated with increased risk of illness and disease, which can be modified to promote sleep health. For sleep to have a restorative effect on the body, it must be of adequate duration and quality; particularly for athletes whose physical and mental recovery needs may be greater due to the high physiological and psychological demands placed on them during training and competition. Sleep has been shown to have a restorative effect on the immune system, the endocrine system, facilitate the recovery of the nervous system and metabolic cost of the waking state and has an integral role in learning, memory and synaptic plasticity, all of which can impact both athletic recovery and performance. Functional food-based interventions designed to enhance sleep quality and quantity or promote general health, sleep health, training adaptations and/or recovery warrant further investigation.

Keywords: sleep; athletes; chrononutrition

---

1. What is Sleep?

Sleep, in humans, is defined as a complex reversible behavioural state where an individual is perceptually disengaged from and unresponsive to their environment [1]. Sleep architecture has two basic states based on physiological parameters: non-rapid eye movement sleep (NREM) and rapid eye movement (REM) sleep [2]. Sleep stages fall along a continuum from fully awake to deep sleep [3]. NREM has been defined as "a relatively inactive yet actively regulating brain in a moveable body" [2], (p.17). In terms of brain activity, the electroencephalogram (EEG) pattern of NREM sleep is commonly described as synchronous (increasing depth of sleep is indicated by progressive dominance of high voltage, low frequency EEG patterns), with characteristic waveforms (sleep spindles, K-complexes and high voltage waves) [2]. NREM is usually associated with minimal or fragmented mental activity. Table 1 shows the traditional four stages of NREM which are associated with differing levels of depth of sleep, with arousal thresholds generally lowest in Stage 1 and highest in Stage 4 sleep [2].

Table 1. Characteristics of NREM Sleep.

Stage	Characteristics
1	Sleep is easily discontinued (e.g., noise, a light touch, etc.) Sleep is easily interrupted Key role in the initial wake to sleep transition Transitional stage throughout the sleep cycle
2	More intense stimuli required to produce arousal (e.g., bright light or loud noise) Indicated by K-complexes or sleep spindles in the EEG High voltage slow wave EEG activity will become apparent
3	High voltage (75 μV) slow wave (two cycles per second [cps]) activity that is $\geq$ 20% but < 50% of EEG activity
4	High voltage slow wave activity is $\geq$ 50% of EEG activity.

(Adapted from: [4]).

In contrast, REM sleep is defined by EEG activation, muscle atonia (paralysis) and episodic bursts of rapid eye movement [2]. REM sleep is associated with cognitive activity, while brain stem mechanisms inhibit spinal motor neurons limiting movement. Hence, REM sleep has been defined as "an activated brain in a paralysed body" [2], (p.16). It should be noted that the American Academy of Sleep Medicine (AASM) have recommended alternative terminology for Sleep staging. Wake is referred to as W, NREM sleep is referred to as N and is divided into three stages: N1 – Stage 1, N2 – Stage 2 and N3 – Slow Wave Sleep or Deep Sleep, i.e., Stage 3 and 4 combined; while REM is referred to as R [5].

Sleep health is a multidimensional pattern of sleep-wakefulness adapted to individual, social and environmental demands, which promotes physical and mental wellbeing [6]. Good sleep health is characterised by satisfaction, appropriate timing, adequate duration, high efficiency and sustained alertness during waking hours [6]. Sleep deprivation adversely affects glucose metabolism and neuroendocrine function which can affect carbohydrate metabolism, appetite, energy intake and protein synthesis [1]. These factors may negatively impact an athlete's nutritional, metabolic and endocrine status impacting athletic performance and recovery [1], (e.g., impaired glucose metabolism could reduce glycogen repletion while impaired protein synthesis could reduce recovery and adaptation from training). This narrative review examines and evaluates the interaction between nutrition and sleep.

How and Why Sleep Occurs

The brain is essentially an electrical system with circuits that switch on and off to promote either wakefulness or sleep. Since the arousal and sleep-promoting systems are mutually inhibitory, a sleep switch or 'flip-flop' model has been proposed [7]. A flip-flop switch contains mutually inhibitory elements where activity in one of the competing sides shuts down inhibitory inputs from the other side producing two discrete states with sharp transitions [8]. Activation of arousal systems inhibits sleep active neurons facilitating sleep while activation of sleep-promoting neurons inhibits arousal-related neurons reinforcing consolidated sleep episodes providing a mechanism for stabilisation of sleep and waking states [9].

The circadian rhythm in humans has been estimated in young males (24.18 ± 0.04 h; PCV 0.54%) and older adults (24.18 ± 0.04 h; PCV 0.58%), low percentage coefficients of variation and no significant difference between the groups indicated a small range variability in circadian rhythms [10]. Humans however, typically display individual differences in their behaviour (e.g., social activities, daytime activities and sleep). Chronotype is the expression of individual circadian rhythmicity and has been categorised as follows: morning types, intermediate types and evening types [11]. Chronotype is, in part, genetic but cultural and environmental factors also affect an individual's sleep pattern. Research in the general population has demonstrated that most people are intermediate types (70%) with the remainder being either morning types (14%) or evening types (16%) [12].

Sleep is a dynamic process largely regulated by two factors; the circadian systems and the sleep homeostat. The Two Process Model for Sleep Regulation was developed to illustrate the interaction of the homeostatic sleep drive (sleep pressure or urge to sleep that accumulates during wakefulness) and the circadian system (endogenous timing system) in the timing and duration of sleep [13,14]. The homeostatic process (S) is a function of sleep and waking, while the circadian process (C) is controlled by a circadian oscillator [10]. S increases during waking and declines during sleep and it interacts with C, which is independent of sleep and waking and receives cues (e.g., light) from the environment [13,14]. The suprachiasmatic nucleus (SCN) in the brain is central to this process but secondary clock systems have been identified throughout the body [14].

Process S is an endogenous mechanism, relying on exogenous cues to regulate it to approximately 24 h. Process S represents sleep debt which increases during waking and reduces during sleep within a range that oscillates within a period that is normally entrained to day and night by process C [14]. When S reaches the lower boundary of the range, awakening is triggered and when S reaches the upper boundary sleep is triggered [14]. In terms of process C, the Two-Process Model focuses on time-of-day effects on sleep propensity, specifically that sleep propensity is minimal near midday and is strongly promoted in the early hours of the morning [13]. This circadian rhythmicity in sleep propensity is combined with S by C dictating the threshold values at which S transitions from sleep to wake, and vice versa [13,14]. Core body temperature and melatonin rhythms are markers of C [11]. The SCN has melatonin receptor cells, as darkness falls, melatonin is secreted by the pineal gland making the individual sleepy [15]. Animal studies have demonstrated that exogenous melatonin and ramelteon (an MT1/MT2 melatonin receptor agonist) function as non-photic entrainers, which phase advance the SCN [16]. A Three-Process Model of Sleep Regulation has also been proposed whereby sleepiness and alertness are stimulated by the combined action of a homeostatic process, a circadian process and sleep inertia process, the model has been extended to include sleep onset latency (the length of time of the transition from wakefulness to sleep), sleep length and performance [17].

Sleep has a restorative effect on the immune system and the endocrine system, facilitates the recovery of the nervous and metabolic cost of the waking state and has an integral role in learning, memory and synaptic plasticity (ability of synapses to strengthen or weaken over time) [18,19]. Sleep, particularly slow wave sleep (or N3) early in the night promotes prolactin release, while the anti-inflammatory actions of cortisol and catecholamines are reduced [18]. Acute sleep deprivation and sleep disturbance (short sleep duration or reduced sleep efficiency) impair adaptive immunity which is associated with reduced response to vaccinations and increased vulnerability to infectious diseases, attributed to reduced growth hormone release during deep sleep and increased sympathetic output [20]. Tumour necrosis factor-α (TNFα) along with other cytokines are considered key to the regulation of sleep in normal physiological conditions [21]. Research has demonstrated that sleep disturbance (i.e. insomnia) and extremes of sleep durations affect risk factors of inflammatory disease and contribute to all-cause mortality [18,22]. Increased levels of circulating inflammatory markers (i.e., C-reactive protein [CRP] and Interleukin-6 [IL-6]) predict body mass gain in older adults [23] and type 2 diabetes [24]. Sleep disturbance is believed to have proximal effects on IL-6, which induces CRP [18], therefore, increases in CRP may be attributed to persistent or severe sleep disturbance. In a recent meta-analysis sleep disturbance (i.e., poor sleep quality, insomnia) was associated with increased levels of IL-6 (ES: 0.20 (0.08–0.31)) and CRP (ES: 0.12 (0.05 0.19)) [18]. Short sleep duration (< 7 h per night) was associated with increased IL-6 (ES: 0.29 (0.05–0.52)), while long sleep duration (> 8 h per night) was also associated with increased IL-6 (ES: 0.11 (0.02–0.20)) but also increased CRP (ES: 0.17 (0.01–0.34)) [18]. Similarly, a meta-analysis of sleep duration and all-cause mortality demonstrated a U-shaped association, whereby long sleep (> 8 h per night) has a 30% (RR: 1.30 (1.22–1.38)) greater risk while short sleep (<7 h per night) has a 12% (RR: 1.12 (1.06 1.18)) greater risk compared to normal sleep reference (7–8 h per night) [25].

Inappropriate timing of lifestyle behaviours can cause disruption to the circadian rhythm, resulting in an altered physiological response (e.g., poor sleep). Lifestyle factors (e.g., caffeine consumption, alcohol consumption and timing of sleep) can cause alterations in environmental cues which may negatively impact circadian rhythms and in turn result in negative physiological consequences [26]. The SCN receives environmental cues such as the light-dark cycle and additional information from other areas of the brain (e.g., when we eat or exercise). Give that Process C can be modified by exogenous cues [27], there is scope for investigation of nutrition interventions to enhance sleep quality and quantity. Similarly, the effect of nutrition interventions that promote athlete recovery on sleep quality and quantity should be investigated.

2. Sleep and Athletes

The classic view of sleep is that it is a recovery process, with the circadian system regulating feelings of sleepiness and wakefulness throughout the day [28]. Cognition, tissue repair and metabolism are critical psychological and physiological factors that contribute to training capacity, recovery and ultimately performance [28]. The relationship between sleep, performance and recovery can be viewed in terms of 3 key factors that affect the recuperative outcome:

1. Sleep length (total sleep duration; hours/night, plus naps)
2. Sleep quality (i.e., the experience and perceived adequacy of sleep)
3. Sleep phase (circadian timing of sleep) [28].

Post-exercise recovery is vital for all athletes. If the balance between training stress and physical recovery is inadequate, performance in subsequent training sessions or competition may be adversely affected [15]. Muscle fatigue or soreness may adversely affect sleep, with inflammatory cytokines linked to disruption of normal sleep [29]. Inadequate recovery can reduce autonomic nervous system (ANS) resources, with an associated reduction in heart rate variability (HRV) and increased resting heart rate [30]. Sleep deprivation is associated with increased catabolic and reduced anabolic hormones which results in impaired muscle protein synthesis [31], blunting training adaptations and recovery.

Sleep disturbances and inadequate sleep duration have been reported in athletic populations. Assessment of the sleep patterns of professional male ice hockey players ($n = 23$) using polysomnography (PSG), demonstrated mean total sleep duration was 6.92 h; 95% CI 6.3–7.5 h [32]. Similarly, sleep was self-reported as the most important recovery modality utilised by South African athletes ($n = 890$; international $n = 183$, national $n = 474$, club $n = 233$) [15]. While a similar study found that 66% ($n = 416$) of elite German athletes ($n = 632$) reported pre-competition insomnia symptomology including difficulty falling asleep, waking during the night and early final waking times [33]. Sleep duration (< 8 h) has been identified as the strongest predictor of injury in adolescent athletes (RR = 2.1; 95% CI: 1.2–3.9) [34]. The Karolinska Athlete Screening Injury Prevention (KASIP) study investigated injury occurrence in Swedish adolescent elite athletes ($n = 340$; 178 males and 162 females) and demonstrated that athletes sleeping >8 h were less likely to suffer an injury (OR: 0.39; 95% CI 0.17–0.96) [35]. The aetiology of sleep disturbances is unclear during periods of intense training, it is unclear whether poor sleep is a symptom of overtraining, or intense training negatively affects sleep and recovery [30]. Sleep also has a pivotal role to play in performance, training adaptations and recovery [1,18]. Given the importance of sleep for athlete recovery, further research is warranted to investigate potential nutritional interventions to promote improved sleep quality and/or duration and recovery in athletes.

2.1. Sleep, Nutrition and Athletes

Nutrition support needs to be periodised in relation to the demands of the athlete's daily training and overall nutritional goals [36]. The focus of 'training' nutrition is to promote adaptations while the focus of 'competition' nutrition is optimal performance [36]. Athletes also have added responsibility to adhere to the World Anti-Doping Agency (WADA) code and are subject to testing for prohibited substances. If an athlete chooses to take any supplement they must do so in a safe and effective manner. Athletes should check that any supplement they take has been tested for banned substances, and independent testing programmes (e.g., Informed Sport and Informed Choice) offer additional protection. Athletes are advised to seek the professional advice of a qualified sports dietician/nutritionist regarding any nutritional supplement. Training adaptations and recovery can be maximised by optimal nutrition practices or impaired by suboptimal nutrition practices [36–38]. Nutrients such as carbohydrate (high glycaemic index evening meal reduced sleep onset latency), protein (consumption of dairy sources may increase sleep duration), ethanol (reduced REM sleep) [37] and caffeine (increased sleep onset latency, reduced total sleep duration and reduced sleep quality) [39], as well as the timing and quantity of meals (large portions and/or meals later in the evening can negatively impact sleep potentially due to the thermogenic effect of digestion) can affect circadian rhythms [40]. Caffeine consumption can lead to poor sleep which, in turn, can lead to increased caffeine consumption. Caffeine increases the state of alertness, antagonising adenosine receptors, which also leads to a reduction in the inclination to sleep [39]. Alcohol consumption has been associated with poorer sleep quality and quantity, reduced REM sleep and increased sleep disturbance in the second half of the sleep bout [41]. Similar to nutrition, sleep disturbances (difficulty initiating or maintain sleep) and sleep deprivation (not getting enough sleep) are risk factors for inflammation [18,42], which can be treated or managed to promote recovery and/or performance. For sleep to have a restorative effect on the body, it must be of adequate duration which is dependent on age [28,42]. Sleep recommendations particularly the amount of sleep required, change over the lifespan from adolescents (8–10 h), adults (7–9 h), and older adults (7–8 h) [42].

2.2. Chrononutrition

Recently the term Chrononutrition has been used to describe the interaction between food and the circadian system [39]. It has been suggested that the internal clock can be altered by changing the timing and nature of food intake [39]. Chrononutrition has been characterised as including two aspects:

1. Timing of food intake or contributions of food components to the maintenance of health; and
2. Timing of food intake or contributions of food components to rapid changes in or resetting of a human's system of internal clocks [39].

Several neurotransmitters are involved with the sleep-wake cycle including 5–hydroxytryptophan (5-HT), GABA, orexin, melanin concentrating hormone, cholinergic, galanin, noradrenaline and histamine [7]. Therefore, nutrition interventions that act on these neurotransmitters could positively impact sleep. Dietary precursors can influence the rate of synthesis and function of neurotransmitters (e.g., serotonin synthesis is dependent on the availability of its precursor tryptophan in the brain) [1]. Tryptophan is transported across the blood brain barrier by a system that shares transporters with several large neutral amino acids (LNAA) [1]. The ratio of tryptophan:LNAA in the blood is vital to the transport of tryptophan into the brain and can be increased through consumption of tryptophan, a high carbohydrate/low protein diet or α-lactalbumin (whey derived protein) [43].

2.3. Carbohydrate

Carbohydrate consumption has been shown to increase plasma tryptophan concentrations [44]. Carbohydrates affect plasma tryptophan:LNAA ratio and may compliment the sleep enhancing effect of consuming tryptophan rich protein [40]. Insulin influences the transport of tryptophan across the blood brain barrier after a carbohydrate rich meal, as it is an anabolic agent it also facilitates the uptake of LNAA by muscle [26]. Consumption of high glycaemic index (GI) carbohydrate increases the ratio of circulating tryptophann:LNAA via direct action of insulin which promotes muscle uptake of LNAA [45]. This increases tryptophan availability for synthesis of serotonin and ultimately melatonin. GI has been shown to affect sleep onset latency (length of time of the transition from wake to sleep) [44]. A high GI meal consumed four hours before bed, significantly ($p = 0.009$) reduced sleep onset latency (9.0 ± 6.2 min) compared to a low GI meal (17.5 ± 6.2 min) and the same meal consumed 1 hour before bed (14.6 ± 9.9 min) [44]. Among a large sample ($n = 4452$) from the National Health and Nutrition Examination survey lower carbohydrate intake (24–h recall and structured interview) has been significantly associated (OR 0.71; 0.55–0.92, p = 0.01) with insomnia symptoms (difficulty maintaining sleep) [46]. Consumption of a high-carbohydrate meal (130 g) when compared to a low-carbohydrate meal (47 g), or a meal containing no carbohydrate, 45 min before bedtime increased REM and decreased light sleep and wakefulness [47]. The timing of carbohydrate evening meals and the carbohydrate content of evening meal on sleep and athlete recovery requires further investigation within athletic populations.

2.4. Melatonin

Melatonin is a hormone secreted by the pineal gland, that has displayed sedative effects [48,49]. Since endogenous melatonin influences core temperature facilitating sleep, increased exogenous melatonin could affect changes in core temperature improving sleep quality [50]. However, the effect is relative to the person's endogenous melatonin levels. In many Western countries, Cow's milk has traditionally been considered a sleep promoting beverage. Melatonin is a naturally-occurring compound in cow's milk, but its concentration increases significantly if cows are milked in darkness at night referred to as 'night time milk' [40]. Increased tryptophan and melatonin concentrations appear to be the property responsible for the sleep promoting effect of night time milk. Consumption of night time milk (melatonin concentration of 39.43 pg/mL) compared to daytime milk (4.03 pg/mL) significantly increased circulating melatonin (26.5%) concentration in rats [51], indicating that high melatonin concentrations are necessary for milk to affect blood melatonin concentrations. When the night time milk was supplemented with tryptophan (2.5 g/L) circulating melatonin concentrations significantly increased further (35.5%) [51].

Melatonin has extremely low toxicity even at relatively high doses and can easily cross physiological barriers due to its optimal size, partial water solubility and high lipid solubility however, it must be noted that there appears to be no added benefit to doses >3 mg [49]. Ingestion of melatonin affects sleep propensity and has hypnotic effects enhancing sleep quality and duration, pharmacological melatonin can be used to manipulate circadian timing [49]. A positive effect of low doses (0.3 mg or 1 mg) of exogenous melatonin (gelatin capsules) on sleep onset latency has been observed in a small group of healthy males (n = 6), when administered at either 6:00 pm (0.3 mg 16.5 ± 19.9 min; 1 mg 12.3 ± 13.6 min; Placebo 23.1 ± 22.7 min) and 8:00 pm (0.3 mg 19.6 ± 14.1 min; 1 mg 20.7 ± 17.7 min; Placebo 53.4 ± 51.9 min) [52]. However, the impact was time dependent as a 0.3 mg dose increased sleep onset latency when consumed at 9:00 pm (0.3 mg 25.1 ± 10.5 min; 1 mg 12.1 ± 7.4 min; Placebo 8.8 ± 4 min) and there was no evidence of an effect when the 1 mg dose was administered at 9:00 pm [52]. The results indicate a low dose of melatonin similar to nocturnal physiological concentrations can elicit a sleep-inducing effect. A dose response relationship was not evident as the 0.3 mg dose, which is similar to endogenous melatonin concentrations, was as effective as the 1 mg dose when administered at 6:00 pm or 8:00 pm.

2.5. Tryptophan Rich Protein

Tryptophan is an essential amino acid that is a precursor to serotonin and melatonin, which can cross the blood-brain barrier by competing for transport with other LNAA [1]. Conversion to serotonin is dependent on sufficient precursor availability in the brain, an increase in brain tryptophan occurs when the ratio of free tryptophan to branched chain amino acids increases, following tryptophan conversion to serotonin, melatonin is produced [1]. Dietary sources of tryptophan include milk, turkey, chicken, fish, eggs, pumpkin seeds, beans, peanuts, cheese, and leafy green vegetables. Dietary tryptophan has been shown to improve sleep, in a comparison of food bars (Food 1: 25 g deoiled butternut squash seed meal and 25 g dextrose, Food 2: 250 mg of pharmaceutical tryptophan and Food 3: 50 g rolled oats [control]) [53]. Food 1 and Food 2 produced significant results ($p \leq 0.05$) for reduction of time awake during the night (19.2% and 22.1%), increased sleep efficiency (% of time spent in bed; asleep) (5.19% and 7.36%) and increased subjective sleep quality (12.2% and

11.8%) [53], indicating that relatively small doses (250 mg) of dietary tryptophan can positively impact sleep. The milk protein, α-lactalbumin has been reported as having the highest natural levels of tryptophan among all protein food sources [54]. Ingestion of α-lactalbumin enriched whey protein, significantly ($p < 0.05$) increased tryptophan:LNAA by 48% compared to a casein enriched diet [54]. In a similar study, healthy adults ($n = 14$) with sleep complaints consumed milkshakes containing either α-lactalbumin (20 g) or a casein placebo. Evening ingestion of α-lactalbumin resulted in a 130% increase in tryptophan:LNAA prior to bed and modest but significant reduction in morning sleepiness and improved alertness the following morning [55]. Tryptophan depletion studies have demonstrated decreased tryptophan plasma concentrations affected sleep fragmentation (arousal index (events/h)), REM sleep latency (the interval between first epoch of stage 2 and the first epoch of REM sleep), and REM density (the cumulated duration of each REM burst divided by the duration of each REM sleep period) compared to baseline and placebo [56,57]. Consumption of tryptophan rich protein (e.g., milk) could affect changes in core temperature improving sleep quality [51]. The effects of tryptophan rich protein (e.g., α-lactalbumin enriched whey and casein) interventions on sleep and recovery, warrant further investigation.

2.6. Antioxidants

Both the general population and athletes can benefit from nutritional and supplementation support to boost immunity and reduce acute and chronic inflammation during periods of increased training load and competition. Antioxidants are any substance that significantly delay or prevent oxidative damage of a target molecule [58]. The fact that exercising muscles produce free radicals has motivated many athletes to consume antioxidant supplements in an attempt to reduce exercise induced free-radical damage and/or muscle fatigue. The antioxidant capacity of several dietary micronutrients is an emerging area of interest to support the endogenous antioxidant defence system of athletes and attenuate the negative effects of oxidative damage due to free radicals. Antioxidant consumption may influence recovery from exercise but may also influence sleep since sleep regulation is influenced by pro-inflammatory cytokines [59]. Dietary antioxidants (e.g., vitamin C and vitamin E) augment endogenous antioxidant content within skeletal muscle [59]. Vitamin E is a fat-soluble vitamin made up of several isoforms known as tocopherols, with α-tocopherol being the most active and abundant [60]. Vitamin E is an important antioxidant due to its abundance within cells, mitochondrial membranes and its ability to act directly on reactive oxygen species [60]. Vitamin E reacts with other antioxidants such as vitamin C, beta-carotene and lipoic acid, which have the capacity to regenerate vitamin E from its oxidised form. However, supplementation with high doses (800 IU/day) of vitamin E did not counteract OS in triathletes ($n = 38$), the intervention group demonstrated significantly ($p \leq 0.05$) higher levels of post-race inflammation and OS (Plasma F_2-isoprostanes increased 181% versus 97% and IL-6 166 ± 28 pg·mL^{-1} versus 88 ± 13 pg·mL^{-1}) than the control group [60]. Interestingly, despite increased markers of OS and inflammation in the intervention group, there was no significant difference between the groups in terms of race performance.

Vitamin A is a fat-soluble vitamin present in many lipid substances, beta-carotene can be converted into vitamin A, when necessary, from within the body [61]. Vitamin C is a water-soluble vitamin and is extremely effective in extracellular fluids, but is also effective in the cytosol [61]. It must be noted that antioxidants are heterogeneous, they function in a distinct manner and do not solely regulate ROS [61]. Consumption of an antioxidant does not guarantee that the compound will act as an antioxidant within the body, therefore positive findings from one antioxidant or combination of antioxidants cannot be generalised [59]. It has been suggested that a high intake of antioxidants could potentially reduce training adaptations. It is accepted that repeated exercise bouts (i.e. training) induce disruption in skeletal muscle homeostasis that regulate training adaptations [62,63]. While it has been reported that high doses of antioxidants could reduce training adaptations of muscle mitochondrial biogenesis and VO_{2max}, not all antioxidant studies have demonstrated negative effects and it has been suggested that the specific antioxidant used, the dose and timing of ingestion all affect outcomes [64]. It must be noted that the majority of studies have been conducted on healthy adults and there is inconsistency in terms of supplementation protocols, duration and also a wide variety of exercise protocols have been utilised. Antioxidants reduce OS, play a key role in immunity and may improve recovery following exercise [58,59]. Further research is necessary to investigate the recovery promoting doses of antioxidants within athletic populations [38,65]. The potential sleep promoting benefits of antioxidant consumption/supplementation should be investigated also.

2.6.1. Tart Cherries

Tart cherries contain high concentrations of melatonin and a range of phenolic compounds that have both antioxidant and anti-inflammatory properties [66,67]. A recent study was conducted to investigate the effect of tart cherry juice (2 × servings of 30 mL concentrate) on sleep enhancement, sleep duration and sleep quality [50]. This was the first investigation to demonstrate that tart cherry juice supplementation increased circulating melatonin levels and improved sleep time and quality in healthy adults. In the intervention group tart cherry juice supplementation resulted in significantly elevated total melatonin content, increased time in bed (+24 min), increased total sleep duration (+34 min) improved sleep efficiency total (82.3%) and a significant reduction in daytime napping (–22%) ($p < 0.05$) [50]. It must be noted that elevated melatonin concentrations may not be only mechanism at work as sleep regulation is also influenced by proinflammatory cytokines [3]. Tart cherries also contain numerous compounds that have antioxidant and anti-inflammatory properties. A similar study demonstrated that tart cherry juice consumption resulted in significantly reduced insomnia severity index scores (13.2 ± 2.8 versus control 14.9 ± 3.6; $p < 0.05$) and wake after sleep onset time (62.1 ± 37.4 min versus control, 79.1 ± 38.6 mins; $p < 0.01$), in older females with insomnia ($n = 7$) compared to a placebo [68].

Indeed, there is evidence that tart cherry juice supplementation post exercise may aid recovery from running a marathon [67]. The intervention group demonstrated a more rapid return of baseline isometric knee extension strength (pre-race 432 ± 114 vs. 48h 435 ± 109), 48 h post-marathon which was not demonstrated in the control group (pre-race 384 ± 112 vs. 48 h 349 ± 96) [67], indicating that consumption of tart cherry juice may blunt the secondary muscle damage response (localised inflammation). Post-race levels of inflammation were significantly reduced in the intervention group (IL-6 41.8 pg/mL) compared to the control group (IL-6 82.1 pg/mL) [67]. Similarly, post-race elevations in CRP and uric acid were significantly reduced in the intervention group ($p < 0.001$) [67]. Total antioxidant capacity was increased in both groups post-race (intervention 124% of baseline and control 112% of baseline, $p < 0.01$) and remained elevated at 24 h in the intervention group (114% of baseline) but not the control group [67]. During recovery athletes can suffer from delayed onset muscle soreness (DOMS) [66], which can reduce sleep quantity and quality. A recent study has demonstrated tart cherry juice supplementation (30 mL, twice per day for seven days) reduced the post-exercise decline in functional performance following intermittent sprint activity (maximal voluntary isometric contractions, 20 m sprint, counter movement jump and 505 agility test), DOMS and inflammatory response (Il-6) [66]. With regards the reduction in both DOMS and the post-exercise inflammatory response, in practice, the researchers suggested that this might be beneficial during periods of high-volume training (e.g., pre-season) or where athletes are required to produce multiple performances in a short space of time (e.g., double training sessions), when recovery periods are short [66]. The range of phenolic compounds in cherries which have anti-inflammatory and antioxidant properties may enhance post exercise recovery as well as sleep [50]. It has been proposed that melatonin may be synthesised in mitochondria, making melatonin and its metabolites available to protect the muscle against oxidative stress. Melatonin also increases the protective effects of glutathione, vitamin C and trolox, through regeneration by electron transfer processes [69].

2.6.2. Kiwifruit

Kiwifruit are nutritionally dense containing a range of nutrients that can benefit sleep, health and recovery including serotonin, vitamin C, vitamin E, vitamin K, folate, anthocyanidins, carotenoids, beta-carotene, lutein, potassium, copper and fibre [70]. Interest in the antioxidant capacity, enzyme, polyphenolic and phytochemical content of kiwifruit has increased steadily over the last decade [70–72]. It has been suggested that the various bioactive components in kiwifruit may act synergistically affecting various physiological and metabolic processes (e.g., inhibition of oxidative and inflammatory responses, improved gastrointestinal tract health and bowel function) [73]. Contemporary research has focused on the health benefits of kiwifruit particularly in relation to antioxidant capacity, digestion, iron nutrition, metabolic health and immune function [73].

Regular consumption of kiwifruit has been found to significantly ($p \leq 0.05$) increase plasma vitamin C [74] vitamin E [71] and lutein/zeaxanthin concentrations [71,74]. A study involving volunteers ($n = 25$) with a self-reported sleep disturbance demonstrated consumption of two kiwifruit one hour before bedtime for four weeks significantly improved actigraphy-measured total sleep duration (+16.9%, baseline 354.5 ± 17.1 min; post-intervention 395.3 ± 17.4 min) and sleep efficiency (+2.4%, baseline 93.9 ± 1.03 min; post-intervention 95.9 ± 0.67 min) ($p \leq 0.005$) [70]. Self-report measures also improved significantly, wake after sleep onset reduced (time awake during sleep period) (–28.9%), sleep onset latency reduced (–35.4%) while sleep efficiency increased (5.4%) ($p \leq 0.002$) [70]. Sleep quality was significantly improved following the four-week kiwifruit intervention however, the lack of a control group must be noted and there was a high level of subject drop out ($n = 5$). The findings may be prone to bias as subjects were recruited based on interest in participation in a dietary intervention study relating to sleep.

Serotonin is the end product of L-tryptophan metabolism and is related to REM [70]. The serotonin content in kiwifruit may contribute to improved sleep while the rich antioxidant content may supress free radical expression and inflammatory cytokines. Folate deficiency has been linked to insomnia (difficulty initiating or maintaining sleep, extended periods of wakefulness and/or insufficient sleep) and restless leg syndrome (repeated movement of or undesirable sensations in legs leading to sleep disruption) [70], the high folate content in kiwifruit may improve folate status and consequently improve sleep [70]. Although folates are widely consumed in the diet, they are destroyed by cooking or processing, however, kiwifruit are typically consumed in their raw form. Further research is warranted within athletic populations to investigate the potential sleep promoting properties of kiwifruit and the effect of kiwifruit consumption on both sleep quality, sleep quantity and recovery.

2.7. B Vitamins and Magnesium

Vitamin B_{12} contributes to melatonin secretion, pyridoxine (vitamin B_6) is involved in the synthesis of serotonin from tryptophan and niacin (vitamin B_3) may elicit a tryptophan sparing effect [40]. Niacin can be synthesised endogenously from tryptophan via the Kynurenine Pathway, therefore, consuming a sufficient amount of niacin is necessary to inhibit 2,3-dioxygenase activity eliciting a tryptophan sparing effect, increasing its availability for synthesis serotonin and melatonin [40]. Folate (vitamin B_9) and pyridoxine are involved in the conversion of tryptophan into serotonin [45]. The reduced form of folate (5-methyltetratrahydrofolate) increases tetrahydrobiopterin is a co-enzyme of tryptophan-5-hydroxylase which converts tryptophan into 5-hydroxytryptamine (5-HT) [45]. Pyridoxine's role in the conversion of tryptophan into serotonin is related to the amino acid decarboxylase which speeds up the conversion rate of 5-HT to serotonin [45]. Mixed effects have been observed, with different doses of cobalamin (vitamin B_{12}) on sleep-wake rhythms and delayed sleep phase syndrome (a significant delay in circadian rhythm), while no effect was observed for sleep duration [41].

Magnesium is also believed to enhance melatonin secretion promoting sleep onset and act as a GABA agonist, the main inhibitory neurotransmitter that acts on the CNS [40]. Magnesium is important for the production of the enzyme N-acetyltransferase which converts 5-HT into

N-acetyl-5-hydroxytryptamine, which can then be converted to melatonin [44]. A placebo controlled double blind study on older adults (*n* = 43), demonstrated a food based supplement containing 5 mg melatonin, 225 mg magnesium and 11.25 mg zinc, significantly (*p* < 0.001) improved subjective sleep quality scores in the intervention group but not the controls (difference between the groups 6.8; 95% CI 5.4–8.3) and total sleep duration (182.18 min; 95% CI 160.02–204.34) assessed via actigraphy [75]. The effects were attributed to the synergy between magnesium, zinc and melatonin [75]. It must be noted that supplementing these nutrients will most likely only have an effect in cases of deficiency or insufficiency.

3. Conclusions

Nutrients such as antioxidants, tryptophan rich protein, carbohydrate, melatonin, micronutrients and fruit can affect sleep [37,39,40]. Sleep can be promoted either by inhibiting wake-promoting mechanisms or by increasing sleep promoting factors through nutritional interventions [40]. Based on this review of the existing scientific literature, there appears to be considerable scope for further investigation of nutrition interventions designed to enhance sleep quality and quantity or promote general health, sleep health, training adaptations and/or recovery in both general and athletic populations.

Conflicts of Interest: The authors declare no conflict of interest.

References

1. Halson, S.L. Sleep in elite athletes and nutritional interventions to enhance sleep. *Sports Med.* **2014**, *44*, 13–23. [CrossRef]
2. Carskadon, M.A.; Dement, W.C. Monitoring and Staging Human Sleep. In *Principles and Practice of Sleep Medicine*, 5th ed.; Kryger, M.H., Roth, R., Dement, W.C., Eds.; Elsevier: Philadelphia, PA, USA, 2011; pp. 16–26.
3. Irwin, M.R.; Opp, M. Sleep health: Reciprocal regulation of sleep and innate immunity. *Neuropsychopharmacology* **2017**, *42*, 129–155. [CrossRef]
4. Kryger, M.H.; Roth, T.; Dement, W.C. *Principles and Practice of Sleep Medicine*, 5th ed.; Elsevier: Philadelphia, PA, USA, 2011; pp. 1–15.
5. Berry, R.B.; Brooks, R.; Gamaldo, C.E.; Harding, S.M.; Lloyd, R.M.; Marcus, C.L.; Vaughn, B.V. *The AASM Manual for the Scoring of Sleep and Associated Events: Rules, Terminology and Technical Specifications: Version 2.3;* American Academy of Sleep Medicine: Darien, IL, USA, 2017.
6. Buysse, D.J. Sleep health: Can we define it? Does it matter? *Sleep* **2014**, *37*, 9–17. [CrossRef]
7. Saper, C.B.; Scammell, T.E.; Lu, J. Hypothalamic regulation of sleep and circadian rhythms. *Nature* **2005**, *437*, 1257–1263. [CrossRef] [PubMed]
8. Saper, C.B.; Chou, T.C.; Scammell, T.E. The sleep switch: Hypothalamic control of sleep and wakefulness. *Trends Neurosci.* **2001**, *24*, 726–731. [CrossRef]
9. McGinty, D.; Szymusiak, R. Sleep-promoting mechanisms in mammals. In *Principles and Practice of Sleep Medicine*, 4th ed.; Kryger, M.H., Roth, R., Dement, W.C., Eds.; Elsevier: Philadelphia, PA, USA, 2005; pp. 169–184.
10. Czeisler, C.A.; Duffy, J.F.; Shanahan, T.L.; Brown, E.N.; Mitchell, J.F.; Rimmer, D.W.; Ronda, J.M.; Silva, E.J.; Allan, J.S.; Emens, J.S.; et al. Stability, precision, and near-24-hour period of the human circadian pacemaker. *Science* **1999**, *284*, 2177–2181. [CrossRef] [PubMed]
11. Rosenthal, L.; Day, R.; Gerhardstein, R.; Meixner, R.; Roth, T.; Guido, P.; Fortier, J. Sleepiness/alertness among healthy evening and morning type individuals. *Sleep Med.* **2001**, *2*, 243–248. [CrossRef]
12. Vitale, J.A.; Weydahl, A. Chronotype, Physical Activity, and Sport Performance: A Systematic Review. *Sports Med.* **2017**, *47*, 1859–1868. [CrossRef]
13. Borbély, AA. A two-process model of sleep regulation. *Hum. Neurobiol.* **1982**, *1*, 195–204.
14. Borbély, A.A.; Daan, S.; Wirz-Justice, A.; Deboer, T. The two-process model of sleep regulation: A reappraisal. *J. Sleep Res.* **2016**, *25*, 131–143. [CrossRef]

15. Venter, R.E. Role of sleep in performance and recovery of athletes: A review article. *South Afr. J. Res. Sport Phys. Educ. Recreat.* **2012**, *34*, 167–184.

16. Rawashdeh, O.; Hudson, R.L.; Stepien, I.; Dubocovich, M.L. Circadian periods of sensitivity for ramelteon on the onset of running-wheel activity and the peak of suprachiasmatic nucleus neuronal firing rhythms in C3H/HeN mice. *Chronobiol. Int.* **2011**, *28*, 31–38. [CrossRef] [PubMed]

17. Åkerstedt, T.; Folkard, S. The three-process model of alertness and its extension to performance, sleep latency, and sleep length. *Chronobiol. Int.* **1997**, *14*, 115–123. [CrossRef] [PubMed]

18. Irwin, M.R.; Olmstead, R.; Carroll, J.E. Sleep disturbance, sleep duration, and inflammation: A systematic review and meta-analysis of cohort studies and experimental sleep deprivation. *Biol. Psychiatry* **2016**, *80*, 40–52. [CrossRef] [PubMed]

19. Frank, M.G. The mystery of sleep function: Current perspectives and future directions. *Rev. Neurosci.* **2006**, *17*, 375–392. [CrossRef]

20. Irwin, M.R. Why sleep is important for health: A psychoneuroimmunology perspective. *Ann. Rev. Psychol.* **2015**, *66*, 143–172. [CrossRef]

21. Dimitrov, S.; Besedovsky, L.; Born, J.; Lange, T. Differential acute effects of sleep on spontaneous and stimulated production of tumour necrosis factor in men. *Brain Behav. Immun.* **2015**, *47*, 201–210. [CrossRef]

22. Vgontzas, A.N.; Fernandez-Mendoza, J.; Liao, D.; Bixler, E.O. Insomnia with objective short sleep duration: The most biologically severe phenotype of the disorder. *Sleep Med. Rev.* **2013**, *17*, 241–254. [CrossRef]

23. Barzilay, J.I.; Forsberg, C.; Heckbert, S.R.; Cushman, M.; Newman, A.B. The association of markers of inflammation with weight change in older adults: The Cardiovascular Health Study. *Int. J. Obes.* **2006**, *30*, 1362–1367. [CrossRef]

24. Irwin, M.R.; Cole, S.W. Reciprocal regulation of the neural and innate immune systems. *Nat. Rev. Immunol.* **2011**, *11*, 625–632. [CrossRef] [PubMed]

25. Cappuccio, F.P.; D'Elia, L.; Strazzullo, P.; Miller, M.A. Sleep duration and all-cause mortality: A systematic review and meta-analysis of prospective studies. *Sleep* **2010**, *33*, 585–592. [CrossRef]

26. Golem, D.L.; Martin-Biggers, J.T.; Koenings, M.M.; Davis, K.F.; Byrd-Bredbenner, C. An integrative review of sleep for nutrition professionals. *Adv. Nutr.* **2014**, *5*, 742–759. [CrossRef] [PubMed]

27. Johnston, J.D. Physiological links between circadian rhythms, metabolism and nutrition. *Exp. Physiol.* **2014**, *99*, 1133–1137. [CrossRef]

28. Samuels, C.; James, L.; Lawson, D.; Meeuwisse, W. The Athlete Sleep Screening Questionnaire: A new tool for assessing and managing sleep in elite athletes. *Br. J. Sports Med.* **2016**, *50*, 418–422. [CrossRef]

29. Hausswirth, C.; Louis, J.; Aubry, A.; Bonnet, G.; Duffield, R.; Le Meur, Y. Evidence of disturbed sleep and increased illness in overreached endurance athletes. *Med. Sci. Sports Exerc.* **2014**, *46*, 1036–1045. [CrossRef]

30. Hynynen, E.S.A.; Uusitalo, A.; Konttinen, N.; Rusko, H. Heart rate variability during night sleep and after awakening in overtrained athletes. *Med. Sci. Sports Exerc.* **2006**, *38*, 313–317. [CrossRef]

31. Dattilo, M.; Antunes, H.K.M.; Medeiros, A.; Neto, M.M.; Souza, H.S.D.; Tufik, S.; De Mello, M.T. Sleep and muscle recovery: Endocrinological and molecular basis for a new and promising hypothesis. *Med. Hypotheses* **2011**, *77*, 220–222. [CrossRef]

32. Tuomilehto, H.; Vuorinen, V.P.; Penttilä, E.; Kivimäki, M.; Vuorenmaa, M.; Venojärvi, M.; Airaksinen, O.; Pihlajamäki, J. Sleep of professional athletes: Underexploited potential to improve health and performance. *J. Sports Sci.* **2017**, *35*, 704–710. [CrossRef]

33. Erlacher, D.; Ehrlenspiel, F.; Adegbesan, O.A.; El-Din, H.G. Sleep habits in German athletes before important competitions or games. *J. Sports Sci.* **2011**, *29*, 859–866. [CrossRef] [PubMed]

34. Milewski, M.D.; Skaggs, D.L.; Bishop, G.A.; Pace, J.L.; Ibrahim, D.A.; Wren, T.A.; Barzdukas, A. Chronic lack of sleep is associated with increased sports injuries in adolescent athletes. *J. Pediatr. Orthop.* **2014**, *34*, 129–133. [CrossRef] [PubMed]

35. Frohm, A.; Kottorp, A.; Fridén, C.; Heijne, A.; Von Rosen, P. Too little sleep and an unhealthy diet could increase the risk of sustaining a new injury in adolescent elite athletes. *Scan. J. Med. Sci. Sports* **2016**, *27*, 1364–1371.

36. Thomas, D.T.; Erdman, K.A.; Burke, L.M. American College of Sports Medicine Joint Position Statement. Nutrition and Athletic Performance. *Med. Sci. Sports Exerc.* **2016**, *48*, 543–568.

37. Jeukendrup, A.E. Periodized Nutrition for Athletes. *Sports Med.* **2017**, *47*, 1–13. [CrossRef] [PubMed]

38. Close, G.L.; Hamilton, D.L.; Philp, A.; Burke, L.M.; Morton, J.P. New strategies in sport nutrition to increase exercise performance. *Free Radic. Biol. Med.* **2016**, *98*, 144–158. [CrossRef]

39. Tahara, Y.; Shibata, S. Chrono-biology, Chrono-pharmacology and Chrono-nutrition. *J. Pharmacol. Sci.* **2014**, *124*, 320–335. [CrossRef] [PubMed]

40. Peukhuri, K.; Sihvola, N.; Korpela, R. Diet promotes sleep duration and quality. *Nutr. Res.* **2012**, *32*, 309–319. [CrossRef] [PubMed]

41. Roehrs, T.; Roth, T. Sleep, sleepiness, and alcohol use. *Alcohol. Res. Health* **2001**, *25*, 101–109. [PubMed]

42. Hirshkowitz, M.; Whiton, K.; Albert, S.M.; Alessi, C.; Bruni, O.; DonCarlos, L.; Hazen, N.; Herman, J.; Katz, E.S.; Kheirandish-Gozal, L.; et al. National Sleep Foundation's sleep time duration recommendations: Methodology and results summary. *Sleep Health* **2015**, *1*, 40–43. [CrossRef] [PubMed]

43. Silber, B.Y.; Schmitt, J.A.J. Effects of tryptophan loading on human cognition, mood, and sleep. *Neurosci. Biobehav. Rev.* **2010**, *34*, 387–407. [CrossRef] [PubMed]

44. Afaghi, A.; O'connor, H.; Chow, C.M. High-glycemic-index carbohydrate meals shorten sleep onset. *Am. J. Clin. Nutr.* **2007**, *85*, 426–430. [CrossRef] [PubMed]

45. Ordóñez, F.M.; Oliver, A.J.S.; Bastos, P.C.; Guillén, L.S.; Domínguez, R. Sleep improvement in athletes: Use of nutritional supplements. *Am. J. Sports Med.* **2017**, *34*, 93–99.

46. Grandner, M.A.; Jackson, N.; Gerstner, J.R.; Knutson, K.L. Sleep symptoms associated with intake of specific dietary nutrients. *J. Sleep Res.* **2014**, *23*, 22–34. [CrossRef]

47. Porter, JM.; Horne, JA. Bed-time food supplements and sleep: Effects of different carbohydrate levels. *Electroencephalogr. Clin. Neurophysiol.* **1981**, *51*, 426–433. [CrossRef]

48. Ramis, M.; Esteban, S.; Miralles, A.; Tan, D.-X.; Reiter, R. Protective effects of melatonin and mitochondria-targeted antioxidants against oxidative stress: A review. *Curr. Med. Chem.* **2015**, *22*, 2690–2711. [CrossRef] [PubMed]

49. Bonnefont-Rousselot, D.; Collin, F. Melatonin: Action as antioxidant and potential applications in human disease and aging. *Toxicology* **2010**, *278*, 55–67. [CrossRef] [PubMed]

50. Howatson, G.; Bell, P.G.; Tallent, J.; Middleton, B.; McHugh, M.P.; Ellis, J. Effect of tart cherry juice (Prunus Cerasus) on melatonin levels and enhanced sleep quality. *Eur. J. Nutr.* **2012**, *51*, 909–916. [CrossRef]

51. Milagres, M.P.; Minim, V.P.; Minim, L.A.; Simiqueli, A.A.; Moraes, L.E.; Martino, H.S. Night milking adds value to cow's milk. *J. Sci. Food Agric.* **2014**, *94*, 1688–1692. [CrossRef] [PubMed]

52. Pires, M.L.N.; Benedito-Silva, A.A.; Pinto, L. Acute effects of low doses of melatonin on the sleep of young healthy subjects. *J. Pineal Res.* **2001**, *31*, 326–332. [CrossRef] [PubMed]

53. Hudson, C.; Hudson, S.P.; Hecht, T.; MacKenzie, J. Protein source tryptophan versus pharmaceutical grade tryptophan as an efficacious treatment for chronic insomnia. *Nutr. Neurosci.* **2005**, *2*, 121–127. [CrossRef] [PubMed]

54. Markus, C.R.; Olivier, B.; Panhuysen, G.E.; Van der Gugten, J.; Alles, M.S.; Tuiten, A.; Westenberg, H.G.; Fekkes, D.; Koppeschaar, H.F.; de Haan, E.E. The bovine protein α-lactalbumin increases the plasma ratio of tryptophan to the other large neutral amino acids, and in vulnerable subjects raises brain serotonin activity, reduces cortisol concentration, and improves mood under stress. *Am. J. Clin. Nutr.* **2000**, *71*, 1536–1544. [CrossRef]

55. Markus, C.R.; Jonkman, L.M.; Lammers, J.H.; Deutz, N.E.; Messer, M.H.; Rigtering, N. Evening intake of α-lactalbumin increases plasma tryptophan availability and improves morning alertness and brain measures of attention. *Am. J. Clin. Nutr.* **2005**, *81*, 1026–1033. [CrossRef]

56. Arnulf, I.; Quintin, P.; Alvarez, J.C.; Vigil, L.; Touitou, Y.; Lèbre, A.S.; Bellenger, A.; Varoquaux, O.; Derenne, J.P.; Allilaire, J.F.; et al. Mid-morning tryptophan depletion delays REM sleep onset in healthy subjects. *Neuropsychopharmacology* **2002**, *27*, 843–851. [CrossRef]

57. Bhatti, T.; Gillin, J.C.; Seifritz, E.; Moore, P.; Clark, C.; Golshan, S.; Stahl, S.; Rapaport, M.; Kelsoe, J. Effects of a tryptophan-free amino acid drink challenge on normal human sleep electroencephalogram and mood. *Biol. Psychiatry* **1998**, *43*, 52–59. [CrossRef]

58. Halliwell, B.; Gutteridge, J.M. *Free Radicals in Biology and Medicine*; Oxford University Press: Oxford, UK, 2015.

59. Nieman, D.C.; Mitmesser, S.H. Potential Impact of Nutrition on Immune System Recovery from Heavy Exertion: A Metabolomics Perspective. *Nutrients* **2017**, *9*, 513. [CrossRef] [PubMed]

60. Nieman, D.C.; Henson, D.A.; Mcanulty, S.R.; Mcanulty, L.S.; Morrow, J.D.; Ahmed, A.; Heward, C.B. Vitamin E and immunity after the Kona triathlon world championship. *Med. Sci. Sports Exerc.* **2004**, *36*, 1328–1335. [CrossRef]
61. Finaud, J.; Lac, G.; Filaire, E. Oxidative stress. *Sports Med.* **2006**, *36*, 327–358. [CrossRef]
62. Cobley, J.N.; Margeritelis, N.V.; Morton, J.P.; Close, G.L.; Nikolaidis, M.G.; Malone, J.K. The basic chemistry of exercise induced DNA oxidation: Oxidative damage, redox signalling, and their interplay. *Front. Physiol.* **2015**, *6*, 182–188. [CrossRef]
63. Cobley, J.N.; McHardy, H.; Morton, J.P.; Nikolaidis, M.G.; Close, G.L. Influence of Vitamin C and Vitamin E on redox signalling: Impications for exercise adaptations. *Free Radic. Biol. Med.* **2015**, *84*, 65–76. [CrossRef] [PubMed]
64. Mankowski, R.T.; Anton, S.D.; Buford, T.W.; Leeuwenburgh, C. Dietary antioxidants as modifiers of physiologic adaptations to exercise. *Med. Sci. Sports Exerc.* **2015**, *47*, 1857–1868. [CrossRef]
65. Pritchett, K.; Moore, A. Food with Benefits: Gain the Competitive Edge With a "Food-First" Approach. *ACSM's Health Fit. J.* **2018**, *22*, 29–33. [CrossRef]
66. Bell, P.; Stevenson, E.; Davison, G.; Howatson, G. The effects of Montmorency tart cherry concentrate supplementation on recovery following prolonged, intermittent exercise. *Nutrients* **2016**, *8*, 441. [CrossRef] [PubMed]
67. Howatson, G.; McHugh, M.P.; Hill, J.A.; Brouner, J.; Jewell, A.P.; Van Someren, K.A.; Shave, R.E.; Howatson, S.A. Influence of tart cherry juice on indices of recovery following marathon running. *Scan. J. Med. Sci. Sports* **2010**, *20*, 843–852. [CrossRef]
68. Pigeon, W.R.; Carr, M.; Gorman, C.; Perlis, M.L. Effects of a tart cherry juice beverage on the sleep of older adults with insomnia: A pilot study. *J. Med. Food* **2010**, *13*, 579–583. [CrossRef] [PubMed]
69. Galano, A.; Castañeda-Arriaga, R.; Pérez-González, A.; Tan, D.X.; Reiter, R. Phenolic Melatonin-Related Compounds: Their Role as Chemical Protectors against Oxidative Stress. *Molecules* **2016**, *21*, 1442. [CrossRef] [PubMed]
70. Lin, H.H.; Tsai, P.S.; Fang, S.C.; Liu, J.F. Effect of kiwifruit consumption on sleep quality in adults with sleep problems. *Asia Pac. J. Clin. Nutr.* **2011**, *20*, 169–174.
71. Hunter, D.C.; Skinner, M.A.; Wolber, F.M.; Booth, C.L.; Loh, J.M.; Wohlers, M.; Stevenson, L.M.; Kruger, M.C. Consumption of gold kiwifruit reduces severity and duration of selected upper respiratory tract infection symptoms and increases plasma vitamin C concentration in healthy older adults. *Br. J. Nutr.* **2012**, *10*, 1235–1245. [CrossRef] [PubMed]
72. Thorpy, M.J. Classification of sleep disorders. *Neurotherapeutics* **2012**, *9*, 687–701. [CrossRef]
73. Singletary, K. Kiwifruit: Overview of potential health benefits. *Nutr. Today* **2012**, *47*, 133–147. [CrossRef]
74. Beck, K.; Conlon, C.A.; Kruger, R.; Coad, J.; Stonehouse, W. Gold kiwifruit consumed with an iron-fortified breakfast cereal meal improves iron status in women with low iron stores: A 16-week randomised controlled trial. *Br. J. Nutr.* **2011**, *105*, 101–109. [CrossRef] [PubMed]
75. Rondanelli, M.; Opizzi, A.; Monteferrario, F.; Antoniello, N.; Manni, R.; Klersy, C. The effect of melatonin, magnesium, and zinc on primary insomnia in long-term care facility residents in Italy: A double-blind, placebo-controlled clinical trial. *J. Am. Geriatr. Soc.* **2011**, *59*, 82–90. [CrossRef]

© 2019 by the authors. Licensee MDPI, Basel, Switzerland. This article is an open access article distributed under the terms and conditions of the Creative Commons Attribution (CC BY) license (http://creativecommons.org/licenses/by/4.0/).

Review

The Role of Mineral and Trace Element Supplementation in Exercise and Athletic Performance: A Systematic Review

Shane Michael Heffernan [1,*], **Katy Horner** [1], **Giuseppe De Vito** [1] **and Gillian Eileen Conway** [2]

[1] School of Public Health, Physiotherapy and Sports Science, University College Dublin, D04 V1W8 Dublin 4, Ireland; katy.horner@ucd.ie (K.H.); giuseppedevito@ucd.ie (G.D.V.)

[2] School of Food Science and Environmental Health, Dublin Institute of Technology, Dublin 8, Ireland; Gillian.Conway@nibrt.ie

* Correspondence: shane.heffernan@ucd.ie; Tel.: +353-(0)1-716-3256

Received: 31 January 2019; Accepted: 19 March 2019; Published: 24 March 2019

Abstract: Minerals and trace elements (MTEs) are micronutrients involved in hundreds of biological processes. Deficiency in MTEs can negatively affect athletic performance. Approximately 50% of athletes have reported consuming some form of micronutrient supplement; however, there is limited data confirming their efficacy for improving performance. The aim of this study was to systematically review the role of MTEs in exercise and athletic performance. Six electronic databases and grey literature sources (MEDLINE; EMBASE; CINAHL and SportDISCUS; Web of Science and clinicaltrials.gov) were searched, in accordance with PRISMA guidelines. Results: 17,433 articles were identified and 130 experiments from 128 studies were included. Retrieved articles included Iron ($n = 29$), Calcium ($n = 11$), Magnesium, ($n = 22$), Phosphate ($n = 17$), Zinc ($n = 9$), Sodium ($n = 15$), Boron ($n = 4$), Selenium ($n = 5$), Chromium ($n = 12$) and multi-mineral articles ($n = 5$). No relevant articles were identified for Copper, Manganese, Iodine, Nickel, Fluoride or Cobalt. Only Iron and Magnesium included articles of sufficient quality to be assigned as 'strong'. Currently, there is little evidence to support the use of MTE supplementation to improve physiological markers of athletic performance, with the possible exception of Iron (in particular, biological situations) and Magnesium as these currently have the strongest quality evidence. Regardless, some MTEs may possess the potential to improve athletic performance, but more high quality research is required before support for these MTEs can be given. PROSPERO preregistered (CRD42018090502).

Keywords: ergogenic aids; nutritional supplements; physical performance; exercise and sport nutrition; muscle function

---

1. Introduction

Minerals and trace elements (MTEs) are inorganic micronutrients found in a variety of plant and animal foods [1–3]. Inadequate MTE intake has been linked to a number of health conditions, such as diabetes, cardiovascular and kidney disease, aging and fracture risk [4–13]. These micronutrients are involved in hundreds of biological processes relevant to exercise and athletic performance, such as energy storage/utilization, protein metabolism, inflammation, oxygen transport, cardiac rhythms, bone metabolism and immune function [14–18]. However, despite the biological importance of MTEs, population data suggests that current RDAs are not being achieved, with Selenium, Magnesium, Calcium, Iron and Zinc of particular concern (60%, 50%, 51%, 30% and 17% reported deficiencies, respectively; [19–21]). While not adhering to RDAs does not strictly result in biological deficiencies, certain dietary and lifestyle choices may introduce additional challenges to RDA adherence and lead to deficiencies. The Western-Type diet (high in animal protein, saturated fats and refined carbohydrates) is the most adopted diet in first-world adult populations and shows deficiencies in Phosphorus (supplemented as Phosphate) and Magnesium [22]. The Atkins for Life diet, South Beach Diet, and DASH diet result in Chromium, Iodine and Molybdenum deficiencies [23]; Eat to Live-Vegan, Aggressive Weight Loss diet in Calcium, Selenium and Zinc; Fast Metabolism Diet in Calcium, Magnesium and Potassium; Eat, Drink and Be Healthy diets in Calcium and Potassium [24]; and a strict Vegan diet shows deficiencies in Iodine and Selenium [25]. However, a Mediterranean-style diet has been suggested to mitigate some of these deficiencies and may be superior to other diets for micronutrient intake [26].

The exposure of many exercisers and athletes to commercially available diets, via the wider and social media, can lead to the adoption of food choice in-line with these diets [27] and result in associated inadequate MTE intakes and deficiencies [28–32]. For example, some athletes (n = 25 male Polish) may be up to 60% deficient in the dietary intake of particular micronutrients [Magnesium (Mg); 33]. Similar to the available population data (for example, 15% deficiency of Mg; [22–25]), there are variabilities in the level of micronutrient deficiency in the diet, depending on the geographical location [31,33–35]. In an Australian population of elite female athletes (n = 72), Calcium (22%), Iron (19%) and Magnesium (15%) intakes were identified as deficient when assessed by a food frequency questionnaire [34]. Conversely, in a large Dutch population of sub-elite athletes (n = 553, female n = 226), the only deficiencies identified were Selenium (11%) in the whole group and Iron in females (38%; [35]), assessed by 24 h recall. These may be due to the mineral constituents of the dietary choices within each population [1,2,36] and the soil environments in different geographical locations [37].

There is also considerable potential for error in validity and reliability at all stages of dietary intake assessment, regardless of the method used [38]. Biomarkers can provide an objective assessment of nutrient status. However, among other limitations, few nutrients have reference ranges for well-trained athletes [38]. In situations of high metabolic demand, such as exercise or athletic training, inadequate circulating and cellular MTEs may impair optimal physiological performance [14,30] and may require supplementation [17]. The exact impact of these deficiencies or supplementation on athletic performance remains generally unclear [39]. However, there may be ergogenic properties of MTEs in achieving or possibly surpassing the RDA, that are specifically designed for the general population health. While there is currently no consensus as to the efficacy of MTE supplementation for exercise and any physiological measure of athletic performance, recent studies show that ~50% of athletes consume some form of micronutrient supplement and between 5–27% are MTEs (Iron, Calcium, Zinc, Selenium or Chromium; [31,40]). Synthesis of some MTE research, in the context of athletic performance, can be found throughout the literature [41–48]; however, there are some MTEs and more recent studies that have yet to be systematically reviewed. Collating the current evidence on the efficacy of MTE supplementation for athletic performance in a single article is warranted and may provide a useful tool for improving the knowledge base of athletes and sport/exercise practitioners; specifically, the effects of MTEs on those phenotypes that benefit general markers of performance, for example, a lower body power output, maximal endurance capacity, maximal and relative muscle mass and strength, and fatigue recovery capacity.

Therefore, the present aim was to systematically review the literature and critically synthesise the available evidence on MTE supplementation for enhancing exercise and physiological aspects of athletic performance. In addition, this review aimed to make recommendations about the efficacy of MTE supplementation for optimising athletic performance, based on the quality of the retrieved research.

2. Methods

The review was conducted in accordance with the Preferred Reporting Items for Systematic Review and Meta-Analysis (PRISMA, checklist in Appendix A) statement [49] and preregistered with PROSPERO (CRD42018090502).

2.1. Search Strategy

A systematic search of six electronic databases and grey literature sources (MEDLINE; EMBASE; CINAHL and SportDISCUS; Web of Science and clinicaltrials.gov) was performed using predefined search terms deduced from the eligibility criteria and PICO guidelines [50] between January and April 2018. The reference lists of identified reviews and included articles were hand searched for potentially relevant articles. Where appropriate, the search was conducted using Medical Subject Headings and Boolean operators of keywords relating to population (athlete etc.), intervention (Calcium etc.), and outcome (athletic performance etc.; Appendix B).

2.2. Study Selection and Data Extraction

Following the initial data search, a complete search of titles and abstracts was performed by two independent reviewers (SH and GC). Titles and abstracts were screened for eligibility according to predefined inclusion and exclusion criteria. In the case of an inclusion discrepancy, the two independent reviewers discussed the merits of selection. If a consensus could not be reached, a third reviewer (KH) was consulted to resolve the issue and an agreement was achieved. Retrieved articles in fulfilment of the inclusion criteria were accessed, in full, and critically appraised, and the article data was extracted using a customised form (content of data extraction was decided by two reviewers) and characterised into methodological (risk of bias etc.), participant characteristics (age etc.) and study characteristics (sample size etc.). In any cases where more than one distinct experiment was performed in a single article, the experimental data for each experiment was extracted and assessed separately (for example, in Shae et al. [51]).

2.3. Eligibility Criteria

Studies of healthy adult and athletic populations (+17 years), English language, both sexes and supplement studies were included. Reports on animals, cells, children, diseased populations, dietary intake only and psychological phenotypes were excluded. Diseased populations where mineral deficiency was part of the diagnosis/prognosis were excluded; however, studies including individuals with deficiencies that were otherwise healthy were included. Interventional and control trials were the target designs for the present review to ensure the capture of all relevant articles. The following article types were excluded: editorials, systematic reviews, letters to the editor, commentaries, duplicated publications and articles that combine minerals with other molecules such as the amino acid aspartic acid combined with Zinc and Magnesium in ZMA [52]. In additionally retrieved (not identified through database searches) articles, the references list(s) were scanned for appropriate original articles.

2.4. Quality Assessment and Risk of Bias Tool

The Effective Public Health Practice Project Quality Assessment Tool (EHPP; [53]) was used to assess study quality and risk of bias independently by two authors (SH and GC). On the occasion of a discrepancy in the global quality, a third independent reviewer (KH) assessed the article(s) and a consensus was achieved. To ensure selection consistency and quality assurance, a random sub-sample of retrieved studies were cross-checked, post-hoc and independently, by KH (blinded to the original review decisions). For studies describing athletes as 'well/highly/trained/elite', selection bias was graded as "Somewhat Likely" (this decision was made in consultation with the EPHPP licence holders at McMaster University).

3. Results

A total of 17,459 articles were identified, and after removing duplicates, 14,144 articles remained. Of these, 13,816 were excluded after title and abstract screening. The remaining 328 articles were screened in full text and 128 studies met the eligibility criteria (Figure 1).

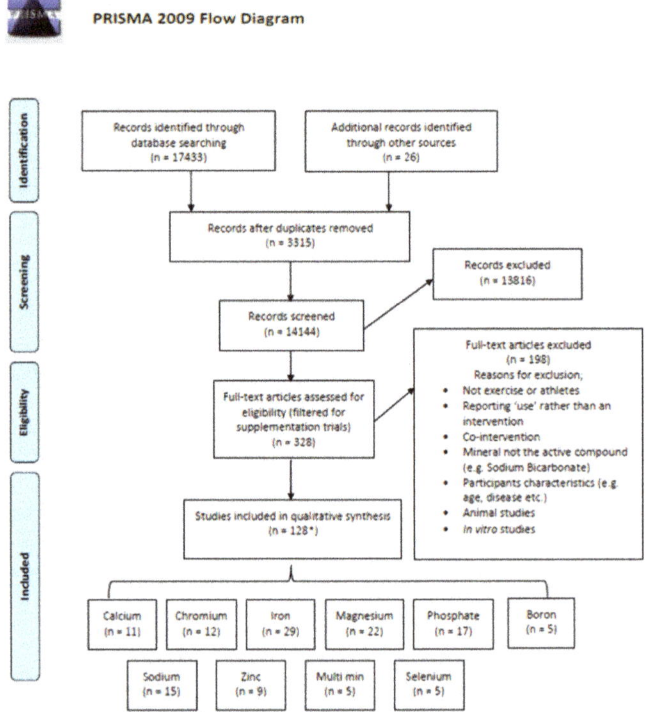

PRISMA 2009 Flow Diagram

Figure 1. PRISMA schematic summarising the search strategy and study selection. * One study assessed both Zn and Se separately and in combination (*n* = 3 groups). Thus, is counted in subsections Zn, Se and multi-minerals (also accounted for in all other cumulative study sample calculations).

3.1. Study Characteristics

The 128 eligible studies consisted of 3643 participants (1387 Females), aged between 17–75 years, and included 24 studies of elite athletes (Supplementary Table S1). The eligible studies related to Iron (Fe; *n* = 29), Calcium (Ca; *n* = 11), Magnesium, (Mg; *n* = 22), Phosphate (P; *n* = 17), Zinc (Zn; *n* = 9), Sodium (Na; *n* = 15), Boron (Br; *n* = 5), Selenium (Se; *n* = 5), Chromium (Cr; *n* = 12) and Multi-mineral articles (*n* = 5). No relevant research articles were identified for Copper, Manganese, Iodine, Nickel, Fluoride or Cobalt.

3.2. Study Quality

Using the quality assessment tool, eight articles were identified as strong, 95 as moderate and 25 as weak. Of these, only Fe and Mg included articles of sufficient quality to be classified as strong (Fe = 6, Mg = 2; Figure 2). Overall, the majority of retrieved articles were assigned a moderate quality (77%, Figure 2 and Supplementary Table S1).

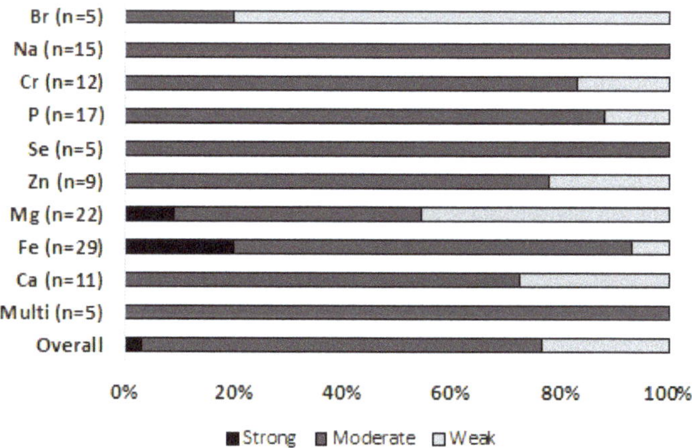

Figure 2. EPHPP global quality rating. Presented as percentage of articles rated as strong, moderate and weak for each mineral.

4. Discussion

This systematic review aimed to synthesise the evidence relating to the effects of MTE supplementation on athletic performance and related physiological phenotypes in the adult population. Quality of evidence investigating MTEs and athletic performance is lacking, with only eight articles classified as strong. Nonetheless, there is limited but growing evidence for potential benefits of some MTEs in relation to athletic performance (Fe and Mg), although the majority of research quality remains moderate-weak (Figure 2). One article was identified that presented evidence for a possible benefit of a particular combination of minerals on athletic performance-related phenotypes (see Section 4.10). Furthermore, the present review retrieved recent, although still limited, evidence for a 'natural' mineral-rich compound *Lithothamnion* and its potential for athletic performance-related haematological phenotypes, although currently no evidence for functional performance.

4.1. Iron

Twenty nine articles fulfilled the inclusion criteria for iron (Fe), including 946 participants (females, $n = 776$). These consisted of 19 randomised control trials (RCT's), 10 of which referred to elite athletes. Iron is the most studied mineral in exercise and athletic performance, with the best quality research (Figure 2). Non-anaemic Fe deficiency (serum ferritin <20.0 $\mu g \cdot L^{-1}$, Hb >115 $\mu g \cdot L^{-1}$) and anaemia (serum ferritin <12.0 $\mu g \cdot L^{-1}$, Hb <115 $\mu g \cdot L^{-1}$) are common at all levels of athletic performance and are thought to affect physiological capacity. Fe deficiency with and without anaemia has been repeatedly shown to be attenuated following both oral and intravenous (IV) Fe supplementation in a variety of sports [54–59]. However, the potential benefits of Fe supplementation on physiological performance may be dependent of baseline ferritin level, Fe dose and route of administration.

Baseline iron status or ferritin level is a major factor that could impact the efficacy of Fe supplementation on Fe status and performance related outcomes. For example, in a group of elite endurance athletes ($n = 178$, 80 females) divided by baseline ferritin levels prior to training at moderate altitude, those with high ferritin levels (>100 $\mu g \cdot L^{-1}$) were given no supplement, mid-ferritin levels (~76 $\mu g \cdot L^{-1}$) were supplemented with 105 mg Fe and low ferritin levels (~25 $\mu g \cdot L^{-1}$) were supplemented with 210 mg for two to four weeks [60]. Haemoglobin mass (HBmass) increased in those with low and mid-baseline ferritin levels supplemented with Fe, but there was no change in non-supplemented athletes. In addition, follow up ferritin levels increased by 37% in the group with the lowest baseline levels of ferritin, whereas in the other groups, ferritin decreased and total erythrocyte Fe uptake increased as a result of moderate altitude [60]. In female distance runners and triathletes with low baseline ferritin levels, high dose Fe (350 mg of ferrous gluconate) over eight weeks following recovery from two weeks of intensive training similarly increased serum ferritin levels, but had no effects on serum Iron, haematocrit or markers of immune function (natural killer cells), compared to calcium carbonate [61]. Hinton and Sinclair [62] also showed that daily 30 mg of oral Fe in 20 NAID males and females (ferritin <16 $\mu g \cdot L^{-1}$) for six weeks did not alter VO_{2peak}, but improved serum ferritin, energetic efficiency during submaximal exercise and attenuated the decline in ventilatory threshold that was observed with the placebo. Elsewhere, in active females with low ferritin, eight weeks of Fe supplementation (100 $mg \cdot d^{-1}$) increased serum ferritin, Hb, VO_{2peak} and reduced blood lactate following submaximal exercise [63]. Others have shown that in male runners with relatively high ferritin (~61 $\mu g \cdot L^{-1}$), assessed throughout a 20-day 500 km road race with low-dose Fe supplementation (36 $mg \cdot day^{-1}$), serum Iron and ferritin did not change significantly [64]. Furthermore, in non-Fe-deplete (ferritin >70 $\mu g \cdot L^{-1}$) elite male boxers, 1335 mg of ferrous-glycine-sulphate (equivalent to 200 mg elementary Fe) had no effect on physiological parameters (VO_{2peak} and other ventilatory markers) measured at moderate altitude (18,000 m) over 18 days [65]. Similarly, in non-deplete (Ferritin >100 $\mu g \cdot L^{-1}$) exercising females, 12 weeks of low dose oral Fe (50 $mg \cdot day^{-1}$) had no effect on VO_{2peak} [66].

It is important to acknowledge, however, that not all studies show improvements in iron status or endurance performance-related outcomes, such as VO_{2peak} or lactate levels, in those with low ferritin levels. For example, in female athletes with low (<26 $\mu g \cdot L^{-1}$) ferritin, no improvement in VO_{2peak}, lactate concentrations or time to exhaustion has also been shown following eight weeks of supplementation (100 $mg \cdot day^{-1}$) compared to the placebo, despite improvements in iron status [67,68]. In an older study using extremely low Fe over eight weeks (9 and 18 $mg \cdot day^{-1}$ - the RDA of the day; [69]) in females with a wide range of baseline ferritin levels, Fe had no effect on serum ferritin, Iron, Hb or haematocrit. Elsewhere, in anaemic and non-anaemic females with a wide range of baseline ferritin levels, somewhat of a high dose (160 $mg \cdot day^{-1}$ ferrous sulphate) during a 42 day period of intense physical training (5–6 h per day, 6 days per week) maintained ferritin and resulted in a significant increase in VO_{2peak} at three weeks, but not at six weeks, compared to the placebo [70]. However, it was not possible to separate the findings by baseline iron status, limiting the interpretation in this regard. Collectively, although some inconsistency in findings are evident, the majority of studies that show benefits of Fe supplementation on measures of iron status and outcomes related to endurance performance are in individuals with low baseline ferritin levels, whereas there is little evidence for beneficial effects when Iron status is not compromised.

In the context of "real-world" athletic performance, Fe has been shown to have positive effects on measures of functional performance. An improvement in a 15 km time trial was demonstrated following four weeks of high intensity endurance training (75–85% max heart rate) with six weeks of supplementation (100 $mg \cdot day^{-1}$) and was accompanied by increases in serum ferritin and Hb in NAID females [71]. In a large study (*n* = 171) of female military recruits undergoing eight weeks of basic combat training, the sample was divided into normal Iron status, NAID and Fe-deficient-anaemia, with each group receiving either a low dose of 100 mg of ferrous sulphate or placebo [72]. The supplemented Fe-deficient anaemia group completed a two mile time trial 110 seconds faster and improved indicators of mood, particularly 'vigour'. Elsewhere, amateur female rowers (Ferritin <29 $\mu g \cdot L^{-1}$) undergoing six weeks of specific training with 100 $mg \cdot day^{-1}$ Fe showed a slower rate of lactate accumulation at the 1 km and 2 km stages of a 4 km time trial and recovered quicker (~7%) at 5 min post exercise [54]. Furthermore, in those with the lowest baseline Fe (Ferritin <20 $\mu g \cdot L^{-1}$), Fe improved 4 km time trial energy efficiency (kcal) more than the placebo, although this was not observed in those with a higher Fe status [54]. Others have shown that in endurance trained females with very low baseline ferritin levels (14–18 $ng \cdot dL^{-1}$), taking 60 mg of $Fe \cdot day^{-1}$ over eight weeks of intense training 3000 m time (~10 $m \cdot min^{-1}$) improved along with running velocity to lactate threshold and onset of blood lactate accumulation improved-but there was no change in VO_{2peak} compared to the placebo [73].

These findings collectively show the benefits of Fe supplementation on some measures of functional performance in individuals with low baseline ferritin levels.

Other performance-related outcomes, including fatigue resistance and strength, have also been shown to improve following Fe supplementation in females with low ferritin levels, and in studies of females with varying Iron status. Brutsaert et al. [74] investigated the effect of six weeks of oral Fe (20 mg·day^{-1} elemental Fe) on maximal voluntary contractions (MVC) following a quadriceps muscle fatiguing protocol in NAID females (ferritin <15 µg·L^{-1}; n = 20) over six weeks. Fe increased post-fatiguing MVC by 15% and showed a significant improvement in fatigue resistance in quadriceps MVC, compared to the placebo. In elite volleyball players who were NAID or had adequate iron stores that were supplemented with a high dose (105 mg·day^{-1} elemental Fe) for 11 weeks during the competitive season, markers of dynamic strength, the clean and jerk, power clean and total mean strength performance (ranging between ~10–40%) were enhanced and Iron loss was prevented compared to controls [75].

Along with dose, the route of Fe administration may also influence the efficacy of supplementation in individuals of varying Iron status. Although used since the 1970's [76], recent advances have shown the rise in the use of parenteral Fe preparations, which appear superior to oral supplementation in enhancing measures of athletic performance [77], but not ubiquitously [42]. Six weeks of IV was superior to high-dose oral Fe for improving VO$_{2max}$ (by 2.5%) and max running time (3.7%) in non-anaemic endurance runners (ferritin ≤65 µg·L^{-1} and [Hb] >12 g·dL^{-1}) [77]. Furthermore, when subdivided into low (ferritin <35 µg·L^{-1} and transferrin saturation <20%, or ferritin <15 µg·L^{-1}) or everyone else (classified as 'suboptimal'), HBmass (~4–5%), VO$_{2max}$ (2%) and max running time (6%) increased more in the low IV group compared to the oral low or suboptimal IV group [77]. Others have shown an improvement in serum ferritin but no performance effect of IV Fe compared to the placebo [78]. However, this may be due to methodological limitations, including that follow up testing was performed ~10 days after the last supplementation period and the testing was not standardised to facilitate the athletes' disciplines. In highly trained distance runners (6 male, 8 female) without clinical Fe deficiency (~60 km·week-training^{-1}; ferritin 30–100 µg·L^{-1}), three Fe injections over four weeks had no effect on 3000 running performance, but improved perceived fatigue and mood compared to saline injections [79]. Therefore, the current evidence on whether Fe IV infusion or injections are superior to oral supplementation is limited.

In conclusion, there appears to be a range of factors, including baseline Iron status, dose and route of administration, that may influence the efficacy of Fe supplementation on Iron status and performance. It is generally accepted that Fe increases HBmass, leading to greater oxygen delivery; however, there may be other mechanisms not related to erythropoiesis and oxygen transport at play (i.e. no change in HBmass in non-Fe deficient athletes [77,79]) that, over longer trial periods this could lead to performance enhancement. The evident general trend, as is also evident in recent systematic reviews, is that Fe supplementation may benefit Iron status and athletic performance in individuals with a compromised Iron status [42,43]. In NAID females, oral elemental Fe supplementation between ~15–60 mg·day^{-1} or 100 mg·day^{-1} of ferrous sulfate over six to eight weeks may be adequate to elect performance adaptations in 3000 m running time, running velocity to lactate threshold, onset of blood lactate accumulation, quadriceps MVC fatigue resistance, post-fatiguing MVC, 4 km rowing time trial energy efficiency (kcal), preventing exercise-induced Fe loss, Hb, VO$_{2peak}$, improving 15 km cycling and two mile running time trial, quicker recovery post exercise and indicators of mood. Approximately 100 mg·day^{-1} elemental Fe over 11 weeks has been shown to result in adaptations in markers of dynamic and absolute strength. Intravenous Fe administration may be beneficial to improving running time and VO$_{2max}$ performance; however, the current evidence on whether Fe IV infusion or injections are superior to oral supplementation is limited and the administration (i.e. via a medical professional) makes it difficult to recommend such approaches at this time. Future studies should focus on recruiting larger samples, including elite athletes, tracking longer term supplementation and considering alternative mechanisms to the potential changes in HBmass. As the majority of individuals supplementing with Fe are not elite athletes, investigations of Fe-associated training adaptations in non-elite individuals are also currently limited and warrant attention.

4.2. Calcium

Eleven articles fulfilled the inclusion criteria for Calcium (Ca), including 311 participants (females, n = 58). These consisted of 19 RCT's, one of which referred to elite athletes. Exercise is known to induce modest Ca loss following non-weight bearing steady-state activities; however, Ca supplementation might mitigate this loss. Ca loss can also lead to hormonal changes [80] that might impair muscle function; however, limited quality data exists for direct markers of functional performance.

An investigation of Ca supplementation on Ca homeostasis during exercise in healthy active premenopausal women demonstrated that Ca supplementation (800 mg above the controlled allowance for placebo) effectively attenuated exercise-induced Ca loss [81]. In fact, Ca supplementation altered the usual Ca loss associated with exercise into positive retention. These data are important because cellular Ca is vital for skeletal muscle function, particularly the calcium-dependent troponin complex, and parathyroid hormone (PTH) homeostasis [82,83]. The precise impact of circulating Ca on muscle function and exercise/athletic performance, however, remains unclear. For example, additional Ca (35 mg·kg·day^{-1}; standard intake ~1800 mg·day^{-1}) did not affect Ca balance or VO_{2peak} performance during normal endurance training (67 km·week^{-1}) or when trained endurance athletes were restricted to 7 km·day^{-1} for 12 months [84]. However, with restricted activity, iPTH (ionised PTH) decreased to a greater extent in the Ca supplemented group compared to the non-supplemented group [84]. In a very similar study design, Zorbras et al. [85] also showed that when athletes considered to be Ca deficient at baseline (~1500 mg·day^{-1} dietary intake) restricted training (0.7 km·day^{-1}) and consumed 55 mg·kg·day^{-1} Ca over 12 months, circulating PTH was reduced.

Continual IV infusion of Ca (156 mg), with the intention of maintaining consistent blood Ca during exercise [86], has also been shown to attenuate the usual exercise-induced increase in PTH by 65%, compared with a saline infusion. In agreement, elite athletes consuming either high (972 mg·L^{-1}) or low (18 mg·L^{-1}) Ca water during exercise found similar effects for PTH (at peak concentrations, Low Ca = 36 pg·mL versus High Ca = 65 pg·mL; [87]). This is potentially important for exercise and athletic performance as high PTH levels have been linked to significant muscle impairments [88,89], PTH replacement in hypothyroid patients can decrease maximal muscle strength by ~30% and can impair neuromuscular motor unit action potentials [90].

Another route to ensure adequate or supplementary Ca intake is through food control and whole food supplementation. Haakonssen et al. [91] provided female road cyclists with a meal consisting of either 46 mg (control group) or 1352 mg (Calcium group) of dairy Ca 90 mins prior to a 90-min cycle trial. Blood samples were taken pre-trial; immediately pre-exercise and at 40 min, 100 min and 190 min post-exercise. Serum iCa decreased with the onset of exercise, but was higher in the Ca supplemented group (by ~0.040 mmol·L) immediately prior to exercise and remained higher post-exercise and at 40min post-exercise [91]. Similarly, iPTH was persistently lower in the Ca supplemented group compared to controls prior to, post-exercise and during recovery - to the largest extent immediately post-exercise (1.55 times lower) and was independent of Ca loss through sweat.

Cinar et al. [92–96] investigated the influence of Ca (~37 mg·kg·day^{-1}) and exhaustive exercise (90 min·day^{-1}, 5 days·week^{-1}) on a number of exercise-related blood markers, in a series of studies in the same sample of amateur athletes (n = 30). The authors concluded that circulating Ca, Potassium (K), Copper (Cu), Testosterone (T), Glucose, Leukocyte and Erythrocyte levels were altered following exhaustive exercise and occurred to a greater extent with exercise plus Ca supplementation [92–96]. However, these conclusions are misleading, as the presented results suggest that it was more likely that exhaustive exercise that influenced these changes, rather than Ca supplementation. For example, the exercise condition (no supplementation) increased circulating Cu by ~0.34 mg·dL and circulating T by ~4 pg·mL; whereas Ca supplementation plus exercise increased Cu by ~0.85 mg·dL and T by ~3 pg·mL and Ca alone resulted in the opposite—a decrease in Cu and T (data derived from [92,93]).

This trend continued for Leukocyte and Erythrocyte levels, showing that Ca supplementation likely had no effect on exercise-related blood biomarkers; rather, exhaustive exercise was the driver for the observerations. Furthermore, the authors found no effect of Ca supplementation and/or exhaustive exercise on plasma adrenocorticotropic hormone (ACTH) and cortisol levels [95]. This data is curious as it has been known for some time that both biomarkers increase following exercise, particularly exhaustive exercise [97].

Although Ca is vitally important for muscle and cardiovascular function [98,99], there is currently no evidence that Ca supplementation has any direct effect on athletic performance (currently, only aerobic capacity has been investigated). Nonetheless, calcium supplementation at oral doses between 800 (over 8 days) to 1352 mg (single meal prior to exercise), or IV infusion at 156 mg (prior to and during exercise), may attenuate post-exercise reductions in serum iCa and Ca loss, with lower doses appear to have no effect. Supplementary Ca may also reduce the exercise-induced increase in iPTH. As mentioned above, the impact on PTH could potentially have implications for muscle strength [88,89]. However, there is currently no direct evidence to support the hypothesis that Ca supplementation may enhance muscle physiological capacity through the actions of PTH. Future research should investigate this using detailed measures such as hormonal changes, muscle fibre characteristics, muscle-derived biochemical effects, action potentials, and functional and specific strength measures.

4.3. Magnesium

Twenty two articles fulfilled the inclusion criteria for magnesium (Mg), including 663 participants (females, $n = 72$). These consisted of ten RCT's, three of which referred to elite athletes. Evidence is growing that Mg may be an important element to maintain muscle mass, power and markers of systemic inflammation [100,101], although the effects of supplementation on these parameters remains ambiguous [102]. There appears to be some inconsistencies in the literature that are likely to be a result of dosing, baseline mineral status/intake, exercise intensity and population. For example, low dose Mg supplementation may be simply elevating Mg to the required physiological levels, rather than being sufficient to have an ergogenic affect. This is reflected in the general trend for higher doses to have more positive results [103,104] than lower doses [105–108].

Exercise is known to effect Mg metabolism and there is an alternating response, depending on exercise intensity [107,109,110]. In elite handball athletes, greater time spent exercising at low-to-moderate intensity was associated with higher plasma Mg levels, whereas a greater time spent training at >80% residual heart rate was associated with lower Mg levels ($r = 0.38$, $p < 0.01$; [109]). Intense training over four weeks similarly resulted in lower Mg levels in elite volleyball players, including in those supplementing with Mg [107]. However, others have shown that supplementing with 400 mg·day^{-1}, in addition to 217 mg per 1000 kcal of dietary Mg, maintained Mg status throughout an elite basketball competitive season, except during the most competitive portion of the season [110]. These data imply that with adequate Mg availability and under conditions of low physiological demand, Mg is released into the circulation, likely from muscle, but not necessarily used. Whereas, when stores are inadequate to supply the demand in conditions of physiological stress, circulating Mg levels decrease, highlighting the potentially important role of muscle in Mg metabolism A responder/non-responder paradigm has also been suggested to play a role in Mg and exercise metabolism, and requires further explanation [111].

The exact relationship between blood and muscle Mg is unclear and also requires further investigation. When well-trained endurance athletes were restricted from running (13.9 km·day^{-1} to 4.7 km·day^{-1}) for 12 months, they entered a negative Mg balance (Mg intake lower than faecal and urinary losses), compared to a positive balance in athletes where exercise was non-restricted. This negative balance resulted in increased serum Mg, irrespective of low dose Mg supplementation (0.5 mg kg day^{-1}; [112]). Interestingly, the negative balance was greater when reduced activity was combined with Mg supplementation and the greatest positive balance was evident in the athletes who combined Mg supplementation with their usual training. These data suggest that when Mg is supplemented during exercise restriction, significant amounts are excreted, likely because lower amounts of Mg are being utilised. This highlights the role of muscle in Mg metabolism.

Two older studies have investigated muscle-derived Mg in the context of supplementation and post-exercise requirement [113,114], with no exercise-associated change in muscle-derived Mg content identified. In one study, Mg supplementation (365 mg day^{-1}) had no effect on muscle or serum Mg concentrations or on 42 km marathon running performance compared to a placebo group [113]. However, the post-exercise muscle biopsies were take 48 h after the marathon and by this time, Mg concentrations are likely to have returned to resting levels. Weller et al. [114] investigated the effect of Mg supplementation (500 mg day^{-1}) over three weeks on muscle, serum, leukocyte and other blood-derived cell Mg in a group of athletes with low-normal serum Mg levels. There was no effect on muscle, serum or blood Mg concentrations and no difference in aerobic capacity, neuromuscular function or exercise haematological parameters between Mg and placebo groups. However, there was a weak correlation between muscle Mg (measured by NMR) and total Mg in mononuclear leukocytes, and an inverse correlation between muscle and red cell Mg. It appears that serum Mg does not necessary reflect muscle Mg (although serum is often measured), however exercise was not well-recorded and muscle Mg was not measured following exercise. These data are interesting but owing to the lack of rigorous methodology, the effect of exercise with Mg supplementation on muscle-derived Mg is currently unknown and warrants future investigation.

The effects of Mg supplementation on functional performance outcomes and related measures in general appear inconsistent. For example, there is currently no evidence to support Mg supplementation enhancing endurance capacity [100,106,107,112,114], despite a logical biological potential [115]. However, positive effects have been shown in some other outcome measures. In elite volleyball athletes, Mg supplementation (350 mg·day^{-1}) over four weeks improved (~6%) countermovement jump, compared to the placebo, although there was no difference in neuromuscular capacity (Isokinetic dynamometry) [107]. Kass et al. found improvements in blood pressure at rest and during recovery following Mg supplementation over two weeks (>300 mg·day^{-1}), but no effect on isometric bench press or endurance performance indicators [105]. However, in a follow up study, Kass and Poeira [116] investigated the effect of acute (300 mg·day^{-1} for one week) and chronic (4 weeks) Mg supplementation compared to a placebo on exercise and recovery from resistance exercise in 13 recreational endurance athletes. These athletes were already consuming ~370 mg·day^{-1} of dietary Mg and the results showed that the effects may depend on duration of supplementation. A 40 km cycling time-trial was carried out to elicit physiological stress, deemed typical training, where blood pressure and 1 repetition maximum (1RM) bench press (following the time-trial) were assessed at baseline and over two consecutive days following the supplementation period. Acute Mg increased 1RM by 17% compared to baseline (prior to commencing supplementation), whereas in the chronic (over 4 weeks) Mg group, there was no change in 1RM bench press. In addition, acute Mg showed no decline in force during repetitions to fatigue the next day following the time-trial, whereas in the chronic Mg group, there was a 32% performance decrement. Blood pressure also showed a greater and more consistent reduction in the acute versus chronic Mg groups [116]. However, changes in neuromuscular strength may be influenced by a higher dose and duration of Mg supplementation with training intensity. In previously untrained participants, eight mg·kg·day^{-1} Mg (achieving ~144% of the RDA compared to 70% in the control group) for seven weeks, together with lower body resistance exercise three times per week, improved absolute (~40 Nm) and relative (~0.9 Nm·kg of lean body mass) strength compared to a control group [103]. In contrast, in an older population (>65 years) undertaking a 'mild fitness program', twelve weeks of 300 mg·day^{-1} ($n = 62$) had no effect on neuromuscular or handgrip strength compared to a control group (no placebo or intervention, $n = 77$; [117]). However, other markers of physical function improved (Short Physical Performance Battery; chair stand, 4m walking test) following Mg supplementation and the improvements were more pronounced in those with a lower Mg intake. Combined, these data suggest (although relatively weakly with the current evidence) that in younger populations, there may be a capacity for high-dose

Mg to elicit improvements in functional markers of athletic performance and with longer trial periods, neuromuscular strength. Whereas in older populations, moderate dose Mg can improve functional markers of health and standard markers of physical function.

There appears to be some evidence that Mg can also positively influence exercise-induced haematological changes. Four weeks of 500 mg·day^{-1} of Mg in an amateur rugby union ameliorated the exercise-induced increase in IL-6 (a marker of systemic inflammation), reduced white blood cell count, neutrophils percentage, post-game cortisol, increased adrenocorticotropic hormone and lymphocytes percentage, compared to controls [104]. These trends were evident at various time-points throughout six days of recovery following a game. Interestingly, a significant reduction in cortisol was observed in the Mg supplemented group on the day prior to and the morning of the game, but not on the day after the game [104]. The same researchers, using a similar design, showed that Mg mitigated exercise-induced DNA damage in the presence of H_2O_2, but not without, following the same amateur rugby game [118]. Recent evidence has shown the importance of some of these molecules in exercise adaptation [119–121], however the implications of these changes for athletic performance require further investigation.

It should also be noted that Cinar et al. presented findings on the effects of Mg (10 mg·kg·day^{-1} over 4 weeks) and exercise training on haematological measures of immune function, and hormonal, insulin and glucose status at rest and following an exhaustive exercise protocol [122–126]. The authors concluded that Mg supplementation in "sportsmen" elicited higher circulating Mg and Zn [123], blood cell count were reduced; erythrocytes were increased [123], and leukocytes, erythrocytes, ACTH, cortisol, glucose and testosterone were increased both at rest and following exercise to exhaustion [122,124–126]. However, as noted with other papers by these authors, these conclusions should be taken with caution as many of the biomarker changes were likely the result of the training conditions rather than Mg supplementation.

In conclusion, the current evidence suggests that 300–500 mg·day^{-1} for short-term supplementation (~1–4 weeks) can have a positive influence on functional dynamic measures of muscle performance (CMJ, 1RM and fatigue resistance) and exercised-induced inflammation, DNA damage, cortisol and immunological blood markers, but no effect on isokinetic performance [107,116,118]. Whereas, longer supplementation trials (~7 weeks) can elicit training-induced adaptive responses in young untrained populations [103]—whether this would be reflected in well-trained athletic populations is yet to be established. Furthermore, Mg (300 mg·day^{-1} over 12 weeks) may improve markers of functional performance in older populations and may be a consideration to maintain functional capacity throughout aging, but may require an even longer treatment period [127]. Lastly, there is currently no evidence to support a benefit of Mg supplementation to improve endurance performance-related outcomes, despite a logical biological potential [115]. However, the literature is limited and further research is required. Mg appears to have some ergogenic potential, but much more evidence is needed in a variety of populations (untrained, elite athletes and elderly) and in response to both aerobic and resistance/dynamic power training and performance. Little is currently known about the direct effect of Mg ingestion on muscle in response to exercise and further investigation is needed to uncover the mechanisms of action for the above physiological responses.

4.4. Phosphate

Seventeen articles fulfilled the inclusion criteria for phosphate (P). These included 247 participants (females, $n = 48$), consisting of 13 RCT's, two of which referred to elite athletes. Phosphate (commonly supplemented as sodium phosphate (SP); the supplementary form of phosphorus) has been shown to improve a range of parameters, including sprint time, cycling power output, VO_{2Peak}, resting HR, biomarkers markers of metabolic demand and measures of cardiac function (echocardiographic).

Several studies have shown improvements in sprint time, total work and power output in a range of athletes. In recreational male ($n = 11$) and female ($n = 12$) team sport athletes, supplementation with 50 mg·kg·FFM^{-1} of SP over six consecutive days was superior for improving sprint performance compared to a placebo using magnitude-based inferences [128,129]. Participants performed a simulated team-game circuit with a 6×20 m repeated sprint set performed before, at half-time and at the end. In males, SP resulted in faster times for all sprints compared to the placebo, caffeine or a combination of SP plus caffeine (Cohen's $d' = 0.5$–0.8), and in females, both SP and combined SP plus caffeine improved repeated sprint ability compared to placebo [128,129]. The same researchers showed that in female team sport athletes ($n = 13$), SP was superior (Cohen's $d' = 0.5$–0.8) in improving total sprint times for all sets and overall, and the best sprint times were improved for all sets, with ~6% improvement after SP compared to the placebo and beetroot juice [130]. This group also demonstrated improvements in total work and power output in trained cyclists in a race simulation on days 1 and 4 following six days SP supplementation, compared to no change in the placebo [131]. Others have also shown improvements in power output by ~30 W and finishing time during a 1 km time trial in well-trained cyclists following six days SP (4 g day^{-1}) supplementation, compared to a placebo ($n = 7$; [132]).

A range of outcome measures related to endurance performance, including VO_{2peak} and blood lactate, have also been shown to improve following SP supplementation. In competitive male cyclists, SP loading significantly improved VO_{2peak} and interestingly, a second loading phase separated by a washout period (15–35 days) resulted in even greater improvements, compared to a placebo ($n = 12$; [133]). An older study comparing two days of potassium phosphate (PP) loading (4 g·day^{-1}) to placebo in highly-trained endurance runners (crossover $n = 8$) showed that PP may reduce the rate of perceived exertion (RPE) during the mid-stages of maximal treadmill running, although no changes in physiological parameters were identified. This may be due to the short loading phase [134]. Others have shown that SP loading (4 g) over three days improved VO_{2peak} in elite endurance athletes by ~9%, increased blood haematocrit levels, glucose and other markers of metabolic demand [135]. SP also enhanced echocardiographic measures of endurance performance by 5–12% (ejection fraction and fractional shortening). These data are an indicator of the possible mechanisms that SP may have on athletic performance (in competitive male endurance athletes, $n = 6$). Particularly as low and high P has been linked to impaired echocardiographic measures in critically diseased populations [136–140], this emphasises the importance of P in cardiac function. Surprisingly, no recent echocardiographic data related to SP and athletic performance was identified in the present review, but would be of interest for future investigations.

Recent evidence suggests that SP appears to maintain the exercise-associated benefits of supplementation when the loading phase (50 mg·kg-FFM·day^{-1} for 6 days) is followed by a lower dose (25 mg·kg-FFM·day^{-1}) for a subsequent three weeks [141,142]. In elite off-road mountain cyclists, the loading phase increased VO_{2peak} (by 5.3%), and reduced resting HR (by 9.6%), max HR (by 2.7%), and HR at lactate threshold (1.7%), all of which were maintained following lower dosing, with no change in placebo [142]. Furthermore, maximal power output did not change due to the loading phase, but increased significantly after the maintenance phase, implying a possible delayed response to the peripheral tissues (as described with Zinc). However, to date, this has not been experimentally shown. This adaption could be a result of increased 2,3-diphosphoglycerate (2,3-DPG) concentrations as 2,3-DPG decreases the affinity of Hb for oxygen, thus resulting in the greater unloading of oxygen to the peripheral tissues [142] and increased 2,3-DPG after SP loading [142–144].

It is interesting to note that "moderately" trained individuals supplemented in a similar way (6-day loading) showed no effect on aerobic capacity or power output [145–148]. This suggests that SP may only improve aerobic performance when individuals are well-trained or elite athletes, indicating that the dietary ingestion of phosphorus may be adequate outside of extreme physiological requirements. This may be reflected in the relatively unchanged blood phosphate levels reported by some authors following supplementation in well-trained athletes [131,133,135]—although this has not been observed by all [134,142]. It is important to note, as suggested by Kreider et al. [135], that serum phosphate concentrations may not accurately assess the effects of phosphate loading on intracellular phosphate levels and oxidative metabolism. However, no investigation of muscle-derived P activity was retrieved in this review.

In conclusion, the current evidence indicates several ergogenic effects of SP supplementation with ~4 g·day^{-1} over three-to-six days on a range of performance-related outcomes, such as sprint time, cycling power output, VO_{2Peak}, resting HR, biomarkers of metabolic demand and measures of cardiac function (echocardiographic). However, these ergogenic effects are limited to highly-trained individuals and may not aid recreational athletes when dietary phosphorus is adequate and outside of the requirements of highly-trained athletes. In addition, benefits in endurance performance-related outcomes can be further maintained with a lower dose of ~2–4 g·day^{-1} [141,142]. As there is some evidence for a delayed response to the adaptation of muscle power [142], this is an interesting area for future investigation, along with determining if this is a central or peripheral physiological adaptation and if it is directly related to changes in 2,3-DPG. Furthermore, Kreider et al. [135] presented some possible mechanistic data in relation to morphological changes to the myocardium in a small sample, but this requires replication with a larger sample in future investigations.

4.5. Zinc

Nine articles fulfilled the inclusion criteria for zinc (Zn). These included 229 participants (females, $n = 50$), consisting of four RCT's, one of which referred to elite athletes. Few studies exist that have directly assessed the effect of Zn (without other molecules) on athletic performance phenotypes and the majority appear to assess physiological variables that indirectly contribute or augment exercises-induced reduction in immunity [149–151].

Nevertheless, six weeks of low-dose Zn (gluconate; 30 mg·day^{-1}) have been shown to improve estimated VO_{2peak} (Bruce protocol), similar to that of HITT training (~4%; additional to team training) and when combined, Zn and HITT further improved estimated VO_{2peak} (by 9.33%) in female futsal athletes [152]. Whereas, two weeks of Zn carnosine at 70 mg·day^{-1} had no effect on 80% VO_{2Peak} performance or lactate accumulation, although this time-period was sufficient for Zn to reduce exercise-induced gut permeability by 71% ("leaky gut"; [153]). Similarly, one week of low-dose Zn gluconate at 20 mg·day^{-1} did not improve estimated VO_{2peak} (Astrand and Ryhming protocol) compared to a placebo in healthy sedentary males [154]. However, Zn supplementation reduced blood viscosity (lower viscosity is associated with aerobic exercise as it promotes oxygen delivery; [155]), which, over time, could potentially lead to improved aerobic performance. Overall, these findings show inconsistent effects of Zn supplementation on VO_{2peak}.

The absence of change in endurance performance parameters may be due to the potential existence of a latent response to Zn. In elite cyclists with Zn deficiency (<11 μmol·L) receiving 22 mg·day^{-1} for 30 days and maintaining their normal training, participants' Zn status improved and following the cessation of Zn supplementation and replacement with a placebo continued for a subsequent 30 days with plasma Zn concentrations improving further in ~1 μmol·L increments (elite cyclists; [156]). It is plausible that a longer trial period may be required to achieve more complex physiological effects with Zn. These results showed that as plasma Zn increased, Cu decreased, resulting in an increase in the Zn:Cu ratio, and this was positively correlated with an increase in insulin and HOMA2-IR (a measure of insulin resistance; 27% and 47%, respectively), indicating that Zn may impair glucose utilisation [156]. These findings could provide some rationale to the relatively inconsistent estimated VO_{2peak} results previously presented [152–154]. It is important to note that these data contrast the findings of improved insulin sensitivity with exercise training [157–159]. Zn also improves insulin sensitivity in individuals with obesity [160] and there is also evidence of high insulin sensitivity in elite athletes [161]. Therefore, caution needs to be adopted when considering these results until there is replication in elite and non-elite athletic populations.

Zn has also been shown to affect the anabolic hormone testosterone (T) in some, but not all, studies. Cinar et al. [162] demonstrated that high-dose Zn (2.5–3 mg·kg·day^{-1}) appeared to increase post-exercise bound and free T (3.1 pg·dL^{-1}). However, six weeks of resistance training with and without Zn induced greater changes (–5.7 pg·dL^{-1}, –4.7 pg·dL^{-1}, respectively) compared to no change in the control group (no Zn or training). Contrary to the results of Cinar et al. [163], 30 days of Zn

(22 mg·day^{-1}; Zn gluconate) in elite Zn deficient male cyclists, during regular training, failed to change the thyroid stimulating hormone thyroxine or the active form triiodothyronine. This further supports the hypothesis that the resistance training stimuli enacted the hormonal changes observed by Cinar et al., [156] not the supplementation. Nonetheless, others have shown that Zn may contribute to exercise-induced hormonal adaptations. In a double-blind placebo controlled trial (over 4 weeks; in Zn sufficient well-trained road cyclists), Zn (30 mg; Zn sulphate) increased free T (by ~4 pg·dL^{-1}) following a bout of exhaustive exercise, but did not affect resting levels [164].

Despite the popularity of Zn supplementation among athletes [31,40], there is little quality evidence that Zn can improve athletic performance and existing evidence relates to short trial periods (1–6 weeks). There is limited evidence that 20-30 mg day^{-1} may improve estimated VO$_{2Peak}$ [152], reduce blood viscosity [154] and result in ~4 pg·dL^{-1} increase in T following high-intensity exercise [164] over a period of one to six weeks. Nonetheless, there were a number of methodological limitations in these studies, including that the VO$_{2peak}$ was estimated and not measured. Therefore, the true impact of Zn on aerobic capacity remains unclear and requires further investigation. There are some curious results pertaining to insulin resistance in Zn deficient elite cyclists following Zn supplementation [156]. Although contrasting our current understanding of insulin during exercise [157–159] and the effect of Zn on insulin in other populations, this Zn-insulin paradigm in elite cyclists needs to be investigated further. Furthermore, little quality research on the effect of Zn supplementation on markers of muscle strength or hormonal responses to exercise in healthy adult populations was retrieved. Thus, in addition to investigating the effects on endurance capacity, research should focus on other markers of athletic performance, such as muscle strength and dynamic power for longer trial periods than have been investigated.

4.6. Sodium

Fifteen articles fulfilled the inclusion criteria for sodium (Na). These included 387 participants (females, $n = 27$), consisting of five RCT's, none of which referred to elite athletes. Na supplementation is an effective strategy for improving exercise-induced changes in sodium balance [165–170], particularly in hot conditions [166,171,172]. However, the effectiveness of Na (in the form of loading) for enhancing athletic performance (and related physiological phenotypes) is controversial [173].

Some controlled laboratory and 'real-world' studies suggest that Na may be beneficial [165,166,174], whereas others show no-benefit [167,175–178] for a variety of reasons, which may include environmental, gender, dose, exercise type and training differences [165,166,179]. In a recent laboratory study (temperature approximately at 21 °C) of trained endurance athletes, capsules containing 360 mg Na and 540 mg chloride (Cl; standard table salt) were provided immediately prior to starting an exercise test (2h treadmill or cycling at 60% HR max) and at 25 min intervals during exercise, but had no impact on time-to-exhaustion following the exercise test [175]. The largest study to investige the effect of Na on athletic performance [176] sampled South African ironman triathlon finishers in 2001 (mixed athletic level; $n = 53$ supplemented with 620 mg sodium chloride, $n = 61$ had placebo and $n = 299$ as a control group) with an air temperature ranging from 16 to 21 °C during the race and found that Na had no effect on race finishing times or other related physiological variables. However, training status was not considered (finishing time standard deviation was ~96 mins) and there are some indications that this may influence the physiological impact of Na [179]. Regardless, when comparing high- (164 mmol·L) to low-dose Na (10 mmol·L) in trained endurance athletes exercising in the heat (32 °C), high-dose Na-supplemented athletes showed a greater exercise tolerance (96.1 min vs. 75.3 min) in men and women [165,166]. This suggests that when exercising in heat, high-dose Na may improve athletic performance, although replication in studies with a placebo group is required [165,166].

Interestingly, there is some evidence from a series of studies by Zorbas and colleagues that suggest during a period of inactivity, supplementing with Na may maintain pre-inactivity VO$_{2peak}$, even following 12 months of significantly reduced activity [169,170,180]. During a period of bed rest (from 74 km·wk^{-1} running), Na increased body mass and body fat, but surprisingly maintained VO$_{2peak}$ (pre-rest = 66 vs. post-rest = 67 mL·kg^{-1}·min^{-1}), compared to a placebo [170]. Furthermore, following 364 days of reduced activity (0.7 km·day^{-1}), Na maintained similar VO$_{2peak}$ results to those of bed rest (pre-rest = 67 vs. post-rest = 68 mL·kg^{-1}·min^{-1}; [169]). This maintenance is proposed to result from an increased plasma volume (also shown elsewhere; [165,166]) and higher arterial blood pleasure [169,180]. This may be a consideration for injured endurance athletes to possibly maintain VO$_{2peak}$ performance. However, given the limited data and as Na intake is linked to cardiovascular disease in the sedentary general population [181], caution and professional supervision should be considered if adopting such strategies.

Others have investigated the effects of low-dose 0.2 g·kg^{-1} Na on mean muscle power in well-trained cyclists and although not traditionally significant ($p = 0.09$), it was shown that there may be some 'positive' effect of Na on peak muscle power, with the Na showing ~30 watts greater than the placebo [174]. However, others have shown that arm ergonometric peak power, lactate or blood PH were not different in supplemented (0.21 g·kg^{-1} of Na) college wrestlers ($n = 8$) were compared to placebo (346 vs. 354 Watts; [178]).

Currently, there is no clear evidence that Na in healthy competitive endurance athletes, competing in temperate conditions, will have an ergogenic effect. However, there is some evidence that when exercising in high temperatures, a high-dose Na (164 mmol·L) beverage ingested prior to exercise may improve exercise tolerance [165,166]. As there is evidence that training status may significantly affect the physiological impact of Na [179], further studies similar to Huw-Butler et al. [176] with a higher dose of Na (>164 mmol·L), considering training status and assessing finishing times, would further establish the real-world impact of Na on athletic performance. Despite this, there is some interesting (although not confirmed) evidence to suggest that during a period of inactivity, supplementing with Na may maintain pre-inactivity VO$_{2peak}$ [169,170,180]. If found to be replicable (and safe), this could be a useful strategy to maintain performance through short-term periods of injury for elite/highly-trained endurance athletes, but warrants considerable investigation regarding its safety as an addition to a rehabilitation program.

4.7. Selenium

Five articles fulfilled the inclusion criteria for selenium (Se). These included 124 participants (females, $n = 0$), consisting of four RCT's, one of which referred to elite athletes. There appears to be no benefit to supplementing with additional Se on athletic performance, although research is limited.

In trained cyclists, Se (200 µg·day^{-1}) over a four-week period had no additional benefit to the exercise-induced (exhaustive) increase in T or lactate accumulation [164]. Furthermore, following 10 weeks of endurance training, in previously untrained participants, consuming 180 µg of Se had no effect on mitochondrial activity, myosin heavy chain expression in muscle fibres or aerobic performance [182]. However, Se has been shown to increase glutathione peroxidase (which protects against oxidative stress) to a greater extent than a placebo, in response to exercise [183]. The absence of Se-induced exercise-adaptation to endurance exercise may not be surprising as the same group also showed that Se dampened the rate of exercise-induced mitochondrial density and overall biogenesis [184]. The authors suggested that this was potentially due to the increased cellular anti-oxidative capacity evidenced by their previous finding of increased glutathione peroxidase following exercise training [183,184]. This mechanism is logical as Se toxicity (excessive Se) can induce excessive mitochondrial oxidative stress, leading to organelle damage and dysfunction [185]. Furthermore, because mitochondrial density is a strong predictor of aerobic capacity [186], excessive Se may in fact be harmful to athletic performance, but inadequate amounts may increase exercise-induced oxidative stress over time [183,187]. However, if Se is inadequate, this may increase exercise-induced oxidative stress over time [183,187]. Savory et al. [187] investigated the effects of three weeks of sodium selenite (200 µg; supplementary form of Se) on exercise-induced (30 min at 70% VO$_{2peak}$) oxidative stress in normal weight (NW) and overweight (OW) individuals. Se had no effect on exercise-induced markers of oxidative stress in NW, but mitigated the increase in lipid hydroperoxide (a marker of fatty acid oxidation) observed in the placebo condition in OW. It is important to note that at baseline, OW had low (according to recent population data; [188–190]) Se plasma concentrations (46 µg·L) compared to NW (68 µg·L) and this may suggest that individuals deficient in Se may experience greater exercise-induced oxidative stress. In this case, Se could be a potentially useful strategy to reduce chronic exercise-induced oxidative stress, considering that 11% of athletes may have deficient Se intakes [35] and that certain dietary choices that are currently growing in popularity can lead to Se deficiency, such as the Vegan diet [25].

While there appears to be no beneficial effect of supplementing with additional Se on athletic performance, current evidence suggests that it may be important to maintain an adequate Se status and that there may be an important limit [191]. Excessive Se may be harmful to athletic performance, but inadequate amounts may increase chronic exercise-induced oxidative stress [183,187]. Nonetheless, currently there are only indications as to the detrimental effects of Se deficiency on athletic performance-related phenotypes [187]. Therefore, significantly more research is needed, particularly considering the potential deficiencies that exist in athletic populations [35].

4.8. Chromium

Twelve articles fulfilled the inclusion criteria for chromium (Cr), including 526 participants (females, n = 249). These consisted of 11 RCT's, none of which referred to elite athletes. Cr has been shown to be beneficial in several diseases [192], with a relatively small effect on reducing body mass in humans [193] and the majority of studies in athletic populations have focused on examining body composition-related outcomes. The most recent and well-conducted investigation (randomized, double-blind placebo controlled) supplemented a group of young (19 years) female swimmers with 400 µg of chromium picolinate over a 26-week competitive season [194]. Cr supplementation had no effect on body composition after 12/13 weeks, but following 26 weeks (training tapered during week 23–26), fat free mass (FFM) increased (~2.2 for Cr vs. 1.0 kg for placebo) and fat mass percentage deceased to a great extent compared to the placebo. In the largest study to investigate Cr and anthropometric changes in 154 adults (non-athletes) who were tracked (no training intervention) over 72 days, both 200 µg and 400 µg of Cr daily reduced percentage body fat (~1.4%), tended to increase FFM (~0.5 kg) and these results were slightly better in the 400 µg group [195]. Nonetheless, the vast majority of other data showed no association with markers of anthropometric or body composition change with [196–203] and without exercise intervention [198], in athletes [196,198,199] or non-athletes [197,198,200,201,203], in a range of doses (200–924 µg) and trial periods (6–13 weeks not including the longer studies of [194,195]).

There appears to be no benefit of Cr muscle strength and power [196–203], aerobic capacity [199], anaerobic capacity [199], fibre type [200,201] or insulin metabolism [199,204,205]. However, it should be noted that the trial periods in these studies were far shorter than studies showing modest body composition changes with Cr supplementation [194] and it may be that longer periods are required. The absence of good evidence for the benefit of Cr on human physiology has prompted some to question its efficacy as an essential mineral [206]. Overall, the current evidence of Cr supplementation for athletic performance is lacking, with most evidence showing no effect (with shorter trial periods <13 weeks), however further longer term studies may be warranted.

4.9. Boron

Five articles fulfilled the inclusion criteria for Boron (Br). These included 122 participants (females, *n* = 84), consisting of two RCT's, none of which referred to elite athletes. Br is an important mineral for human health [207], however little attention has been given to the context of athletic performance. The limited evidence that exists suggests that seven weeks of Br supplementation had no effect on total-T, FFM, 1RM squat or bench press in bodybuilders [208,209] and when athletes and sedentary controls are supplemented with Br (3 mg·day^{-1} for 10 months), serum phosphorus levels were lowered [210,211]. However, this effect is diminished with exercise training [211]. There is also some evidence that Br can negatively affect Mg in athletes, but not ubiquitously [210,211].

Currently, there is no evidence to support the use of Br for any athletic performance phenotype. To the authors' knowledge, there is no evidence in humans that Br has any potential in improving physiological athletic performance, although there is some evidence in animal models for improved bone metabolism [207], but not BMD in humans [210,212].

4.10. Multi Minerals

Five articles which fulfilled the inclusion criteria involved multi-minerals. These included 152 participants (females, $n = 23$), consisting of three RCT's, one of which referred to elite athletes. The articles investigated exercise/athletic performance-related variables with combinations of minerals (in a number of biological forms) inadvertently or deliberately to investigate their cumulative effect, with mixed results [51,164,213–215].

There is some evidence of improvements in 'real-world' endurance performance with multi-mineral supplementation. Del Coso et al. [213] loaded 13 triathletes with an electrolyte combination of 2580 mg of Na, 3979 mg of Cl, 756 mg of K, and 132 mg of Mg, divided into three dosing intervals, during a half ironman race and compared the results to a placebo group of 13 athletes. The supplemented group had a faster cycle speed and tendency towards faster running speed along with a quicker finishing time (by ~25 min), compared to the placebo group. Others have investigated the effects of a combination of Zn-Se supplementation on blood lactate and testosterone, compared to a placebo and Zn or Se alone over four weeks in road cyclists (with 3–4 years' experience, $n = 32$; [164]), showing no effect of Zn-Se combined on blood lactate or total and free-T pre- or post-exhaustive exercise.

Marine multi-minerals naturally enriched in calcium have also been the focus of some more recent studies. Barry et al. [214], Shea et al. [51] and Sherk et al. [215] investigated the acute effects of a mineral-rich algae compound (*Lithothamnion* species) in a number of exercise contexts. Barry et al. [214] showed that this marine-derived MTE (~12% Ca, ~1% Mg and >70 trace elements) consumed before and during a 35 km cycling time trial attenuated the exercise-induced increase in PTH in amateur athletes ($n = 35$) compared to the placebo (potentially beneficial to muscle physiology, see Ca section; [88,89]). However, there was no effect on biomarkers of bone resorption or 35 km cycling time trial performance. In a larger follow up study ($n = 51$) investigating the same supplementation and dose (1000 mg), but in a chewable form 30 minutes prior to exercise, Sherk et al. [215] found an attenuation of the exercise-induced decrease in iCa following supplementation compared to the placebo. In addition, there was a trend towards an attenuated PTH response to exercise following supplementation, but there were no effects on bone resorption. Similar results were seen in unfit postmenopausal women, where exercise (60 min of walking at 75% VO_{2peak}) resulted in a decrease of iCa and an increased PTH, but this was attenuated when supplementation commenced 60 minutes before exercise (8 vs. 26 pg·mL; [51]).

There is currently little evidence for the beneficial effects of mineral combinations from either metal element combinations or natural complexes on athletic performance. Nonetheless, there is some evidence that a combination of Na, K and Mg loading may improve aspects of half ironman performance [213]. However, real-world replication is needed with complementary controlled laboratory evidence before this strategy can be recommended. Furthermore, while there is currently some evidence of attenuation of an exercise-induced increase in PTH and decrease in iCa, there is no evidence for performance effects of mineral-rich algae compounds (*Lithothamnion* species).

5. Limitations

The present study was conducted in accordance with the PRISMA and PICO guidelines and was written and conducted with reference to the AMSTAR 2 systematic review assessment tool [216]. Nonetheless, there are some unavoidable limitations within the present review. Due to the lack of retrieved experimental data, only Iron, Calcium, Magnesium, Phosphate, Zinc, Sodium, Selenium, Boron and Chromium MTEs were reviewed. Both controlled trials and cross-sectional studies of either athletes or non-athletic populations were included. The rationale for this decision was to incorporate as much evidence as possible of athletic performance-related phenotypes. The sample consisted of only adult non-diseased populations, which may have resulted in the elimination of a limited number of semi-relevant articles involving less serious diseases or involving younger populations. Furthermore, it was not possible to perform meta-analysis due to the wide range of outcome measures investigated in studies.

6. Conclusions

Currently, there is not sufficient evidence to suggest specific guidelines to assist in formulating mineral specific dietary recommendations to improve athletic performance, other than to assess baseline mineral insufficiency and ensure adherence to RDAs (while there are currently no athlete specific guidelines). In general, the scientific evidence to support the use of mineral supplementation for sports performance is lacking in volume and quality. However, there are some notable exceptions that may be better utilised under particular physiological states (deficiencies, thermoregulatory stress etc.) rather than as general ergogenic aids. Iron and Magnesium supplementation remain the minerals with the most and highest quality research, although greater advances in randomised control trials are required. Furthermore, there is a need for the replication of some key, good quality studies investigating the efficacy of particular minerals for athletic performance. In general, the conclusion and recommendations of the present review are in agreement with the International Society of Sports Nutrition [16,18], however the present review adds to the existing literature and current knowledge of MTEs and athletic performance with a systematic review, including all recent up-to-date articles of MTEs implicated to potentially have a role in athletic performance.

7. Key Points

- Iron and Magnesium supplementation have the best quality evidence for improvements to markers and outcomes related to exercise capacity and athletic performance.
- In NAID females, oral supplementation of 100 mg day^{-1} ferrous sulfate or providing elemental Fe between 15–60 mg·day^{-1} over six-to-eight weeks may be adequate to elect performance adaptations in 3000 m running time, running velocity to lactate threshold, onset of blood lactate accumulation, quadriceps MVC fatigue resistance, post-fatiguing MVC, 4 km rowing time trial energy efficiency (kcal), preventing exercise-induced Fe loss, Hb, VO$_{2peak}$, improving 15 km cycling and two mile running time trial, quicker recovery post exercise and indicators of mood. Approximately 100 mg·day^{-1} elemental Fe over 11 weeks has been shown to result in adaptations in markers of dynamic and absolute strength.
- 300–500 mg·day^{-1} of Magnesium in the short-term (~1–4 weeks) may have a positive influence on functional dynamic measures of muscle performance (CMJ, 1RM, fatigue resistance) and longer-term (>7 weeks) benefits on quadricep torque measurements.

Supplementary Materials: The following are available online at http://www.mdpi.com/2072-6643/11/3/696/s1, Table S1: All reviewed manuscript details.

Author Contributions: Study concept and design: S.M.H., G.D.V., and G.E.C. Data extraction and analysis: S.M.H., G.E.C., and K.H. Manuscript drafting: S.M.H. and K.H. Manuscript revision and approval S.M.H., K.H., G.D.V., and G.E.C.

Funding: This research received no external funding.

Conflicts of Interest: S.H. is in receipt of grant funding from Marigot Ltd. (V1253). K.H., G.D.V., and G.C. have no conflicts of interest to declare.

Appendix A

Table A1. PRISMA check list.

Section/Topic	#	Checklist Item	Reported on Page #
TITLE			
Title	1	Identify the report as a systematic review, meta-analysis, or both.	1
ABSTRACT			
Structured summary	2	Provide a structured summary including, as applicable: background; objectives; data sources; study eligibility criteria, participants, and interventions; study appraisal and synthesis methods; results; limitations; conclusions and implications of key findings; systematic review registration number.	1
INTRODUCTION			
Rationale	3	Describe the rationale for the review in the context of what is already known.	2 onwards
Objectives	4	Provide an explicit statement of questions being addressed with reference to participants, interventions, comparisons, outcomes, and study design (PICOS).	2 onwards
METHODS			
Protocol and registration	5	Indicate if a review protocol exists, if and where it can be accessed (e.g., Web address), and, if available, provide registration information including registration number.	2–4
Eligibility criteria	6	Specify study characteristics (e.g., PICOS, length of follow-up) and report characteristics (e.g., years considered, language, publication status) used as criteria for eligibility, giving rationale.	4
Information sources	7	Describe all information sources (e.g., databases with dates of coverage, contact with study authors to identify additional studies) in the search and date last searched.	3–4
Search	8	Present full electronic search strategy for at least one database, including any limits used, such that it could be repeated.	3–4
Study selection	9	State the process for selecting studies (i.e., screening, eligibility, included in systematic review, and, if applicable, included in the meta-analysis).	4
Data collection process	10	Describe method of data extraction from reports (e.g., piloted forms, independently, in duplicate) and any processes for obtaining and confirming data from investigators.	4
Data items	11	List and define all variables for which data were sought (e.g., PICOS, funding sources) and any assumptions and simplifications made.	NA
Risk of bias in individual studies	12	Describe methods used for assessing risk of bias of individual studies (including specification of whether this was done at the study or outcome level), and how this information is to be used in any data synthesis.	4

Table A1. *Cont.*

Section/Topic	#	Checklist Item	Reported on Page #
Summary measures	13	State the principal summary measures (e.g., risk ratio, difference in means).	NA
Synthesis of results	14	Describe the methods of handling data and combining results of studies, if done, including measures of consistency (e.g., I^2) for each meta-analysis.	NA
Risk of bias across studies	15	Specify any assessment of risk of bias that may affect the cumulative evidence (e.g., publication bias, selective reporting within studies).	4
Additional analyses	16	Describe methods of additional analyses (e.g., sensitivity or subgroup analyses, meta-regression), if done, indicating which were pre-specified.	NA
RESULTS			
Study selection	17	Give numbers of studies screened, assessed for eligibility, and included in the review, with reasons for exclusions at each stage, ideally with a flow diagram.	5
Study characteristics	18	For each study, present characteristics for which data were extracted (e.g., study size, PICOS, follow-up period) and provide the citations.	5
Risk of bias within studies	19	Present data on risk of bias of each study and, if available, any outcome level assessment (see item 12).	Figure 2
Results of individual studies	20	For all outcomes considered (benefits or harms), present, for each study: (a) simple summary data for each intervention group (b) effect estimates and confidence intervals, ideally with a forest plot.	Supplementary Table S1
Synthesis of results	21	Present results of each meta-analysis done, including confidence intervals and measures of consistency.	NA
Risk of bias across studies	22	Present results of any assessment of risk of bias across studies (see Item 15).	Supplementary Table S1
Additional analysis	23	Give results of additional analyses, if done (e.g., sensitivity or subgroup analyses, meta-regression [see Item 16]).	NA
DISCUSSION			
Summary of evidence	24	Summarize the main findings including the strength of evidence for each main outcome; consider their relevance to key groups (e.g., healthcare providers, users, and policy makers).	5–25
Limitations	25	Discuss limitations at study and outcome level (e.g., risk of bias), and at review-level (e.g., incomplete retrieval of identified research, reporting bias).	25
Conclusions	26	Provide a general interpretation of the results in the context of other evidence, and implications for future research.	25–26
FUNDING			
Funding	27	Describe sources of funding for the systematic review and other support (e.g., supply of data); role of funders for the systematic review.	26

From: Moher D, Liberati A, Tetzlaff J, Altman DG, The PRISMA Group (2009). Preferred Reporting Items for Systematic Reviews and Meta-Analyses: The PRISMA Statement. PLoS Med 6(7): e1000097. doi:10.1371/journal.pmed1000097 For more information, visit: www.prisma-statement.org.

Appendix B

Example search terms (PubMed)

P

Athletes OR "sports people" OR "Athletes" [Majr] OR "Sports" [Mesh]

Calcium

I

Calcium[Title/Abstract] OR "Calcium" [Majr:NoExp] OR "Calcium, Dietary"[Majr]

C

Controls, placebo etc.

O

"athletic performance" [Title/Abstract] OR "Athletic Performance/physiology" [Majr] OR Exercise OR "Exercise/physiology" [Majr] OR "physical fitness" OR "Physical Fitness/physiology" [Majr] OR "physical function" [Title/Abstract] OR "Skeletal Muscle" [Title/Abstract] OR "Muscle, Skeletal/physiology" [Majr]

Magnesium

Magnesium [Title/Abstract] OR "Magnesium" [Majr:NoExp] OR " Magnesium, Dietary" [Majr]

Sodium

Sodium [Title/Abstract] OR "Sodium" [Majr:NoExp] OR "Sodium, Dietary" [Majr]

Potassium

Potassium [Title/Abstract] OR "Potassium" [Majr:NoExp] OR "Potassium, Dietary" [Majr]

Selenium

Selenium [Title/Abstract] OR "Selenium" [Majr:NoExp] OR "Selenium, Dietary" [Majr]

Iron

Iron [Title/Abstract] OR "Iron" [Majr:NoExp] OR "Iron, Dietary" [Majr]

Phosphorus

Phosphorus [Title/Abstract] OR "Phosphorus" [Majr:NoExp] OR "Phosphorus, Dietary" [Majr]

Phosphate

Phosphate [Title/Abstract] OR "Phosphate" [Majr:NoExp] OR "Phosphate, Dietary" [Majr]

Zinc

Zinc [Title/Abstract] OR "Zinc" [Majr:NoExp] OR "Zinc, Dietary" [Majr]

Copper

Copper [Title/Abstract] OR "Copper" [Majr:NoExp] OR "Copper, Dietary" [Majr]

Manganese

Manganese [Title/Abstract] OR "Manganese" [Majr:NoExp] OR "Manganese, Dietary" [Majr]

Iodine

Iodine [Title/Abstract] OR "Iodine" [Majr:NoExp] OR "Iodine, Dietary" [Majr]

Nickel

Nickel [Title/Abstract] OR "Nickel" [Majr:NoExp] OR "Nickel, Dietary" [Majr]

Boron

Boron [Title/Abstract] OR "Boron" [Majr:NoExp] OR "Boron, Dietary" [Majr]

Fluoride

Fluoride [Title/Abstract] OR "Fluoride" [Majr:NoExp] OR "Fluoride, Dietary" [Majr]

Cobalt

Cobalt [Title/Abstract] OR "Cobalt" [Majr:NoExp] OR "Cobalt, Dietary" [Majr]

References

1. Lazarte, C.E.; Carlsson, N.G.; Almgren, A.; Sandberg, A.S.; Granfeldt, Y. Phytate, zinc, iron and calcium content of common Bolivian food, and implications for mineral bioavailability. *J. Food Compos. Anal.* **2015**, *39*, 111–119. [CrossRef]
2. Gibson, R.S.; Bailey, K.B.; Gibbs, M.; Ferguson, E.L. A review of phytate, iron, zinc, and calcium concentrations in plant-based complementary foods used in low-income countries and implications for bioavailability. *Food Nutr. Bull.* **2010**, *31*, S134–S146. [CrossRef] [PubMed]
3. Gupta, U.; Gupta, S. Sources and deficiency diseases of mineral nutrients in human health and nutrition: A review. *Pedosphere* **2014**, *24*, 13–38. [CrossRef]
4. Asemi, Z.; Jamilian, M.; Mesdaghinia, E.; Esmaillzadeh, A. Effects of selenium supplementation on glucose homeostasis, inflammation, and oxidative stress in gestational diabetes: Randomized, double-blind, placebo-controlled trial. *Nutrition (Burbank, Los Angeles County, Calif.)* **2015**, *31*, 1235–1242. [CrossRef] [PubMed]
5. Oropeza-Moe, M.; Wisloff, H.; Bernhoft, A. Selenium deficiency associated porcine and human cardiomyopathies. *J. Trace Elem. Med. Biol.* **2015**, *31*, 148–156. [CrossRef]
6. Del Gobbo, L.C.; Imamura, F.; Wu, J.H.Y.; de Oliveira-Otto, M.C.; Chiuve, S.E.; Mozaffarian, D. Circulating and dietary magnesium and risk of cardiovascular disease: A systematic review and meta-analysis of prospective studies. *Am. J. Clin. Nutr.* **2013**, *98*, 160–173. [CrossRef]
7. Cheungpasitporn, W.; Thongprayoon, C.; Mao, M.A.; Srivali, N.; Ungprasert, P.; Varothai, N.; Sanguankeo, A.; Kittanamongkolchai, W.; Erickson, S.B. Hypomagnesaemia linked to depression: A systematic review and meta-analysis. *Intern. Med. J.* **2015**, *45*, 436–440. [CrossRef] [PubMed]
8. Fang, X.; Han, H.; Li, M.; Liang, C.; Fan, Z.; Aaseth, J.; He, J.; Montgomery, S.; Cao, Y. Dose-response relationship between dietary magnesium intake and risk of type 2 diabetes mellitus: A systematic review and meta-regression analysis of prospective cohort studies. *Nutrients* **2016**, *8*, 739. [CrossRef]
9. Joosten, M.M.; Gansevoort, R.T.; Bakker, S.J. Low plasma magnesium and risk of developing chronic kidney disease: Results from the PREVEND Study. *Kidney Int.* **2015**, *87*, 1262–1263. [CrossRef] [PubMed]
10. Shaikh, M.N.; Malapati, B.R.; Gokani, R.; Patel, B.; Chatriwala, M. Serum magnesium and vitamin D levels as indicators of asthma severity. *Pulm. Med.* **2016**, *2016*, 1–5. [CrossRef]
11. Rude, R.K.; Singer, F.R.; Gruber, H.E. Skeletal and hormonal effects of magnesium deficiency. *J. Am. Coll. Nutr.* **2009**, *28*, 131–141. [CrossRef]
12. Kunutsor, S.K.; Whitehouse, M.R.; Blom, A.W.; Laukkanen, J.A. Low serum magnesium levels are associated with increased risk of fractures: A long-term prospective cohort study. *Eur. J. Epidemiol.* **2017**, *32*, 593–603. [CrossRef]
13. Zhang, Y.; Xun, P.; Wang, R.; Mao, L.; He, K. Can Magnesium Enhance Exercise Performance? *Nutrients* **2017**, *9*, 946. [CrossRef]
14. Speich, M.; Pineau, A.; Ballereau, F. Minerals, trace elements and related biological variables in athletes and during physical activity. *Clin. Chim. Acta* **2001**, *312*, 1–11. [CrossRef]
15. Williams, M.H. Dietary supplements and sports performance: Minerals. *J. Int. Soc. Sport Nutr.* **2005**, *2*, 43–49. [CrossRef] [PubMed]
16. Kreider, R.B.; Wilborn, C.D.; Taylor, L.; Campbell, B.; Almada, A.L.; Collins, R.; Cooke, M.; Earnest, C.P.; Greenwood, M.; Kalman, D.S. ISSN exercise and sport nutrition review: Research and recommendations. *J. Int. Soc. Sports Nutr.* **2010**, *7–50*, 7. [CrossRef]
17. Misner, B. Food alone may not provide sufficient micronutrients for preventing deficiency. *J. Int. Soc. Sports Nutr.* **2006**, *3*, 51–55. [CrossRef]
18. Kerksick, C.M.; Wilborn, C.D.; Roberts, M.D.; Smith-Ryan, A.; Kleiner, S.M.; Jager, R.; Collins, R.; Cooke, M.; Davis, J.N.; Galvan, E.; et al. ISSN exercise & sports nutrition review update: Research & recommendations. *J. Int. Soc. Sports Nutr.* **2018**, *15*, 38. [PubMed]
19. Bailey, R.L.; West, K.P.; Black, R.E. The epidemiology of global micronutrient deficiencies. *Ann. Nutr. Metab.* **2015**, *66*, 22–33. [CrossRef] [PubMed]
20. Kumssa, D.B.; Joy, E.J.; Ander, E.L.; Watts, M.J.; Young, S.D.; Walker, S.; Broadley, M.R. Dietary calcium and zinc deficiency risks are decreasing but remain prevalent. *Sci. Rep.* **2015**, *5*, 10974. [CrossRef]

21. Gröber, U.; Schmidt, J.; Kisters, K. Magnesium in Prevention and Therapy. *Nutrients* **2015**, *7*, 8199–8226. [CrossRef] [PubMed]

22. Moshfegh, A.; Goldman, J.; Ahuja, J.; Rhodes, D.; LaComb, R. *What We Eat in America, NHANES 2005–2006: Usual Nutrient Intakes from Food and Water Compared to 1997 Dietary Reference Intakes for Vitamin D, Calcium, Phosphorus, and Magnesium*; USDA: Washington, DC, USA, 2009. Available online: http://www.ars.usda.gov/ba/bhnrc/fsrg (accessed on 18 February 2018).

23. Calton, J.B. Prevalence of micronutrient deficiency in popular diet plans. *J. Int. Soc. Sports Nutr.* **2010**, *7*, 24. [CrossRef]

24. Engel, M.; Kern, H.; Brenna, J.T.; Mitmesser, S. Micronutrient gaps in three commercial weight-loss diet plans. *Nutrients* **2018**, *10*, 108. [CrossRef] [PubMed]

25. Kristensen, N.B.; Madsen, M.L.; Hansen, T.H.; Allin, K.H.; Hoppe, C.; Fagt, S.; Lausten, M.S.; Gobel, R.J.; Vestergaard, H.; Hansen, T.; et al. Intake of macro- and micronutrients in Danish vegans. *Nutr. J.* **2015**, *14*, 115. [CrossRef]

26. Castro-Quezada, I.; Roman-Vinas, B.; Serra-Majem, L. The Mediterranean diet and nutritional adequacy: A review. *Nutrients* **2014**, *6*, 231–248. [CrossRef]

27. Birkenhead, K.L.; Slater, G. A Review of Factors Influencing Athletes' Food Choices. *Sports Med.* **2015**, *45*, 1511–1522. [CrossRef] [PubMed]

28. Volpe, S.L. Magnesium and the athlete. *Curr. Sports Med. Rep.* **2015**, *14*, 279–283. [CrossRef] [PubMed]

29. Maynar, M.; Llerena, F.; Bartolome, I.; Alves, J.; Robles, M.C.; Grijota, F.J.; Munoz, D. Seric concentrations of copper, chromium, manganesum, nickel and selenium in aerobic, anaerobic and mixed professional sportsmen. *J. Int. Soc. Sports Nutr.* **2018**, *15*, 8. [CrossRef]

30. Maynar, M.; Munoz, D.; Alves, J.; Barrientos, G.; Grijota, F.J.; Robles, M.C.; Llerena, F. Influence of an Acute Exercise Until Exhaustion on Serum and Urinary Concentrations of Molybdenum, Selenium, and Zinc in Athletes. *Biol. Trace Elem. Res.* **2018**, *186*, 1327–1329. [CrossRef]

31. Wardenaar, F.C.; Ceelen, I.J.; Van Dijk, J.W.; Hangelbroek, R.W.; Van Roy, L.; Van der Pouw, B.; De Vries, J.H.; Mensink, M.; Witkamp, R.F. Nutritional supplement use by dutch elite and sub-elite athletes: Does receiving dietary counseling make a difference? *Int. J. Sport Nutr. Exerc. Metab.* **2017**, *27*, 32–42. [CrossRef]

32. Harrington, J.; Perry, I.; Lutomski, J.; Morgan, K.; McGee, H.; Shelley, E.; Watson, D.; Barry, M. LÁN 2007: Survey of Lifestyle, A itudes and Nutrition in Ireland. Dietary Habits of the Irish Population. Dep Health Child 2008. Available online: http://epubs.rcsi.ie/psycholrep/6 (accessed on 18 February 2018).

33. Wierniuk, A.; Wlodarek, D. Estimation of energy and nutritional intake of young men practicing aerobic sports. *Roczniki Panstwowego Zakladu Higieny* **2013**, *64*, 143–148.

34. Heaney, S.; O'Connor, H.; Gifford, J.; Naughton, G. Comparison of strategies for assessing nutritional adequacy in elite female athletes' dietary intake. *Int. J. Sport Nutr. Exer. Metab.* **2010**, *20*, 245–256. [CrossRef]

35. Wardenaar, F.; Brinkmans, N.; Ceelen, I.; Van Rooij, B.; Mensink, M.; Witkamp, R.; De Vries, J. Micronutrient Intakes in 553 Dutch Elite and Sub-Elite Athletes: Prevalence of Low and High Intakes in Users and Non-Users of Nutritional Supplements. *Nutrients* **2017**, *9*, 142. [CrossRef]

36. Mir-Marques, A.; Cervera, M.L.; de la Guardia, M. Mineral analysis of human diets by spectrometry methods. *TrAC Trend Anal. Chem.* **2016**, *82*, 457–467. [CrossRef]

37. Freeland-Graves, J.H.; Sanjeevi, N.; Lee, J.J. Global perspectives on trace element requirements. *J. Trace Elem. Med. Biol.* **2015**, *31*, 135–141. [CrossRef]

38. Larson-Meyer, D.E.; Woolf, K.; Burke, L. Assessment of Nutrient Status in Athletes and the Need for Supplementation. *Int. J. Sport Nutr. Exerc. Metab.* **2018**, *28*, 139–158. [CrossRef] [PubMed]

39. Lee, N. A review of magnesium, iron, and zinc supplementation effects on athletic performance. *Korean J. Phys. Edu* **2017**, *56*, 797–806. [CrossRef]

40. Knapik, J.J.; Steelman, R.A.; Hoedebecke, S.S.; Austin, K.G.; Farina, E.K.; Lieberman, H.R. Prevalence of dietary supplement use by athletes: Systematic review and meta-analysis. *Sports Med.* **2016**, *46*, 103–123. [CrossRef] [PubMed]

41. Chu, A.; Holdaway, C.; Varma, T.; Petocz, P.; Samman, S. Lower serum zinc concentration despite higher dietary zinc intake in athletes: A systematic review and meta-analysis. *Sports Med.* **2018**, *48*, 327–336. [CrossRef] [PubMed]

42. Burden, R.J.; Morton, K.; Richards, T.; Whyte, G.P.; Pedlar, C.R. Is iron treatment beneficial in, iron-deficient but non-anaemic (IDNA) endurance athletes? a systematic review and meta-analysis. *Br. J. Sports Med.* **2015**, *49*, 1389–1397. [CrossRef] [PubMed]

43. Rubeor, A.; Goojha, C.; Manning, J.; White, J. Does iron supplementation improve performance in iron-deficient nonanemic athletes? *Sports Health* **2018**. Ahead of print. [CrossRef]

44. Peeling, P.; Binnie, M.J.; Goods, P.S.R.; Sim, M.; Burke, L.M. Evidence-based supplements for the enhancement of athletic performance. *Int. J. Sport Nutr. Exerc. Metab.* **2018**, *28*, 178–187. [CrossRef]

45. Newhouse, I.J.; Finstad, E.W. The effects of magnesium supplementation on exercise performance. *Clin. J. Sport Med.* **2000**, *10*, 195–200. [CrossRef]

46. Sieja, K.; von Mach-Szczypińska, J.; Kois, N.; Ler, P.; Piechanowska, K.; Stolarska, M. Influence of selenium on oxidative stress in athletes. *Cent. Eur. J. Sport Sci. Med.* **2016**, *14*, 87–92.

47. Rayssiguier, Y.; Guezennec, C.Y.; Durlach, J. New experimental and clinical data on the relationship between magnesium and sport. *Magn. Res.* **1990**, *3*, 93–102.

48. Armstrong, L.E.; Maresh, C.M. Vitamin and mineral supplements as nutritional aids to exercise performance and health. *Nutr. Rev.* **1996**, *54*, S149–S158. [CrossRef] [PubMed]

49. Liberati, A.; Altman, D.G.; Tetzlaff, J.; Mulrow, C.; Gotzsche, P.C.; Ioannidis, J.P.; Clarke, M.; Devereaux, P.J.; Kleijnen, J.; Moher, D. The PRISMA statement for reporting systematic reviews and meta-analyses of studies that evaluate healthcare interventions: Explanation and elaboration. *PLos Med.* **2009**, *6*, e1000100. [CrossRef] [PubMed]

50. O'Connor, D.; Green, S.; Higgins, J.P. *Defining the Review Question and Developing Criteria for Including Studies*; Cochrane Book Series; Wiley Online Library: Hoboken, NJ, USA, 2008; pp. 81–94.

51. Shea, K.L.; Barry, D.W.; Sherk, V.D.; Hansen, K.C.; Wolfe, P.; Kohrt, W.M. Calcium supplementation and parathyroid hormone response to vigorous walking in postmenopausal women. *Med. Sci. Sports Exerc.* **2014**, *46*, 2007–2013. [CrossRef] [PubMed]

52. Moëzzi, N.; Peeri, M.; Matin, H. Effects of zinc, magnesium and vitamin B6 supplementation on hormones and performance in weightlifters. *Ann. Biol. Res.* **2013**, *4*, 163–168.

53. Thomas, B.; Ciliska, D.; Dobbins, M.; Micucci, S. A process for systematically reviewing the literature: Providing the research evidence for public health nursing interventions. *Worldv. Evid. Based Nurs.* **2004**, *1*, 176–184. [CrossRef]

54. DellaValle, D.M.; Haas, J.D. Iron supplementation improves energetic efficiency in iron-depleted female rowers. *Med. Sci. Sports Exerc.* **2014**, *46*, 1204–1215. [CrossRef] [PubMed]

55. Villanueva, J.; Soria, M.; Gonzalez-Haro, C.; Ezquerra, L.; Nieto, J.L.; Escanero, J.F. Oral iron treatment has a positive effect on iron metabolism in elite soccer players. *Biol. Trace Elem. Res.* **2011**, *142*, 398–406. [CrossRef] [PubMed]

56. Blee, T.; Goodman, C.; Dawson, B.; Stapff, A. The effect of intramuscular iron injections on serum ferritin levels and physical performance in elite netballers. *J. Sci. Med. Sport* **1999**, *2*, 311–321. [CrossRef]

57. Powell, P.D.; Tucker, A. Iron supplementation and running performance in female cross-country runners. *Int. J. Sports Med.* **1991**, *12*, 462–467. [CrossRef] [PubMed]

58. Ashenden, M.J.; Pyne, D.B.; Parisotto, R.; Dobson, G.P.; Hahn, A.G. Can reticulocyte parameters be of use in detecting iron deficient erythropoiesis in female athletes? *J. Sports Med. Phys. Fit.* **1999**, *39*, 140–146.

59. Haymes, E.M.; Puhl, J.L.; Temples, T.E. Training for cross-country skiing and iron status. *Med. Sci. Sports Exerc.* **1986**, *18*, 162–167. [CrossRef] [PubMed]

60. Govus, A.D.; Garvican-Lewis, L.A.; Abbiss, C.R.; Peeling, P.; Gore, C.J. Pre-altitude serum ferritin levels and daily oral iron supplement dose mediate iron parameter and hemoglobin mass responses to altitude exposure. *PLoS ONE* **2015**, *10*, e0135120. [CrossRef] [PubMed]

61. Flynn, M.G.; Mackinnon, L.; Gedge, V.; Fahlman, M.; Brickman, T. Influence of iron status and iron supplements on natural killer cell activity in trained women runners. *Int. J. Sports Med.* **2003**, *24*, 217–222. [CrossRef] [PubMed]

62. Hinton, P.S.; Sinclair, L.M. Iron supplementation maintains ventilatory threshold and improves energetic efficiency in iron-deficient nonanemic athletes. *Eur. J. Clin. Nutr.* **2007**, *61*, 30–39. [CrossRef]

63. LaManca, J.J.; Haymes, E.M. Effects of iron repletion on VO_{2max}, endurance, and blood lactate in women. *Med. Sci. Sports Exerc.* **1993**, *25*, 1386–1392. [CrossRef]

64. Dressendorfer, R.H.; Keen, C.L.; Wade, C.E.; Claybaugh, J.R.; Timmis, G.C. Development of runner's anemia during a 20-day road race: Effect of iron supplements. *Int. J. Sports Med.* **1991**, *12*, 332–336. [CrossRef] [PubMed]

65. Friedmann, B.; Jost, J.; Rating, T.; Weller, E.; Werle, E.; Eckardt, K.U.; Bartsch, P.; Mairbaurl, H. Effects of iron supplementation on total body hemoglobin during endurance training at moderate altitude. *Int. J. Sports Med.* **1999**, *20*, 78–85. [CrossRef]

66. Jensen, C.A.; Weaver, C.M.; Sedlock, D.A. Iron supplementation and iron status in exercising young women. *J. Nutr. Biochem.* **1991**, *2*, 368–373. [CrossRef]

67. Fogelholm, M.; Jaakkola, L.; Lampisjarvi, T. Effects of iron supplementation in female athletes with low serum ferritin concentration. *Int. J. Sports Med.* **1992**, *13*, 158–162. [CrossRef] [PubMed]

68. Klingshirn, L.A.; Pate, R.R.; Bourque, S.P.; Davis, J.M.; Sargent, R.G. Effect of iron supplementation on endurance capacity in iron-depleted female runners. *Med. Sci. Sports Exerc.* **1992**, *24*, 819–824. [CrossRef]

69. Hegenauer, J.; Strause, L.; Saltman, P.; Dann, D.; White, J.; Green, R. Transitory hematologic effects of moderate exercise are not influenced by iron supplementation. *Eur. J. Appl. Physiol. Occup. Physiol.* **1983**, *52*, 57–61. [CrossRef]

70. Magazanik, A.; Weinstein, Y.; Abarbanel, J.; Lewinski, U.; Shapiro, Y.; Inbar, O.; Epstein, S. Effect of an iron supplement on body iron status and aerobic capacity of young training women. *Eur. J. Appl. Physiol. Occup. Physiol.* **1991**, *62*, 317–323. [CrossRef]

71. Hinton, P.S.; Giordano, C.; Brownlie, T.; Haas, J.D. Iron supplementation improves endurance after training in iron-depleted, nonanemic women. *J. Appl. Physiol.* **2000**, *88*, 1103–1111. [CrossRef]

72. McClung, J.P.; Karl, J.P.; Cable, S.J.; Williams, K.W.; Nindl, B.C.; Young, A.J.; Lieberman, H.R. Randomized, double-blind, placebo-controlled trial of iron supplementation in female soldiers during military training: Effects on iron status, physical performance, and mood. *Am. J. Clin. Nutr.* **2009**, *90*, 124–131. [CrossRef]

73. Yoshida, T.; Udo, M.; Chida, M.; Ichioka, M.; Makiguchi, K. Dietary iron supplement during severe physical training in competitive female distance runners. *Res. Sport Med.* **1990**, *1*, 279–285. [CrossRef]

74. Brutsaert, T.D.; Hernandez-Cordero, S.; Rivera, J.; Viola, T.; Hughes, G.; Haas, J.D. Iron supplementation improves progressive fatigue resistance during dynamic knee extensor exercise in iron-depleted, nonanemic women. *Am. J. Clin. Nutr.* **2003**, *77*, 441–448. [CrossRef]

75. Mielgo-Ayuso, J.; Zourdos, M.C.; Calleja-Gonzalez, J.; Urdampilleta, A.; Ostojic, S. Iron supplementation prevents a decline in iron stores and enhances strength performance in elite female volleyball players during the competitive season. *Appl. Physiol. Nutr. Metab. Physiol. Appl. Nutr. Metab.* **2015**, *40*, 615–622. [CrossRef]

76. Ohira, Y.; Edgerton, V.R.; Gardner, G.W.; Senewiratne, B.; Barnard, R.J.; Simpson, D.R. Work capacity, heart rate and blood lactate responses to iron treatment. *Br. J. Haematol.* **1979**, *41*, 365–372. [CrossRef]

77. Garvican, L.A.; Saunders, P.U.; Cardoso, T.; Macdougall, I.C.; Lobigs, L.M.; Fazakerley, R.; Fallon, K.E.; Anderson, B.; Anson, J.M.; Thompson, K.G.; et al. Intravenous iron supplementation in distance runners with low or suboptimal ferritin. *Med. Sci. Sports Exerc.* **2014**, *46*, 376–385. [CrossRef] [PubMed]

78. Peeling, P.; Blee, T.; Goodman, C.; Dawson, B.; Claydon, G.; Beilby, J.; Prins, A. Effect of iron injections on aerobic-exercise performance of iron-depleted female athletes. *Int. J. Sport Nutr. Exerc. Metab.* **2007**, *17*, 221–231. [CrossRef] [PubMed]

79. Woods, A.; Garvican-Lewis, L.A.; Saunders, P.U.; Lovell, G.; Hughes, D.; Fazakerley, R.; Anderson, B.; Gore, C.J.; Thompson, K.G. Four weeks of IV iron supplementation reduces perceived fatigue and mood disturbance in distance runners. *PLoS ONE* **2014**, *9*, e108042. [CrossRef]

80. Townsend, R.; Elliott-Sale, K.J.; Pinto, A.J.; Thomas, C.; Scott, J.P.; Currell, K.; Fraser, W.D.; Sale, C. Parathyroid hormone secretion Is controlled by both ionized calcium and phosphate during exercise and recovery in men. *J. Clin. Endocrinol. Metab.* **2016**, *101*, 3231–3239. [CrossRef] [PubMed]

81. Martin, B.R.; Davis, S.; Campbell, W.W.; Weaver, C.M. Exercise and calcium supplementation: Effects on calcium homeostasis in sportswomen. *Med. Sci. Sports Exerc.* **2007**, *39*, 1481–1486. [CrossRef] [PubMed]

82. Frontera, W.R.; Ochala, J. Skeletal muscle: A brief review of structure and function. *Calcif. Tissue Int.* **2015**, *96*, 183–195. [CrossRef] [PubMed]

83. Pu, F.; Chen, N.; Xue, S. Calcium intake, calcium homeostasis and health. *Food Sci. Hum. Well* **2016**, *5*, 8–16. [CrossRef]

84. Zorbas, Y.G.; Petrov, K.L.; Kakurin, V.J.; Kuznetsov, N.A.; Charapakhin, K.P.; Alexeyev, I.D.; Denogradov, S.D. Calcium supplementation effect on calcium balance in endurance-trained athletes during prolonged hypokinesia and ambulatory conditions. *Biol. Trace Elem. Res.* **2000**, *73*, 231–250. [CrossRef]

85. Zorbas, Y.G.; Kakurin, V.J.; Kuznetsov, N.A.; Yarullin, V.L.; Andreyev, I.D.; Charapakhin, K.P. Measurements in calcium-supplemented athletes during and after hypokinetic and ambulatory conditions. *Clin. Biochem.* **2000**, *33*, 393–404. [CrossRef]

86. Kohrt, W.M.; Wherry, S.J.; Wolfe, P.; Sherk, V.D.; Wellington, T.; Swanson, C.M.; Weaver, C.M.; Boxer, R.S. Maintenance of serum ionized calcium during exercise attenuates parathyroid hormone and bone resorption responses. *J. Bone Miner. Res. Off. J. Am. Soc. Bone Miner. Res.* **2018**, *33*, 1326–1334. [CrossRef] [PubMed]

87. Guillemant, J.; Accarie, C.; Peres, G.; Guillemant, S. Acute effects of an oral calcium load on markers of bone metabolism during endurance cycling exercise in male athletes. *Calcif. Tissue Int.* **2004**, *74*, 407–414. [CrossRef]

88. Visser, M.; Deeg, D.J.; Lips, P. Low vitamin D and high parathyroid hormone levels as determinants of loss of muscle strength and muscle mass (sarcopenia): The Longitudinal Aging Study Amsterdam. *J. Clin. Endocrinol. Metab.* **2003**, *88*, 5766–5772. [CrossRef] [PubMed]

89. Passeri, E.; Bugiardini, E.; Sansone, V.A.; Valaperta, R.; Costa, E.; Ambrosi, B.; Meola, G.; Corbetta, S. Vitamin D, parathyroid hormone and muscle impairment in myotonic dystrophies. *J. Neurol. Sci.* **2013**, *331*, 132–135. [CrossRef]

90. Sikjaer, T.; Rolighed, L.; Hess, A.; Fuglsang-Frederiksen, A.; Mosekilde, L.; Rejnmark, L. Effects of PTH(1-84) therapy on muscle function and quality of life in hypoparathyroidism: Results from a randomized controlled trial. *Osteoporos. Int.* **2014**, *25*, 1717–1726. [CrossRef]

91. Haakonssen, E.C.; Ross, M.L.; Knight, E.J.; Cato, L.E.; Nana, A.; Wluka, A.E.; Cicuttini, F.M.; Wang, B.H.; Jenkins, D.G.; Burke, L.M. The effects of a calcium-rich pre-exercise meal on biomarkers of calcium homeostasis in competitive female cyclists: A randomised crossover trial. *PLoS ONE* **2015**, *10*, e0123302. [CrossRef] [PubMed]

92. Cinar, V.; Baltaci, A.K.; Mogulkoc, R.; Kilic, M. Testosterone levels in athletes at rest and exhaustion: Effects of calcium supplementation. *Biol. Trace Elem. Res.* **2009**, *129*, 65–69. [CrossRef]

93. Cinar, V.; Baltaci, A.K.; Mogulkoc, R. Effect of exhausting exercise and calcium supplementation on potassium, magnesium, copper, zinc and calcium levels in athletes. *Pak. J. Med. Sci.* **2009**, *25*, 238–242.

94. Cinar, V.; Mogulkoc, R.; Baltaci, A.K.; Bostanci, O. Effects of calcium supplementation on glucose and insulin levels of athletes at rest and after exercise. *Biol. Trace Elem. Res.* **2010**, *133*, 29–33. [CrossRef]

95. Cinar, V.; Mogulkoc, R.; Baltaci, A.K. Calcium supplementation and 4-week exercise on blood parameters of athletes at rest and exhaustion. *Biol. Trace Elem. Res.* **2010**, *134*, 130–135. [CrossRef]

96. Cinar, V.; Cakmakci, O.; Mogulkoc, R.; Baltaci, A.K. Effects of exhaustion and calcium supplementation on adrenocorticotropic hormone and cortisol levels in athletes. *Biol. Trace Elem. Res.* **2009**, *127*, 1–5. [CrossRef] [PubMed]

97. Farrell, P.A.; Garthwaite, T.L.; Gustafson, A.B. Plasma adrenocorticotropin and cortisol responses to submaximal and exhaustive exercise. *J. Appl. Physiol. Respir. Environ. Exerc. Physiol.* **1983**, *55*, 1441–1444. [CrossRef] [PubMed]

98. Burt, M.G.; Mangelsdorf, B.L.; Srivastava, D.; Petersons, C.J. Acute effect of calcium citrate on serum calcium and cardiovascular function. *J. Bone Miner. Res. Off. J. Am. Soc. Bone Miner. Res.* **2013**, *28*, 412–418. [CrossRef]

99. Rios, E. Calcium-induced release of calcium in muscle: 50 years of work and the emerging consensus. *J. Gener. Physiol.* **2018**, *150*, 521–537. [CrossRef] [PubMed]

100. Welch, A.A.; Kelaiditi, E.; Jennings, A.; Steves, C.J.; Spector, T.D.; MacGregor, A. Dietary magnesium Is positively associated with skeletal muscle power and indices of muscle mass and may attenuate the association between circulating C-Reactive Protein and muscle mass in women. *J. Bone Miner. Res. Off. J. Am. Soc. Bone Miner. Res.* **2016**, *31*, 317–325. [CrossRef]

101. Welch, A.A.; Skinner, J.; Hickson, M. Dietary magnesium may be protective for aging of bone and skeletal muscle in middle and younger older age men and women: Cross-sectional findings from the UK biobank cohort. *Nutrients* **2017**, *9*, 1189. [CrossRef] [PubMed]

102. Wang, R.; Chen, C.; Liu, W.; Zhou, T.; Xun, P.; He, K.; Chen, P. The effect of magnesium supplementation on muscle fitness: A meta-analysis and systematic review. *Magn. Res.* **2017**, *30*, 120–132.

103. Brilla, L.R.; Haley, T.F. Effect of magnesium supplementation on strength training in humans. *J. Am. Coll. Nutr.* **1992**, *11*, 326–329. [CrossRef]

104. Dmitrasinovic, G.; Pesic, V.; Stanic, D.; Plecas-Solarovic, B.; Dajak, M.; Ignjatovic, S. ACTH, cortisol and IL-6 Levels in athletes following magnesium supplementation. *J. Med. Biochem.* **2016**, *35*, 375–384. [CrossRef]

105. Kass, L.S.; Skinner, P.; Poeira, F. A pilot study on the effects of magnesium supplementation with high and low habitual dietary magnesium intake on resting and recovery from aerobic and resistance exercise and systolic blood pressure. *J. Sports Sci. Med.* **2013**, *12*, 144–150. [PubMed]

106. Finstad, E.W.; Newhouse, I.J.; Lukaski, H.C.; McAuliffe, J.E.; Stewart, C.R. The effects of magnesium supplementation on exercise performance. *Med. Sci. Sports Exerc.* **2001**, *33*, 493–498. [CrossRef] [PubMed]

107. Setaro, L.; Santos-Silva, P.R.; Nakano, E.Y.; Sales, C.H.; Nunes, N.; Greve, J.M.; Colli, C. Magnesium status and the physical performance of volleyball players: Effects of magnesium supplementation. *J. Sports Sci.* **2014**, *32*, 438–445. [CrossRef]

108. Zorbas, Y.G.; Kakurin, V.J.; Afonin, V.B.; Charapakhin, K.P.; Denogradov, S.D. Magnesium supplements' effect on magnesium balance in athletes during prolonged restriction of muscular activity. *Kidney Blood Press. Res.* **1999**, *22*, 146–153. [CrossRef]

109. Molina-Lopez, J.; Molina, J.M.; Chirosa, L.J.; Florea, D.; Saez, L.; Millan, E.; Planells, E. Association between erythrocyte concentrations of magnesium and zinc in high-performance handball players after dietary magnesium supplementation. *Magn. Res.* **2012**, *25*, 79–88.

110. Cordova Martinez, A.; Fernandez-Lazaro, D.; Mielgo-Ayuso, J.; Seco Calvo, J.; Caballero Garcia, A. Effect of magnesium supplementation on muscular damage markers in basketball players during a full season. *Magn. Res.* **2017**, *30*, 61–70.

111. Porta, S.; Gell, H.; Pichlkastner, K.; Porta, J.; von Ehrlich, B.; Vormann, J.; Stossier, H.; Kisters, K. A system of changes of ionized blood Mg through sports and supplementation. *Trace Elem. Electrolytes* **2013**, *30*, 105–107. [CrossRef]

112. Zorbas, Y.G.; Kakurin, A.G.; Kuznetsov, N.K.; Federov, M.A.; Yaroshenko, Y.Y. Magnesium loading effect on magnesium deficiency in endurance-trained subjects during prolonged restriction of muscular activity. *Biol. Trace Elem. Res.* **1998**, *63*, 149–166. [CrossRef] [PubMed]

113. Terblanche, S.; Noakes, T.D.; Dennis, S.C.; Marais, D.; Eckert, M. Failure of magnesium supplementation to influence marathon running performance or recovery in magnesium-replete subjects. *Int. J. Sport Nutr.* **1992**, *2*, 154–164. [CrossRef] [PubMed]

114. Weller, E.; Bachert, P.; Meinck, H.M.; Friedmann, B.; Bartsch, P.; Mairbaurl, H. Lack of effect of oral Mg-supplementation on Mg in serum, blood cells, and calf muscle. *Med. Sci. Sports Exerc.* **1998**, *30*, 1584–1591. [CrossRef]

115. Darooghegi Mofrad, M.; Djafarian, K.; Mozaffari, H.; Shab-Bidar, S. Effect of magnesium supplementation on endothelial function: A systematic review and meta-analysis of randomized controlled trials. *Atherosclerosis* **2018**, *273*, 98–105. [CrossRef] [PubMed]

116. Kass, L.S.; Poeira, F. The effect of acute vs chronic magnesium supplementation on exercise and recovery on resistance exercise, blood pressure and total peripheral resistance on normotensive adults. *J. Int. Soc. Sports Nutr.* **2015**, *12*, 19. [CrossRef]

117. Veronese, N.; Berton, L.; Carraro, S.; Bolzetta, F.; De Rui, M.; Perissinotto, E.; Toffanello, E.D.; Bano, G.; Pizzato, S.; Miotto, F.; et al. Effect of oral magnesium supplementation on physical performance in healthy elderly women involved in a weekly exercise program: A randomized controlled trial. *Am. J. Clin. Nutr.* **2014**, *100*, 974–981. [CrossRef]

118. Petrovic, J.; Stanic, D.; Dmitrasinovic, G.; Plecas-Solarovic, B.; Ignjatovic, S.; Batinic, B.; Popovic, D.; Pesic, V. Magnesium supplementation diminishes peripheral blood lymphocyte DNA oxidative damage in athletes and sedentary young man. *Oxid. Med. Cell. Longev.* **2016**, *2016*, 2019643. [CrossRef] [PubMed]

119. Margaritelis, N.V.; Theodorou, A.A.; Paschalis, V.; Veskoukis, A.S.; Dipla, K.; Zafeiridis, A.; Panayiotou, G.; Vrabas, I.S.; Kyparos, A.; Nikolaidis, M.G. Adaptations to endurance training depend on exercise-induced oxidative stress: Exploiting redox interindividual variability. *Acta Physiol.* **2018**, *222*, e12898. [CrossRef]

120. Hawley, J.A.; Lundby, C.; Cotter, J.D.; Burke, L.M. Maximizing cellular adaptation to endurance exercise in skeletal muscle. *Cell Metab.* **2018**, *27*, 962–976. [CrossRef]

121. Jaimovich, E.; Casas, M. Evaluating the essential role of RONS in vivo in exercised human muscle. *Acta Physiol.* **2018**, *222*, e12972. [CrossRef] [PubMed]

122. Cinar, V.; Polat, Y.; Baltaci, A.K.; Mogulkoc, R. Effects of magnesium supplementation on testosterone levels of athletes and sedentary subjects at rest and after exhaustion. *Biol. Trace Elem. Res.* **2011**, *140*, 18–23. [CrossRef]

123. Cinar, V.; Mogulkoc, R.; Baltaci, A.K.; Nizamlioglu, M. Effect of magnesium supplementation on some plasma elements in athletes at rest and exhaustion. *Biol. Trace Elem. Res.* **2007**, *119*, 97–102. [CrossRef] [PubMed]

124. Cinar, V.; Nizamlioglu, M.; Mogulkoc, R.; Baltaci, A.K. Effects of magnesium supplementation on blood parameters of athletes at rest and after exercise. *Biol. Trace Elem. Res.* **2007**, *115*, 205–212. [CrossRef]

125. Cinar, V.; Mogulkoc, R.; Baltaci, A.K.; Polat, Y. Adrenocorticotropic hormone and cortisol levels in athletes and sedentary subjects at rest and exhaustion: Effects of magnesium supplementation. *Biol. Trace Elem. Res.* **2008**, *121*, 215–220. [CrossRef] [PubMed]

126. Cinar, V.; Polat, Y.; Mogulkoc, R.; Nizamlioglu, M.; Baltaci, A.K. The effect of magnesium supplementation on glucose and insulin levels of tae-kwan-do sportsmen and sedentary subjects. *Pak. J. Pharm. Sci.* **2008**, *21*, 237–240. [PubMed]

127. Parazzini, F.; Di Martino, M.; Pellegrino, P. Magnesium in the gynecological practice: A literature review. *Magn. Res.* **2017**, *30*, 1–7.

128. Kopec, B.J.; Dawson, B.T.; Buck, C.; Wallman, K.E. Effects of sodium phosphate and caffeine ingestion on repeated-sprint ability in male athletes. *J. Sci. Med. Sport* **2016**, *19*, 272–276. [CrossRef]

129. Buck, C.; Guelfi, K.; Dawson, B.; McNaughton, L.; Wallman, K. Effects of sodium phosphate and caffeine loading on repeated-sprint ability. *J. Sports Sci.* **2015**, *33*, 1971–1979. [CrossRef] [PubMed]

130. Buck, C.L.; Henry, T.; Guelfi, K.; Dawson, B.; McNaughton, L.R.; Wallman, K. Effects of sodium phosphate and beetroot juice supplementation on repeated-sprint ability in females. *Eur. J. Appl. Physiol.* **2015**, *115*, 2205–2213. [CrossRef]

131. Brewer, C.P.; Dawson, B.; Wallman, K.E.; Guelfi, K.J. Effect of sodium phosphate supplementation on repeated high-intensity cycling efforts. *J. Sports Sci.* **2015**, *33*, 1109–1116. [CrossRef] [PubMed]

132. Folland, J.P.; Stern, R.; Brickley, G. Sodium phosphate loading improves laboratory cycling time-trial performance in trained cyclists. *J. Sci. Med. Sport* **2008**, *11*, 464–468. [CrossRef]

133. Brewer, C.P.; Dawson, B.; Wallman, K.E.; Guelfi, K.J. Effect of repeated sodium phosphate loading on cycling time-trial performance and VO$_{2peak}$. *Int. J. Sport Nutr. Exerc. Metab.* **2013**, *23*, 187–194. [CrossRef]

134. Goss, F.; Robertson, R.; Riechman, S.; Zoeller, R.; Dabayebeh, I.D.; Moyna, N.; Boer, N.; Peoples, J.; Metz, K. Effect of potassium phosphate supplementation on perceptual and physiological responses to maximal graded exercise. *Int. J. Sport Nutr. Exerc. Metab.* **2001**, *11*, 53–62. [CrossRef]

135. Kreider, R.B.; Miller, G.W.; Schenck, D.; Cortes, C.W.; Miriel, V.; Somma, C.T.; Rowland, P.; Turner, C.; Hill, D. Effects of phosphate loading on metabolic and myocardial responses to maximal and endurance exercise. *Int. J. Sport Nutr.* **1992**, *2*, 20–47. [CrossRef] [PubMed]

136. Cubbon, R.M.; Thomas, C.H.; Drozd, M.; Gierula, J.; Jamil, H.A.; Byrom, R.; Barth, J.H.; Kearney, M.T.; Witte, K.K. Calcium, phosphate and calcium phosphate product are markers of outcome in patients with chronic heart failure. *J. Nephrol.* **2015**, *28*, 209–215. [CrossRef] [PubMed]

137. Ye, M.; Tian, N.; Liu, Y.; Li, W.; Lin, H.; Fan, R.; Li, C.; Liu, D.; Yao, F. High Serum Phosphorus Level Is Associated with Left Ventricular Diastolic Dysfunction in Peritoneal Dialysis Patients. *PLoS ONE* **2016**, *11*, e0163659. [CrossRef] [PubMed]

138. Keskek, S.O.; Sagliker, Y.; Kirim, S.; Icen, Y.K.; Yildirim, A. Low serum phosphorus level in massry's phosphate depletion syndrome may be one of the causes of acute heart failure. *J. Nutr. Sci. Vitaminol.* **2015**, *61*, 460–464. [CrossRef]

139. Amin, M.; Fawzy, A.; Hamid, M.A.; Elhendy, A. Pulmonary hypertension in patients with chronic renal failure: Role of parathyroid hormone and pulmonary artery calcifications. *Chest* **2003**, *124*, 2093–2097. [CrossRef]

140. Kamiyama, Y.; Suzuki, H.; Yamada, S.; Kaneshiro, T.; Takeishi, Y. Serum phosphate levels reflect responses to cardiac resynchronization therapy in chronic heart failure patients. *J. Arrhythm.* **2015**, *31*, 38–42. [CrossRef]

141. Czuba, M.; Zając, A.; Poprzecki, S.; Cholewa, J. The influence of sodium phosphate supplementation on VO$_{2max}$, serum 2, 3-diphosphoglycerate level and heart rate in off-road cyclists. *J. Hum. Kinet.* **2008**, *19*, 149–164. [CrossRef]

142. Czuba, M.; Zajac, A.; Poprzecki, S.; Cholewa, J.; Woska, S. Effects of sodium phosphate loading on aerobic power and capacity in off road cyclists. *J. Sports Sci. Med.* **2009**, *8*, 591–599. [PubMed]

143. Cade, R.; Conte, M.; Zauner, C.; Mars, D.; Peterson, J.; Lunne, D.; Hommen, N.; Packer, D. Effects of phosphate loading on 2,3-diphosphoglycerate and maximal oxygen uptake. *Med. Sci. Sports Exerc.* **1984**, *16*, 263–268. [CrossRef] [PubMed]

144. Stewart, I.; McNaughton, L.; Davies, P.; Tristram, S. Phosphate loading and the effects on VO_{2max} in trained cyclists. *Res. Q. Exerc. Sport* **1990**, *61*, 80–84. [CrossRef] [PubMed]

145. West, J.S.; Ayton, T.; Wallman, K.E.; Guelfi, K.J. The effect of 6 days of sodium phosphate supplementation on appetite, energy intake, and aerobic capacity in trained men and women. *Int. J. Sport Nutr. Exerc. Metab.* **2012**, *22*, 422–429. [CrossRef] [PubMed]

146. Buck, C.L.; Dawson, B.; Guelfi, K.J.; McNaughton, L.; Wallman, K.E. Sodium phosphate supplementation and time trial performance in female cyclists. *J. Sports Sci. Med.* **2014**, *13*, 469–475. [PubMed]

147. Duffy, D.J.; Conlee, R.K. Effects of phosphate loading on leg power and high intensity treadmill exercise. *Med. Sci. Sports Exerc.* **1986**, *18*, 674–677. [CrossRef] [PubMed]

148. Galloway, S.D.; Tremblay, M.S.; Sexsmith, J.R.; Roberts, C.J. The effects of acute phosphate supplementation in subjects of different aerobic fitness levels. *Eur. J. Appl. Physiol. Occup. Physiol.* **1996**, *72*, 224–230. [CrossRef] [PubMed]

149. Singh, A.; Failla, M.L.; Deuster, P.A. Exercise-induced changes in immune function: Effects of zinc supplementation. *J. Appl. Physiol.* **1994**, *76*, 2298–2303. [CrossRef]

150. Kilic, M.; Baltaci, A.K.; Gunay, M. Effect of zinc supplementation on hematological parameters in athletes. *Biol. Trace Elem. Res.* **2004**, *100*, 31–38. [CrossRef]

151. Polat, Y. Effects of zinc supplementation on hematological parameters of high performance athletes. *Afr. J. Pharm. Pharmacol.* **2011**, *5*, 1436–1440. [CrossRef]

152. Saeedy, M.; Bijeh, N.; Moazzami, M. The effect of six weeks of high-intensity interval training with and without zinc supplementation on aerobic power and anaerobic power in female futsal players. *Int. J. Appl. Exerc. Phys.* **2016**, *5*, 1–10.

153. Davison, G.; Marchbank, T.; March, D.S.; Thatcher, R.; Playford, R.J. Zinc carnosine works with bovine colostrum in truncating heavy exercise-induced increase in gut permeability in healthy volunteers. *Am. J. Clin. Nutr.* **2016**, *104*, 526–536. [CrossRef]

154. Khaled, S.; Brun, J.F.; Cassanas, G.; Bardet, L.; Orsetti, A. Effects of zinc supplementation on blood rheology during exercise. *Clin. Hemorheol. Microcirc.* **1999**, *20*, 1–10. [PubMed]

155. Smith, M.M.; Lucas, A.R.; Hamlin, R.L.; Devor, S.T. Associations among hemorheological factors and maximal oxygen consumption. Is there a role for blood viscosity in explaining athletic performance? *Clin. Hemorheol. Microcirc.* **2015**, *60*, 347–362. [CrossRef] [PubMed]

156. Marques, L.F.; Donangelo, C.M.; Franco, J.G.; Pires, L.; Luna, A.S.; Casimiro-Lopes, G.; Lisboa, P.C.; Koury, J.C. Plasma zinc, copper, and serum thyroid hormones and insulin levels after zinc supplementation followed by placebo in competitive athletes. *Biol. Trace Elem. Res.* **2011**, *142*, 415–423. [CrossRef]

157. Bajpeyi, S.; Tanner, C.J.; Slentz, C.A.; Duscha, B.D.; McCartney, J.S.; Hickner, R.C.; Kraus, W.E.; Houmard, J.A. Effect of exercise intensity and volume on persistence of insulin sensitivity during training cessation. *J. Appl. Physiol.* **2009**, *106*, 1079–1085. [CrossRef]

158. Fisher, G.; Brown, A.W.; Bohan Brown, M.M.; Alcorn, A.; Noles, C.; Winwood, L.; Resuehr, H.; George, B.; Jeansonne, M.M.; Allison, D.B. High intensity interval- vs moderate intensity- training for improving cardiometabolic health in overweight or obese males: A randomized controlled trial. *PLoS ONE* **2015**, *10*, e0138853. [CrossRef]

159. McGarrah, R.W.; Slentz, C.A.; Kraus, W.E. The effect of vigorous versus moderate-intensity aerobic exercise on insulin action. *Curr. Cardiol. Rep.* **2016**, *18*, 117. [CrossRef]

160. Cruz, K.J.; Morais, J.B.; de Oliveira, A.R.; Severo, J.S.; Marreiro, D.D. The effect of zinc supplementation on insulin resistance in obese subjects: A systematic review. *Biol. Trace Elem. Res.* **2017**, *176*, 239–243. [CrossRef]

161. Perseghin, G.; Burska, A.; Lattuada, G.; Alberti, G.; Costantino, F.; Ragogna, F.; Oggionni, S.; Scollo, A.; Terruzzi, I.; Luzi, L. Increased serum resistin in elite endurance athletes with high insulin sensitivity. *Diabetologia* **2006**, *49*, 1893–1900. [CrossRef] [PubMed]

162. Cınar, V.; Talaghir, L.G.; Akbulut, T.; Turgut, M.; Sarıkaya, M. The effects of the zinc supplementation and weight trainings on the testosterone levels. *Hum. Sport Med.* **2017**, *17*, 58–63.

163. Cinar, V.; Akbulut, T.; Sarikaya, M. Effect of zinc supplement and weight lifting exercise on thyroid hormone levels. *Indian J. Physiol. Pharmacol.* **2017**, *61*, 232–236.

164. Shafiei-Neek, L.; Gaeini, A.A.; Choobineh, S. Effect of zinc and selenium supplementation on serum testosterone and plasma lactate in cyclist after an exhaustive exercise bout. *Biol. Trace Elem. Res.* **2011**, *144*, 454–462. [CrossRef]

165. Sims, S.T.; Rehrer, N.J.; Bell, M.L.; Cotter, J.D. Preexercise sodium loading aids fluid balance and endurance for women exercising in the heat. *J. Appl. Physiol.* **2007**, *103*, 534–541. [CrossRef]

166. Sims, S.T.; van Vliet, L.; Cotter, J.D.; Rehrer, N.J. Sodium loading aids fluid balance and reduces physiological strain of trained men exercising in the heat. *Med. Sci. Sports Exerc.* **2007**, *39*, 123–130. [CrossRef]

167. Speedy, D.B.; Thompson, J.M.; Rodgers, I.; Collins, M.; Sharwood, K.; Noakes, T.D. Oral salt supplementation during ultradistance exercise. *Clin. J. Sport Med. Off. J. Can. Acad. Sport Med.* **2002**, *12*, 279–284. [CrossRef]

168. Sanders, B.; Noakes, T.D.; Dennis, S.C. Sodium replacement and fluid shifts during prolonged exercise in humans. *Eur. J. Appl. Physiol.* **2001**, *84*, 419–425. [CrossRef] [PubMed]

169. Zorbas, Y.G.; Petrov, K.L.; Yarullin, V.L.; Kakurin, V.J.; Popov, V.K.; Deogeneov, V.A. Effect of fluid and salt supplementation on body hydration of athletes during prolonged hypokinesia. *Acta Astronaut.* **2002**, *50*, 641–651. [CrossRef]

170. Zorbas, Y.G.; Kakurin, V.J.; Kuznetsov, N.A.; Yarullin, V.L. Fluid and salt supplementation effect on body hydration and electrolyte homeostasis during bed rest and ambulation. *Acta Astronaut.* **2002**, *50*, 765–774. [CrossRef]

171. Anastasiou, C.A.; Kavouras, S.A.; Arnaoutis, G.; Gioxari, A.; Kollia, M.; Botoula, E.; Sidossis, L.S. Sodium replacement and plasma sodium drop during exercise in the heat when fluid intake matches fluid loss. *J. Athl. Train.* **2009**, *44*, 117–123. [CrossRef]

172. Barr, S.I.; Costill, D.L.; Fink, W.J. Fluid replacement during prolonged exercise: Effects of water, saline, or no fluid. *Med. Sci. Sports Exerc.* **1991**, *23*, 811–817. [CrossRef] [PubMed]

173. Hew-Butler, T.; Loi, V.; Pani, A.; Rosner, M.H. Exercise-associated hyponatremia: 2017 Update. *Front. Med.* **2017**, *4*, 21. [CrossRef] [PubMed]

174. Driller, M.; Williams, A.; Bellinger, P.; Howe, S.; Fell, J. The effects of NaHCO3 and NaCl loading on hematocrit and high-intensity cycling performance. *J. Exerc. Phys.* **2012**, *15*, 47–57.

175. Earhart, E.L.; Weiss, E.P.; Rahman, R.; Kelly, P.V. Effects of oral sodium supplementation on indices of thermoregulation in trained, endurance athletes. *J. Sports Sci. Med.* **2015**, *14*, 172–178. [PubMed]

176. Hew-Butler, T.D.; Sharwood, K.; Collins, M.; Speedy, D.; Noakes, T. Sodium supplementation is not required to maintain serum sodium concentrations during an Ironman triathlon. *Br. J. Sports Med.* **2006**, *40*, 255–259. [CrossRef] [PubMed]

177. Cosgrove, S.D.; Black, K.E. Sodium supplementation has no effect on endurance performance during a cycling time-trial in cool conditions: A randomised cross-over trial. *J. Int. Soc. Sports Nutr.* **2013**, *10*, 30. [CrossRef]

178. Aschenbach, W.; Ocel, J.; Craft, L.; Ward, C.; Spangenburg, E.; Williams, J. Effect of oral sodium loading on high-intensity arm ergometry in college wrestlers. *Med. Sci. Sports Exerc.* **2000**, *32*, 669–675. [CrossRef]

179. Hamouti, N.; Del Coso, J.; Ortega, J.F.; Mora-Rodriguez, R. Sweat sodium concentration during exercise in the heat in aerobically trained and untrained humans. *Eur. J. Appl. Physiol.* **2011**, *111*, 2873–2881. [CrossRef] [PubMed]

180. Zorbas, Y.G.; Federenko, Y.F.; Naexu, K.A. Effect of daily hyperhydration on fluid-electrolyte changes in endurance-trained volunteers during prolonged restriction of muscular activity. *Biol. Trace Elem. Res.* **1995**, *50*, 57–78. [CrossRef]

181. Strazzullo, P.; D'Elia, L.; Kandala, N.B.; Cappuccio, F.P. Salt intake, stroke, and cardiovascular disease: Meta-analysis of prospective studies. *BMJ* **2009**, *339*, b4567. [CrossRef]

182. Margaritis, I.; Tessier, F.; Prou, E.; Marconnet, P.; Marini, J.F. Effects of endurance training on skeletal muscle oxidative capacities with and without selenium supplementation. *Trace Elem. Med. Biol.* **1997**, *11*, 37–43. [CrossRef]

183. Tessier, F.; Margaritis, I.; Richard, M.J.; Moynot, C.; Marconnet, P. Selenium and training effects on the glutathione system and aerobic performance. *Med. Sci. Sports Exerc.* **1995**, *27*, 390–396. [CrossRef]

184. Zamora, A.J.; Tessier, F.; Marconnet, P.; Margaritis, I.; Marini, J.F. Mitochondria changes in human muscle after prolonged exercise, endurance training and selenium supplementation. *Eur. J. Appl. Physiol. Occup. Physiol.* **1995**, *71*, 505–511. [CrossRef]

185. Sun, H.J.; Rathinasabapathi, B.; Wu, B.; Luo, J.; Pu, L.P.; Ma, L.Q. Arsenic and selenium toxicity and their interactive effects in humans. *Environ. Int.* **2014**, *69*, 148–158. [CrossRef]

186. Nielsen, J.; Gejl, K.D.; Hey-Mogensen, M.; Holmberg, H.C.; Suetta, C.; Krustrup, P.; Elemans, C.P.H.; Ortenblad, N. Plasticity in mitochondrial cristae density allows metabolic capacity modulation in human skeletal muscle. *J. Physiol.* **2017**, *595*, 2839–2847. [CrossRef]

187. Savory, L.A.; Kerr, C.J.; Whiting, P.; Finer, N.; McEneny, J.; Ashton, T. Selenium supplementation and exercise: Effect on oxidant stress in overweight adults. *Obesity* **2012**, *20*, 794–801. [CrossRef] [PubMed]

188. Alfthan, G.; Eurola, M.; Ekholm, P.; Venalainen, E.R.; Root, T.; Korkalainen, K.; Hartikainen, H.; Salminen, P.; Hietaniemi, V.; Aspila, P.; et al. Effects of nationwide addition of selenium to fertilizers on foods, and animal and human health in Finland: From deficiency to optimal selenium status of the population. *J. trace Elem. Med. Biol.* **2015**, *31*, 142–147. [CrossRef] [PubMed]

189. Hughes, D.J.; Duarte-Salles, T.; Hybsier, S.; Trichopoulou, A.; Stepien, M.; Aleksandrova, K.; Overvad, K.; Tjonneland, A.; Olsen, A.; Affret, A.; et al. Prediagnostic selenium status and hepatobiliary cancer risk in the European Prospective Investigation into Cancer and Nutrition cohort. *Am. J. Clin. Nutr.* **2016**, *104*, 406–414. [CrossRef]

190. Alehagen, U.; Johansson, P.; Bjornstedt, M.; Rosen, A.; Post, C.; Aaseth, J. Relatively high mortality risk in elderly Swedish subjects with low selenium status. *Eur. J. Clin. Nutr.* **2016**, *70*, 91–96. [CrossRef] [PubMed]

191. Pingitore, A.; Lima, G.P.; Mastorci, F.; Quinones, A.; Iervasi, G.; Vassalle, C. Exercise and oxidative stress: Potential effects of antioxidant dietary strategies in sports. *Nutrition (Burbank, Los Angeles County, Calif.)* **2015**, *31*, 916–922. [CrossRef] [PubMed]

192. Marmett, B.; Nunes, R.B. Effects of chromium picolinate supplementation on control of metabolic variables: A systematic review. *J. Food Nutr. Res.* **2016**, *4*, 633–639.

193. Pittler, M.H.; Stevinson, C.; Ernst, E. Chromium picolinate for reducing body weight: Meta-analysis of randomized trials. International journal of obesity and related metabolic disorders. *J. Int. Assoc. Study Obes.* **2003**, *27*, 522–529. [CrossRef] [PubMed]

194. Edwards, W.; Pringle, D.; Palfrey, T.; Anderson, D. Effects of chromium picolinate supplementation on body composition in in-season division I intercollegiate female swimmers. *Med. Sport.* **2012**, *16*, 99–103. [CrossRef]

195. Kaats, G.R.; Blum, K.; Fisher, J.A.; Adelman, J.A. Effects of chromium picolinate supplementation on body composition: A randomized, double-masked, placebo-controlled study. *Curr. Ther. Res.* **1996**, *57*, 747–756. [CrossRef]

196. Clancy, S.P.; Clarkson, P.M.; DeCheke, M.E.; Nosaka, K.; Freedson, P.S.; Cunningham, J.J.; Valentine, B. Effects of chromium picolinate supplementation on body composition, strength, and urinary chromium loss in football players. *Int. J. Sport Nutr.* **1994**, *4*, 142–153. [CrossRef] [PubMed]

197. Hallmark, M.A.; Reynolds, T.H.; DeSouza, C.A.; Dotson, C.O.; Anderson, R.A.; Rogers, M.A. Effects of chromium and resistive training on muscle strength and body composition. *Med. Sci. Sports Exerc.* **1996**, *28*, 139–144. [CrossRef] [PubMed]

198. Lukaski, H.C.; Bolonchuk, W.W.; Siders, W.A.; Milne, D.B. Chromium supplementation and resistance training: Effects on body composition, strength, and trace element status of men. *Am. J. Clin. Nutr.* **1996**, *63*, 954–965. [CrossRef]

199. Walker, L.S.; Bemben, M.G.; Bemben, D.A.; Knehans, A.W. Chromium picolinate effects on body composition and muscular performance in wrestlers. *Med. Sci. Sports Exerc.* **1998**, *30*, 1730–1737. [CrossRef]

200. Campbell, W.W.; Joseph, L.J.; Davey, S.L.; Cyr-Campbell, D., Anderson, R.A.; Evans, W.J. Effects of resistance training and chromium picolinate on body composition and skeletal muscle in older men. *J. Appl. Physiol.* **1999**, *86*, 29–39. [CrossRef]

201. Campbell, W.W.; Joseph, L.J.; Anderson, R.A.; Davey, S.L.; Hinton, J.; Evans, W.J. Effects of resistive training and chromium picolinate on body composition and skeletal muscle size in older women. *Int. J. Sport Nutr. Exerc. Metab.* **2002**, *12*, 125–135. [CrossRef] [PubMed]

202. Livolsi, J.M.; Adams, G.M.; Laguna, P.L. The effect of chromium picolinate on muscular strength and body composition in women athletes. *J. Strength Cond. Res.* **2001**, *15*, 161–166. [PubMed]

203. Hasten, D.L.; Rome, E.P.; Franks, B.D.; Hegsted, M. Effects of chromium picolinate on beginning weight training students. *Int. J. Sport Nutr.* **1992**, *2*, 343–350. [CrossRef] [PubMed]
204. Lefavi, R.G.; Wilson, G.D.; Keith, R.E.; Anderson, R.A.; Blessing, D.L.; Hames, C.G.; McMillan, J.L. Lipid-lowering effect of a dietary chromium (III)—Nicotinic acid complex in male athletes. *Nutr. Res.* **1993**, *13*, 239–249. [CrossRef]
205. Volek, J.S.; Silvestre, R.; Kirwan, J.P.; Sharman, M.J.; Judelson, D.A.; Spiering, B.A.; Vingren, J.L.; Maresh, C.M.; Vanheest, J.L.; Kraemer, W.J. Effects of chromium supplementation on glycogen synthesis after high-intensity exercise. *Med. Sci. Sports Exerc.* **2006**, *38*, 2102–2109. [CrossRef] [PubMed]
206. Vincent, J.B. New Evidence against Chromium as an Essential Trace Element. *J. Nutr.* **2017**, *147*, 2212–2219. [CrossRef]
207. Nielsen, F.H. Update on human health effects of boron. Journal of trace elements in medicine and biology. *Organ Soc. Miner. Trace Elem. (GMS)* **2014**, *28*, 383–387. [CrossRef]
208. Ferrando, A.A.; Green, N.R. The effect of boron supplementation on lean body mass, plasma testosterone levels, and strength in male bodybuilders. *Int. J. Sport Nutr.* **1993**, *3*, 140–149. [CrossRef] [PubMed]
209. Green, N.R.; Ferrando, A.A. Plasma boron and the effects of boron supplementation in males. *Environ. Health Perspect.* **1994**, *102*, 73–77.
210. Meacham, S.L.; Taper, L.J.; Volpe, S.L. Effects of boron supplementation on bone mineral density and dietary, blood, and urinary calcium, phosphorus, magnesium, and boron in female athletes. *Environ. Health Perspect.* **1994**, *102*, 79–82. [PubMed]
211. Meacham, S.L.; Taper, L.J.; Volpe, S.L. Effect of boron supplementation on blood and urinary calcium, magnesium, and phosphorus, and urinary boron in athletic and sedentary women. *Am. J. Clin. Nutr.* **1995**, *61*, 341–345. [CrossRef]
212. Volpe-Snyder, S.L.; Taper, L.J.; Meacham, S.L. The effect of boron supplementation on bone mineral density and hormonal status in college female athletes. *Med. Exerc. Nutr. Health* **1993**, *2*, 323–330.
213. Del Coso, J.; Gonzalez-Millan, C.; Salinero, J.J.; Abian-Vicen, J.; Areces, F.; Lledo, M.; Lara, B.; Gallo-Salazar, C.; Ruiz-Vicente, D. Effects of oral salt supplementation on physical performance during a half-ironman: A randomized controlled trial. *Scand. J. Med. Sci. Sports* **2016**, *26*, 156–164. [CrossRef]
214. Barry, D.W.; Hansen, K.C.; van Pelt, R.E.; Witten, M.; Wolfe, P.; Kohrt, W.M. Acute calcium ingestion attenuates exercise-induced disruption of calcium homeostasis. *Med. Sci. Sports Exerc.* **2011**, *43*, 617–623. [CrossRef] [PubMed]
215. Sherk, V.D.; Wherry, S.J.; Barry, D.W.; Shea, K.L.; Wolfe, P.; Kohrt, W.M. Calcium supplementation attenuates disruptions in calcium homeostasis during exercise. *Med. Sci. Sports Exerc.* **2017**, *49*, 1437–1442. [CrossRef] [PubMed]
216. Shea, B.J.; Reeves, B.C.; Wells, G.; Thuku, M.; Hamel, C.; Moran, J.; Moher, D.; Tugwell, P.; Welch, V.; Kristjansson, E.; et al. AMSTAR 2: A critical appraisal tool for systematic reviews that include randomised or non-randomised studies of healthcare interventions, or both. *BMJ* **2017**, *358*, J4008. [CrossRef] [PubMed]

© 2019 by the authors. Licensee MDPI, Basel, Switzerland. This article is an open access article distributed under the terms and conditions of the Creative Commons Attribution (CC BY) license (http://creativecommons.org/licenses/by/4.0/).

MDPI

St. Alban-Anlage 66

4052 Basel

Switzerland

Tel. +41 61 683 77 34

Fax +41 61 302 89 18

www.mdpi.com

Nutrients Editorial Office

E-mail: nutrients@mdpi.com

www.mdpi.com/journal/nutrients

www.ingramcontent.com/pod-product-compliance
Lightning Source LLC
LaVergne TN
LVHW071512180726

843512LV00014B/1066